Denver International Airport

Other McGraw-Hill Books of Interest

Aircraft Safety: Accident Investigations, Analyses, and Applications
by *Shari Stanford Krause*

Airport Operations, 2d ed
by *Norman Ashford, H. P. Martin, and Clifton A. Moore*

Airport Planning and Management, 3d ed by *Alexander T. Wells*

Becoming a Better Pilot by *Paul A. Craig*

Flying Jets by *Linda D. Pendleton*

Kitplane Construction, 2d ed by *Ronald J. Wanttaja*

Optimizing Jet Transport Efficiency: Performance, Operations, and Economics
by *Carlos E. Padilla*

Piloting for Maximum Performance by *Lewis Bjork*

Redefining Airmanship by *Tony Kern*

They Called It Pilot Error by *Robert Cohn*

Denver International Airport

Lessons Learned

Paul Stephen Dempsey
Andrew R. Goetz
Joseph S. Szyliowicz

McGraw-Hill
New York San Francisco Washington, D.C. Auckland Bogotá
Caracas Lisbon London Madrid Mexico City Milan
Montreal New Delhi San Juan Singapore
Sydney Tokyo Toronto

Library of Congress Cataloging-in-Publication Data
Dempsey, Paul Stephen
Denver International Airport : lessons learned / Paul Stephen Dempsey, Andrew R. Goetz, Joseph S. Szyliowicz
p. cm.
Includes bibliographical references and index.
ISBN 0-07-158184-7 (alk. paper)
1. Denver International Airport. I. Goetz, Andrew R. II. Szyliowicz, Joseph S. III. Title.
TL726.4.D58D46 1996
387.7'36'0978883—dc21 96-48396
CIP

McGraw-Hill

A Division of The ***McGraw-Hill*** *Companies*

2 3 4 5 6 7 8 9 10 FGRFGR 9 9 8 7

ISBN 0-07-158184-7

The sponsoring editor for this book was Shelley IC. Chevalier, the editing supervisor was Ruth W. Mannino, and the production supervisor was Suzanne W.B. Rapcavage. It was set in Garamond by McGraw-Hill's desktop publishing department in Hightstown, N.J.

This book is printed on recycled, acid-free paper containing a minimum of 50% recycled, de-inked fiber.

Contents

Preface

The story of why and how Denver International Airport (DIA) was conceived, planned, designed, financed, and constructed is one of enormous importance and of great potential interest to the world's aviation community. More than a thousand airports around the planet serve international commercial aviation operations, and thousands more serve the domestic air transport market, but the demand for increased capacity seems to grow inexorably.

The need to build new airports and expand existing ones will continue to press communities to address many of the same difficult questions that Denver faced. The emergence of a global economy has made public and private airport and air transport infrastructure more important than ever. Transportation infrastructure is a crucial catalyst for economic growth, creating employment opportunities and stimulating trade and commerce. The ability to transport people and commodities expeditiously, efficiently, and economically is a fundamental element for enhanced trade, income, and prosperity.

Airports and airlines are primary components of the broader tour-and-travel industry. Accounting for $3.5 trillion and employing 127 million people, it is arguably the largest industry in the world. Transportation corridors are the veins and arteries of commerce, communications, and national defense. As an integral part of the infrastructure upon which economic growth is built, a safe, healthy, and efficient system of airports and airways is crucially important for the vitality of the nation it serves. Progress and development in the transport sector often serve as catalysts for broader economic prosperity.

As a consequence, an unprecedented boom in airport construction is under way, particularly in the Pacific Rim. New airports are under construction in Seoul, Macao, Hong Kong, Kuala Lumpur, Bangkok, and many cities throughout China and Indonesia, and airports have recently been completed in Osaka, Japan, and Munich, Germany. Worldwide, more than $250 billion is projected to be spent on new airports and expansion of existing airports through 2010, according to the United Nations' International Civil Aviation Organization (ICAO). DIA, the world's newest, largest, and most technologically advanced airport, can serve as a model of what to do, and what to avoid, in airport planning, location, design, finance, construction, and implementation. It is the purpose of this book to identify, as fairly and candidly as possible, what Denver did well and what it did poorly in building Denver International Airport. To provide context, we include comparative data from major new international airports being built around the world.

This story manifestly needs to be told. It is an interesting drama in its own right, replete with crucial decisions of monumental impact, colorful actors, fame, fortune, deceit, and despair. But this book goes beyond merely describing what happened to provide an evaluation of why these developments occurred and whether they were the most appropriate steps that could have been taken. Such an analysis is essential if the study of DIA is to be of more than historical interest. Knowing what happened can help airport executives, planners, consultants, designers, engineers, architects, contractors, construction executives, managers, other aviation executives, and governmental officials understand the problems involved in planning and implementing airports and other megaprojects. We hope that understanding why certain events transpired will assist professionals in their quest to plan, design, and construct better, more efficient airports. It also permits us to develop an enhanced theoretical understanding of the process.

To do so, it is necessary to place the DIA story within a broader context, because it is impossible to draw generalizations from a single case unless it is based on a more comprehensive analysis. That is our second primary purpose—to analyze these developments from existing theoretical perspectives on planning, decision-making, political science, economic development, and public policy. We also consider the explanatory power of existing theories and the ways in which the DIA case forces us to modify our understanding of how such projects are planned and implemented.

Such an orientation inevitably suggests a strong prescriptive element. We are concerned not only with the theoretical implications of the case, but with its practical, pragmatic applications as well. The lessons that can be derived from Denver's experience are germane to the planning and implementation of airport projects in many countries. These lessons are not limited to specific aspects, such as terminal design or environmental impact assessments, but extend to the fundamental issues of how to plan and implement a project in an environment inevitably marked by great uncertainties and multifaceted political considerations.

Writing this book represents a logical step for the authors, who have lived with this project since its inception. As academics who have analyzed the airline and aviation industries and the relationship between transportation and economic growth, we observed DIA's evolution with more than casual interest, taught courses focusing on it, and, when asked, proffered our observations to the local and national broadcast and print media. We have been gathering materials and evaluating the project for many years.

While working on this project we realized that the literature on airport planning, design, and implementation was devoid of detailed case studies. This book attempts to meet this need, while placing the DIA case within a larger theoretical framework.

The work is organized as follows:

Chapter 1: The world's largest airport

At 53 square miles, DIA is the largest piece of real estate dedicated to commercial aviation on earth. For the foreseeable future, DIA has virtually infinite capacity. As air transport continues to expand, proliferate, and mature, the question of capacity and airport infrastructure remains crucial. Some cities around the world are building new airports; others are expanding existing ones. The current worldwide boom in airport construction does not mean that the problem of capacity provision will be solved easily. In fact, attempts to build new facilities or expand them tend to arouse heated opposition that forces long delays or even outright abandonment. How Denver, a metropolitan area with a population of only 2 million, became the city that decided to build the largest airport in the world is one of the fundamental questions that this book addresses. Chapter 1 places the project in empirical and theoretical perspective and examines its evolution.

Chapter 2: Three hubs to one: The folly of forecasting in the dynamic airline industry

In building DIA, Denver had to confront many problems that any community seeking to expand its aviation capacity must face. Of fundamental importance to the issue of whether a new airport is even needed is the accuracy of forecasting future passenger and freight demand. Forecasting is inevitably more of an art than a science, and the artists in the aviation sector appear to be impressionists rather than realists.

The traffic projections that were made for Denver proved ultimately to have been wildly optimistic, based on the earlier competitive experience of deregulation rather than the later phenomenon of monopoly fortress hubs. In the late-1980s, the FAA projected that Denver would have more than 55 million passengers by 1995; the actual number was less than 31 million. As a consequence, the airport that finally opened was significantly smaller than originally contemplated, declining from 120 planned gates to 88 (compared to Stapleton's 111) as the number of major hub airlines dwindled from three to one.

Accordingly, in this chapter we seek to answer questions such as the following:

- How are forecasts of passenger demand generated?
- Why are these so erroneous?
- How have recent policy changes, such as airline deregulation, affected airport development?
- Given the shortcomings of existing forecasting practices, what kind of planning process is the most effective?

Chapter 3: The politics of DIA's development

Like any other megaproject, new airports are inherently political because they require the support of powerful political forces. The nature of the politics involved varies widely depending on the kind of political system involved. In centralized states, the decision by the ruler suffices; in democratic societies, it is necessary to build a powerful coalition, because without strong political and public support, no project can be implemented. Moreover, airport projects inevitably encounter heated opposition; recent history is littered with projects that have been delayed, postponed, or canceled because of public hostility. How and why opposition was overcome in Denver is an important issue for airport planners and community leaders everywhere, as are such questions as:

- How effective are existing mechanisms for public participation in the planning process?
- What responsibility do the media, elected officials, and planners have to ensure that the public is accurately informed about the issues?
- How does a city secure cooperation with other relevant organizations in airport development or expansion?

Chapter 4: The economic impact of airports

Airports are widely touted as growth generators. The growing interactions between nations and the globalization of trade and finance have important implications for any city seeking to function effectively in this new era:

- To what extent are these implications linked to air travel?
- What might a city hope to realize in terms of new jobs and economic growth, short and long term, with new airport development?
- What are the long-term opportunities for airport expansion and private hotel and other commercial development on annexed and adjacent property?
- What are the advantages of closing an old airport and using it for other developmental purposes?

Chapter 5: Financing the field of dreams

Every airport project requires financing from some source. In this chapter, we discuss the ways in which airports around the world are financed. In the United States, most of the financing comes from private sources. In Denver, it involved the sale by Wall Street investment houses of bonds not backed by the "full faith and credit" of the city and thus of higher risk and enjoying a higher interest rate than municipal bonds.

- How does a city finance an airport or its expansion with private capital?

- What role do bond houses play in the design and implementation process?
- What sources of revenue are available to an airport?
- What is the political role of financing institutions?

Chapter 6: Location, location, location: Site selection, environmental impacts, and ground access

Perhaps the most fundamental issue facing airport planners is where to locate the facility. Denver's decision was to build an entirely new facility on a 53-square-mile parcel of land located 24 miles from downtown Denver rather than expand existing Stapleton Airport (about 7 miles from downtown) onto adjacent federal land (the Rocky Mountain Arsenal). That Denver had such options placed it in an enviable position; few cities have such a vast amount of open space available. Other cities are forced to work within existing constraints and expand only marginally or build a new facility even further away from the community, thus creating additional problems. We consider how alternative airport sites are generated, evaluated, and selected; how effectively governments address environmental impact issues; and what arrangements were made for ground access to DIA.

Chapter 7: Airport layout, design, and technologies

Denver International Airport was equipped with the most modern technologies available and is purported to be the most efficient airport in the world. It finally opened on February 28, 1995, in the midst of a winter snowstorm. Since its opening, the airport has been operating quite smoothly, although its baggage system has been dysfunctional on occasion. The relevant questions discussed include:

- What technologies are utilized at DIA?
- How did DIA address airport security concerns?
- Which runway and terminal configurations are most efficient?
- What cost and design problems are presented in employing "cutting-edge" rather than "off-the-shelf" aviation technologies?

Chapter 8: Architecture and aesthetics

Airports are powerful symbols, and planners and designers everywhere are concerned with the aesthetic dimensions of their projects. Often international design competitions are held. An important criterion is how the design reflects the local culture. Aesthetics were an important element in the DIA project, both in terms of architecture and art. Considerable monies were spent on commissions to sculptors, painters, and other artists. Accordingly, the following questions deserve consideration:

- How were aesthetic considerations integrated into the design?
- What is the quality of the artworks that are displayed throughout the facility?
- How did the art selection process work?
- To what degree has Denver acquired a symbol marking it as a city of the future?

Chapter 9: How did a $1.5 billion airport become a $5.3 billion airport?

DIA came in over budget by hundreds of millions of dollars. As in most megaprojects, the original projections proved to be highly optimistic and the cost escalated constantly. Why was this the case? Although a highly sophisticated management system was supposedly put into place, difficulties of all kinds constantly plagued the project.

- How large were the actual cost overruns?
- What are the advantages and disadvantages of different management strategies used for airport planning?
- Why did the management strategies used for DIA produce such mixed results?
- To what extent were the cost overrun difficulties the result of corruption and incompetence?

Chapter 10: Airport planning theory in perspective

Denver International Airport's many travails have broad implications for those interested specifically in current and future airport projects and generally in megaproject planning and decision-making. They underline the degree to which existing planning practices leave much to be desired.

- What improvements can be made in designing and implementing large airport projects?
- Are there models and theories of decision-making that suggest ways in which the process might be improved?
- If so, how might these be applied to megaprojects in general and airports specifically?

Chapter 11: What have we learned?

Finally, we return to the specific case of Denver International Airport and analyze the successes and failures of the project and its costs and benefits. We summarize the specific reasons for these outcomes and conclude by answering the following basic question: What does Denver's experience teach us?

This then is the journey that awaits you. Hopefully, it will culminate with an understanding of how planning, decision-making, and implementation occurred in the specific case of the new Denver International Airport and the outcomes that resulted. Our goal is to provide our readers not only with the details of what happened in this interesting and important case, but also with lessons and insights that are of relevance to their particular situations. In short, we hope that everyone throughout the world who is struggling with the challenge of how to create the capacity to meet the needs of future generations of travelers will find the journey worthwhile.

Paul Stephen Dempsey

Andrew R. Goetz

Joseph Szyliowicz

Acknowledgments

The need for a book such as this one crystallized at the University of Denver's Conference on Aviation and Airport Infrastructure held in Denver in December 1993. This conference brought together many key actors. We have been fortunate to know many individuals who played a key role in the project and are grateful for their willingness to share with us the details of what has been a unique experience and an education. Many other experts also generously lent their time and provided us with valuable insight.

Accordingly, we wish to thank Robert Albin, Peter Boyles, Michael Boyd, George Brewer, Richard Boulware, Chuck Cannon, Jim De Long, George Doughty, Ginger Evans, Richard Fleming, Kevin Flynn, Tom Gougeon, Steve Kaplan, Gene Levin, Mike McKee, Steven Paulson, Steven Rhodes, James (Skip) Spensley, Genniffer Sussman, Dick Veazey, Gordon Yale, and Richard Young.

Nor can we ignore the many reporters of the *Denver Post*, *Rocky Mountain News*, and *Westword*, who chronicled the tale as the story unfolded, because their articles helped us develop what we hope is an accurate portrait of a highly complicated subject. We would like to thank Denver International Airport and the City and County of Denver for many of the photographs and other visual images that appear in this book.

We are also grateful for the help that we have received from our graduate research assistants, Greg Hall and Sam Schinta of the College of Law, and Perin Arkun and Dan Wessner of the Graduate School of International Studies. We are particularly grateful to those who took the time to read and comment on various sections of the manuscript.

Portions of Chapters 1 and 10 were earlier published in "Getting Realistic about Megaproject Planning: The Case of the New Denver

International Airport," by J. Szyliowicz and A.R. Goetz, in *Policy Sciences* (Volume 28, 1995, pp. 347 to 367, Kluwer Academic Publishers) and are reprinted by permission.

Finally, we must thank our wives, Kerry, Andrea, and Irene, who learned more about DIA than they probably ever wanted to know, and who patiently supported us during this undertaking.

* * *

The cover of this book is a composite of three separate photographs—one of the golden fields in the Great Plains, one of the Rocky Mountains against a blue sky, and the third of the main terminal building of Denver International Airport. There is no venue from which such a photograph can be taken. The composite was prepared by Denver International Airport as a signature for an airport that sits at the junction of North America's vast plains and dynamic mountains.

The authors chose this photograph because it symbolizes the dreams and illusions of DIA's planners and decision-makers and the different perspectives from which the airport continues to be viewed.

1

The world's largest airport

Denver International Airport (DIA) is one of the most impressive airports in the world. It is among the world's most technologically sophisticated, with state-of-the-art navigational, weather, people-moving, baggage-handling, airfield, communications, and security facilities. Terminal and runway configurations allow expeditious hub rotations, enhancing airline labor and aircraft utilization and productivity. DIA's distinctive architecture leaves its mark on Denver the same way the Opera House emblazons itself on Sydney.

At 53 square miles, DIA is twice the size of Manhattan Island. DIA is the largest piece of real estate dedicated to commercial aviation on Earth. For the foreseeable future, DIA has virtually infinite capacity, with the ability to expand from its initial five runways and 88 gates to 12 runways and more than 200 gates and to accommodate up to 200 million passengers per year, something no other airport in the world can yet hope to replicate. Whether or not the airport is needed to handle these traffic needs, DIA is positioned to become the preeminent airport for the 21st century, one that can serve in many ways as a model for other airports being built around the world. Delegations from dozens of countries have already visited DIA to learn more about its technologies and operating systems.

But DIA was not without its problems. Originally scheduled to open in October 1993, DIA finally opened on February 28, 1995, after four postponements, principally because of difficulties with its automated baggage system. Originally estimated to cost $1.5 billion, its price escalated to more than $5 billion. Originally designed to serve three major hubbing airlines, the airport is dominated today only by United Airlines. Originally estimated to serve a total of 56 million passengers by 1995, the facility served just over 30 million passengers in 1995. En route to its belated opening, the airport encountered numerous difficulties with the major tenant airlines, the air cargo carriers, and the rental car companies. And, immediately after its opening, it began to receive complaints from irate residents about aircraft noise, even though the airport was built on a 53-square-mile site that had been carefully selected to minimize noise impact.

Nonetheless, Denver accomplished what had eluded the grasp of so many cities whose dreams of new or expanded airports were thwarted by environmental, financial, and political opposition or the unavailability of sufficient real estate. Such dreams were made possible in the first place only because of one of humanity's most profound technological achievements—aviation. From the Wright brothers' inaugural flight at Kitty Hawk in 1903 to the Apollo landing on the Moon in 1969, the speed of progress in our ability to traverse the skies and space is unprecedented. Like no other industry, air transport shrinks the planet, promotes time-space convergence, and cross-fertilizes disparate cultures and economies. Through technological developments such as the jet engine, the wide-bodied commercial aircraft, and the supersonic transport, even the most remote corners of the world can now be reached easily. Today, large jumbo jets routinely take off, travel thousands of miles, and land, connecting people and places from opposite sides of the world in only a few hours. Aviation is mobility for the human race.

The importance of air transport is likely to increase in the future as national economies become ever more inextricably intertwined. Global trade and tourism require efficient and expeditious systems through which to pump raw materials, consumer goods, and business and pleasure travelers. In the 21st century, the world's population will continue to grow robustly, as will income and the ability and desire of individuals to see the planet and experience diverse cultures. Furthermore, structural changes in the global economy, from reliance on traditional sectors and industries (raw material production and heavy manufacturing) toward light manufacturing, producer services, computing and information processing, and high-technology industries, favor air transport.

As air transport continues to expand, proliferate, and mature, the question of capacity and airport infrastructure remains crucial. Ever since the Wright brothers and Charles Lindbergh first demonstrated the feasibility of flight, the question of the best places to take off and land has been of concern to aviators, city planners, military officers, and political leaders. Originally, airports were simply open fields, but they soon acquired structures and, for the sake of safety and efficiency, simple navigational aids. Over time they became more sophisticated as air transport developed and more and more aircraft took to the air. Furthermore, developments in aircraft technology toward heavier, faster, and larger planes powered by more complex engines forced the aviation community to grapple more seriously with the problematic airport question.

As burgeoning numbers of people and goods are transported through the air, existing facilities become ever more strained. Accommodating this explosive growth has proven to be no easy matter. Nearly every country has been confronted with the challenge of airport planning. Recently, Munich, Germany, and Osaka, Japan, opened new facilities, and, at present, plans are under way to build major new airports in China, Thailand, Hong Kong, Macau, Malaysia, and South Korea. Their significant features, including current passenger totals at the existing city airports, are listed in Table 1-1.

We compare and contrast each of these new airport projects with DIA in the ensuing chapters. Other cities are also building new airports. Athens, Greece, is planning a new airport at Spada to open in 1997, and Oslo, Norway, is building a new airport at Gardermoen to open in 1998. The new airport is Norway's largest development project ever. It is located 25 miles north of Oslo. Its capacity will be 12 million passengers annually by the year 2000 (Hill 1995). Guangzhou, China, is building a new airport, to be completed in the year 2005.

Many others are expanding existing airports. London's Heathrow Airport is planning a new terminal 5, while Milan is planning a new terminal as well (Trautmann 1993). A new runway has been built at Sydney's Kingsford Smith Airport, and a second airport west of Sydney at Badgery's Creek is under development (Meredith 1995). Vancouver is building a new international terminal building and runway. Vancouver's new 1.13 million square foot international terminal will have 15 gates, with sufficient capacity for 17 million passengers by 2005 (VIAA 1995). Pittsburgh built an entirely new terminal building. Many other airports are undergoing expansion, particularly those in the Asia-Pacific Region.[1] These include Tokyo (Haneda), Tokyo

1. For a comprehensive summary of airport projects in Asia, see Mecham (1994) and Neilan (1994).

Table 1-1. Major new airports of the 1990s

Airport	Opening date	Projected cost (billion $)	Runways (capacity)	Passengers (millions, 1994)
Munich	1992	$ 5	2	14
Osaka (Kansai)	1994	$14.4	1 (3)	20
Denver (International)	1995	$ 5.3	5 (12)	33
Macau	1995	$ 0.9	1	2
Hong Kong (Chek Lap Kok)	1997	$20.2	1 (2)	25
Kuala Lumpur (Sepang)	1998	$ 3.5	1 (2)	NA
Seoul (Yongjong)	1999	$ 7	2	18
Bangkok (Nong Ngu Hao)	2000	$ 3.9	2 (4)	21

(Narita), Bangkok (Don Muang), Beijing, Shanghai, Wuhan, Liangjiang, Guangzhou, Haikou, Hangzhou, Nanchang, Shenzhen, Zhuhai, Hanoi, Ho Chi Minh City, Danang, Penang, Bombay, Calcutta, New Delhi, and Madras. Table 1-2 summarizes airport projects around the world.

How many of these airport projects will meet their planners' expectations remains to be seen. It is not at all clear, for example, that China's southern region, where five new airports are being built (Hong Kong, Zhuhai, Macao, Shenzhen, and Guangzhou), will need all this additional capacity (Liden 1995). Although the region has been growing rapidly, these five airports can each handle 75 million passengers per year—almost China's entire aviation traffic in 1994. Numerous forecasts suggest that these facilities will be needed. But as the DIA case suggests, forecasting future airport requirements has not yet achieved the accuracy of weather forecasts.

Demand forecasts, however, are not the only factor that drives the decision to build airports—political, economic, and symbolic considerations are also important. (We shall discuss this point in detail in Chapter 3.) Airports are more than just aviation facilities; they are places where people, goods, and planes converge. They are major economic institutions, with significant direct and indirect impacts. They provide employment for many, are a major source of local revenue, and stimulate regional economic growth. As part of the national infrastructure, they facilitate the flow of commerce and people, domestically and internationally. An airport also conveys an image of the place that it serves and thus has a very powerful symbolic appeal. The first impression that many visitors today have of a city, region, or a nation is through its airport. In earlier periods, ocean piers and train stations were the portals through which people entered and exited the gateway of travel to distant lands. Today, the commercial airport is the door through which we must pass to reach the far corners of our planet. Airports also reflect the increasing importance of air transportation to our highly interconnected global society. They are monumental accomplishments, symbols of progress and technological prowess through which nations and cities are able to express their connection to the modern world.

Unfortunately, airports also have significant negative externalities and are not universally welcomed. There are many reasons for public opposition to airports. They are costly, consume vast amounts of

Table 1-2. Examples of major airport development and expansion projects

Country	Costs (U.S.$)	Airport	Development
Abu Dhabi	$200 million	Ai Ain International	New airport
Australia	$300 million	Kingsford-Smith	New parallel runway
	A$250 million	Brisbane International	Expansion project
Bahrain	$100 million	Bahrain	Expansion project
Bulgaria	$200 million	Sophia International	Redevelopment
Canada	$700 million	Pearson International	Redevelopment
	$350 million	Vancouver International	Expansion
China	$280 million	Fuzhou	New airport
	$400 million	Beijing International	New terminal
	$161 million	Guilin International	New airport
	$ 1 billion	Zhuhai	New airport
	$ 10 billion	Shanghai (Pudong)	New airport
	$130 million	Nanjing Luko	New airport
Cyprus	$300 million	Larnaca International	Redevelopment
Ethiopia	$240 million	Addis Ababa	Redevelopment
Germany	$ 1.1 billion	Frankfurt Airport	New terminal
	$ 7.1 billion	Munich	New airport
Greece	$ 2.3 billion	Sparta (Athens)	New airport

Hong Kong	$20.2 billion	Chek Lap Kok	New development
Italy	$ 3 billion	Rome	Upgrade of facilities
Japan	$ 1.5 billion	Tokyo (Haneda)	Redevelopment
	$ 800 million	Narita Airport	Reconstruction
	$14.4 billion	Kansai International	New airport
	$ 362 million	Okinawa (Naha International)	Domestic terminal
Korea	$ 5 billion	Yongjong Island (Seoul)	New airport
Lebanon	$ 400 million	Beruit International	Refurbishment
Macau	$ 950 million	Macau	New airport
Malaysia	$ 3.6 billion	Sepang	New airport
New Zealand	$ 200 million	Christchurch International	Terminal facilities
Norway	$1.85 billion	Gardermoen (Oslo)	New airport
Pakistan	$ 200 million	New Lahore International	New airport
Russia	$ 100 million	Anadyr	Refurbishment
Singapore	$ 545 million	Changi	Expansion project
Thailand	$ 3.9 billion	Bangkok	New airport
Turkmenistan	$ 100 million	Turkmenistan	New terminal

Table 1-2. *(continued)*

Country	Costs (U.S.$)	Airport	Development
United Kingdom	$ 2.1 billion	London Heathrow	New terminal
United States	$ 2.5 billion	Miami International	Expansion program
	$760 million	Orlando International	Redevelopment
	$ 4.3 billion	John F. Kennedy	Redevelopment
	$600 million	Bergstrom International	Conversion, civil airport
	$375 million	Cincinnati International 3	Reconstruction
	$ 5.3 billion	Denver International	New airport
	$ 1 billion	San Francisco International	New terminal
	$ 4 billion	Boston	New tunnel
	$ 2 billion	Washington	New, expanded terminals
	$ 1.3 billion	Philadelphia	Expansion program
	$350 million	Dallas	New runway
	$690 million	Pittsburgh	New terminal
Vietnam	$200 million	Hanoi (Noi Bai International)	Upgrading
	$200 million	Ho Chi Minh City	New terminal

space, are grievous noise polluters, and contribute to urban sprawl by encouraging dispersed land-use patterns. Sometimes public preferences conflict; the desire to have an airport within reasonably close driving distance might be antithetical to the desire to lie in one's backyard hammock unmolested by the blast of jet engines. Given these characteristics and their high visibility—many people are affected by an airport's location—news that a community is considering new construction or even expansion quickly becomes politically controversial. The pattern of controversy is, of course, profoundly influenced by the culture and the nature of the political system.

Under these conditions, it is not surprising that airport capacity often remains an unresolved issue. The current worldwide boom in airport construction does not mean that the problem of capacity provision will be solved easily. In fact, attempts to build new facilities or expand them tend to arouse heated opposition that forces long delays or even outright abandonment. Munich Airport, for example, took more than 15 years to complete; London's third airport took even longer.

This problem is especially evident in the United States. Though the United States has built more than 16,000 mostly small, local, and private airports, a clear need exists for enhanced commercial capacity at the major cities. Given the phenomenal growth of domestic and international aviation and the fact that the United States is the world's largest aviation market, it is surprising that no new airports had been built there since Dallas/Fort Worth in 1974, more than two decades ago, even though numerous cities, such as Chicago, Boston, Philadelphia, Seattle, Miami, and San Diego, tried to do so. This hiatus in new airport construction is even more alarming considering that the total number of U.S. air passengers increased approximately two and a half times from 1974 to 1994 (FAA 1974, 1994). At the same time, serious congestion and delays have been experienced throughout the air traffic system, and numerous cities have continued to encounter serious capacity limitations. Although some cities, such as Atlanta and Pittsburgh, were able to renovate and expand their existing facilities, only one U.S. city built a major new airport during this 20-year period—Denver, Colorado.

How did Denver accomplish that which so many other cities aspire to? What exactly did Denver accomplish? What kind of airport did it build, what technologies did it incorporate, and how well have these worked? How did Denver find ample real estate within 25 miles of its central business district? How was it able to assemble billions of dollars in private and public capital to finance one of the largest public works projects in the United States? How was the vision of a new airport sold to the political constituencies, the business community, the press, and ultimately, the public? What lessons emerge from this experience that can be of use to airport planners and designers throughout the world who are struggling with the need to expand capacity? These are the fundamental questions to be addressed in this book. To provide perspective, we begin with a brief sketch of its evolution.

The birth of Denver International Airport

The decision to build Denver International Airport was part of a process that dates back at least to 1974. Stapleton International Airport (Fig. 1-1), owned and operated by the City and County of Denver (hereinafter City of Denver), had undergone periodic expansions to accommodate increased air traffic since its opening in 1929. By the mid-1970s, local planners agreed that Stapleton, which had grown to become one of the busiest airports in the world, had to be expanded again to meet projected future demand. Furthermore, both pairs of runways were located so close together that, during periods of inclement weather, only one could be used, creating delays at Stapleton and many other airports across the nation (Szyliowicz and Goetz 1995).

With the outward expansion of the Denver metropolitan area, Stapleton's location, only 7 miles from downtown (Fig. 1-2), was very convenient for most people. Stapleton, however, had become bounded on two sides by residential communities (whose inhabitants later started complaining of noise pollution from the airport), and on a third by expanding commercial development. To its north, the U.S. Army manufactured and stored nerve gas and its by-products at the Rocky Mountain Arsenal, which was eventually shut down. After decades of chemical weapons production and inadequate pollution control, the Arsenal had become one of the most contaminated parcels of land in the country—a condition which led to its being designated a Superfund site.[2] North, east, and west of the Arsenal lie various communities that form part of neighboring

2. A Superfund site is a heavily polluted location that has been accorded a priority for cleanup. For a description of early Arsenal cleanup plans, see Wiley and Rhodes (1987).

Fig. 1-1. *Stapleton International Airport had served Denver since the 1920s but had insufficient capacity for projected demand.*

Adams County. These geographic factors were to greatly influence future developments. (See Fig. 6-2.)

In 1974, the Denver Regional Council of Governments (DRCOG) recommended that a Regional Airport Systems Plan be developed around Stapleton. Two years later, the City of Denver hired the consulting firm of Peat Marwick Mitchell to prepare a master plan for Stapleton's expansion. In September 1978, the Denver Chamber of Commerce convened a special Airport Task Force. In January 1979, DRCOG began formal inquiries with the Army over whether it would approve the airport's expansion onto the Arsenal. Adams County officials reacted negatively because they had their own plans for the 24 square miles of land on which the Arsenal was sited.

The U.S. Federal Aviation Administration (FAA) strongly supported efforts to increase airport capacity in Denver either through expansion or new construction and agreed to pay 75 percent of the costs of a more comprehensive, full-scale site study that DRCOG (whose members include representatives from all the regional cities and local governments, including Adams County) was to undertake beginning in 1979. Concerned with retaining control over the future of its

Fig. 1-2. *View of the Denver skyline from Stapleton International Airport.*

airport, the City of Denver insisted that two Denver city council members be added to DRCOG's executive committee. Concomitantly, the City of Denver was completing work on its own master plan, and in June 1982, Mayor Bill McNichols announced a $1.4 billion Stapleton expansion plan that called for building new runways on the Arsenal. In March 1983, the city hired a Washington law firm to handle negotiations with the Army over the Arsenal land.

Meanwhile, the DRCOG study had analyzed six possible sites for a new facility as part of its alternatives analysis. It concluded that expansion was the cheapest alternative but that if Arsenal cleanup expenses were included, an entirely new airport could be built on a site 40 miles east of Denver for the same price. Confronted with a difficult choice, DRCOG concluded that the decision was a political, not a technical, one and chose expansion on June 20, 1983, by a vote of 28 to 11 (*Denver Post* 1983).

Adams County, which had led the opposition to expanding Stapleton onto the Arsenal, continued to work to prevent the plan's implementation. It had retained its own Washington law firm and threatened to fight expansion in the courts. It could mobilize three powerful arguments—violations of federal regulations concerning aircraft noise, the additional environmental risks that would be associated with the construction and operation of airport facilities on severely contaminated land, and safety due to the proximity of the proposed runways to existing development.

Also in 1983, Mayor Bill McNichols was defeated in his bid for re-election by a young, former state legislator, Federico Peña. In February 1983, Peña strongly criticized the notion of building a new airport: "In terms of access, convenience, and land-use impacts, development of a new regional airport represents an inferior choice. . . . At present, the commitment and financial resources required to build such a facility do not exist." He believed expansion of Stapleton onto the Arsenal "represent[ed] the best long-term option available" (Kowalski 1994).

Shortly after Peña's inauguration in June 1983, the new administration decided to carry out an internal review to reconsider all aspects of the airport project. The review questioned the viability of the expansion because of such factors as the difficulties presented by Adams County's adamant opposition, the feasibility and costs of toxic cleanups (an estimated $6 billion in 1982), and the risks of future litigation related to the Arsenal's Superfund designation and to problems of increased noise pollution on surrounding residential areas. Was the expansion simply a short-term solution that would not meet future needs? Was another solution possible, one that would promote Denver's interests and be acceptable to Adams County?

Negotiations were begun in February 1984, and by January 1985, Denver and Adams County officials announced that they had signed a Memorandum of Understanding (MOU) whereby Adams County would cede (pending voter approval) approximately 20 square miles of virtually uninhabited land east of the Arsenal on which Denver would build a new airport, and Stapleton would be closed.

This agreement met with a mixed reception. The local business community supported it enthusiastically. On the other hand, many local citizens were not convinced of the need for a new facility, and the airlines were unenthusiastic about a new airport because their costs of operation in Denver would undoubtedly rise.

The role of the airlines was crucial because they were the organizations that would bear most of the financial burden of the new facility. Traditionally, airlines had always played an important part in airport development decisions, but the Airline Deregulation Act of

1978 brought about important changes in the relationship between airlines and airports. Whereas airlines had been subject to the Civil Aeronautics Board (CAB) routing and pricing regulations, under deregulation they were free to act on the basis of their perceptions of market demand. Thus airlines could serve or abandon any routes at any price that pleased them, given gate access and the FAA's safety regulations. As a result, hub and spoke networks emerged. Airlines could concentrate banks of flights from various departure points into a central transferring facility, enabling a single airline to service a wider range of markets than had been possible with the previous linear route system. Some airports lost their importance and became peripheral spokes, while others emerged as major connecting hubs. Increasingly, individual airlines came to dominate key airports by creating "fortress hubs" (e.g., TWA in St. Louis, Northwest in Minneapolis, and USAir in Pittsburgh).

Denver, however, was initially in the unique and fortunate position of serving as a hub for three airlines—Frontier, Continental, and United—which brought it unprecedented low fares and a sharp increase in passenger traffic. In 1980, the number of total passengers arriving, connecting, or departing in Denver had expanded to more than 20 million, and by 1986, it had jumped to more than 34 million. These developments were perceived by Denver planners, consultants, and the FAA as underscoring the need for a new airport, since Stapleton was experiencing serious congestion and delay problems under the increased load. Furthermore, their projections suggested that the situation would worsen dramatically, and that Stapleton, being unable to handle the projected traffic load by the early 1990s, would become a serious bottleneck in the national air-traffic system. Although these projections were questioned by some analysts, they remained sacrosanct with the FAA and the City of Denver.

By the late 1980s, however, Frontier had disappeared (absorbed by People Express, which, in turn, was consumed by Texas Air's Continental Airlines), Continental was suffering severe financial and organizational instability, and passenger enplanements at Denver had actually started to decline. Although both Continental and United representatives participated in early planning for the new airport, the airlines were now hostile to the idea of a new facility. With Frontier gone, United and Continental controlled the Denver market and feared that the new facility, with its massive expansion of capacity, would allow other airlines to enter their market.

From the airlines' perspective, the existing airport facility was adequate, though they desired some additional improvements that Denver agreed to undertake in exchange for support of the new airport. By the summer of 1987, tensions were so high that the airlines flatly stated that the new airport was not needed and stopped paying some of their taxes. The City of Denver responded by halting work at Stapleton on the airlines' projects and reaffirming its commitment to the new project.

Meanwhile, in September 1986, the Colorado Forum produced an economic impact study that attributed 10 percent of the state's earned income, 21,000 direct jobs, and 140,000 indirect jobs to Stapleton International Airport. It estimated a new, replacement airport would generate another 90,000 new jobs and would require 10,000 construction workers to build (Albin 1994). It also projected the new airport would generate $8.2 billion annually in business revenue by the year 2010 and $206 million in state and local taxes. Planners then spoke of an airport with more than 100 gates (Kowalski 1994).

Building a new facility, however, was possible only if Adams County voters approved the annexation by Denver of land for a new airport in the election scheduled for May 17, 1988. The campaign was heated due to the City of Denver's unpopularity with Adams County voters and opposition by many groups to various aspects of the agreement. The proponents, however, marshaled impressive resources. Led by Governor Roy Romer (who campaigned actively via the breakfast "oatmeal circuit"[3]), the entire Colorado Congressional delegation, and an organization of remarkable effectiveness (funded by a war chest of more than $1.5 million), they won 56 percent of the vote.

Although most of the legal obstacles had been overcome and the planning process was well under way, Denver residents began to voice opposition to the project. Critics were demanding that the city hold its own referendum on the airport issue, and they eventually formed into two groups. The first, "Vote NO," organized in February 1989, consisted of persons who were not convinced that the new airport was needed or that it would promote Denver's development. They noted that the projections of future growth at Stapleton were

3. Governor Romer made about 50 early-morning visits to Adams County diners and restaurants to convince local residents to vote for the annexation that would permit Denver to build a new airport.

not being met because the rate of increase had slowed dramatically and developed analyses that purported to show that the project was not financially viable. They argued that the people of Denver should have the right to pass judgment on the largest public works project ever proposed in Colorado. The second, "Save Our Stapleton," was composed of owners of hotels around the existing airport who would obviously suffer if Stapleton were replaced by a new airport, planning for which included at least one new hotel on site.

The proponents of the airport were confronted with a difficult decision: whether to hold an election, which was not legally required, and run the risk of seeing the project defeated, or to just push ahead with the new airport. Mayor Peña and his advisors perceived the risk of losing to be high, and even though the vote would only be advisory, a negative outcome would make it difficult for the city to proceed with the project. Still, they decided that the mayor, who had been elected on a populist platform, could not deny citizens the right to vote on the largest public works project in the city's history.

The election was held on May 16, 1989. The supporters of the project organized with the same effectiveness that had characterized their campaign in the Adams County vote. Since they included the political establishment, the business community, and the media, they had access to ample resources and the skills to use them effectively. The opposition, on the other hand, was weak, divided, and possessed only limited resources. The result was an overwhelming victory (63 percent) for the new airport.

The election had important repercussions in Washington, D.C. Now Congress recognized that the project had solid local endorsement, and, with strong support from Transportation Secretary Sam Skinner, who wanted a national demonstration project, as well as much effort by the state's congressional delegation, they agreed to help finance the project in the amount of a $500 million Airport and Airways Trust Fund grant. By late summer 1989, the first federal funding installment ($60 million) was received, the FAA approved the final environmental impact statement, and groundbreaking on the project occurred on September 28, 1989 (Fig. 1-3). The major part of the funding, however, would still have to come from the sale of bonds, which, upon completion, amounted to more than $4 billion sold since the first $704 million issue on May 8, 1990 (Kowalski 1994).

For the bonds to sell, however, the project had to be financially viable. This meant that Denver had to secure the cooperation of the two major hubbing airlines, most importantly by persuading them to sign agreements for significant gate utilization. This led to difficult negotiations, because the airlines were still not convinced that a new airport would serve their interests. Although faced with a difficult decision—to halt the project or to continue building it with no guarantee that it would ever be financially viable—Denver's policymakers decided to proceed. They believed that Denver was such a strategic location that Continental and United would have to agree, eventually, to use the new facility. And, if one of them chose to move their hub operations elsewhere, other airlines would seize the opportunity to establish a hub in Denver, or so they believed.

Shortly thereafter, Denver's gamble seemed to pay off, though at a very high cost. Hoping to increase its market share, Continental Airlines decided to support the project and signed an agreement in March 1990 to lease 30 gates for five years at the new airport. In return, it secured concessions from the city valued at $58 million, as well as several modifications, the most important of which was a major change in the design—a pedestrian bridge from the terminal to its concourse so that its passengers would not have to use the

Fig. 1-3. *Construction of Denver International Airport began in 1989. The new airport was completed in 1995.*

underground train. United Airlines finally reached an agreement with the city in June 1991, committing itself to 45 gates in return for even greater design changes and concessions that totaled $204 million. Moreover, it was agreed that the airlines would pay the city no more than $20 per enplaned passenger (in 1991 dollars) to use the new facility.

United's other demands, however, would profoundly affect the project's implementation. In addition to various concourse modifications, United insisted on a fully automated, high-speed baggage system that would move the passengers' luggage rapidly between the terminal and its concourse through a system of tunnels. A few weeks later, the city decided that, rather than having separate baggage systems for each airline, the automated system should be extended to the entire airport (Russell 1994).

By 1990, further design changes in the new airport were required due to the realization that passenger totals at Stapleton had stopped growing and had actually been declining since 1986. As late as 1987, the FAA was still predicting that by 2000, Denver would have the second busiest airport in the nation, and city consultants were predicting DIA would soon be serving 100 million passengers annually (Kowalski 1994). But these incredibly rosy forecasts belied the reality of four straight years of declining traffic in the late 1980s. In recognition of these declines, Mayor Peña decided to scale back the new airport, originally from 120 gates to 94, then later from 94 gates to 85, while the number of runways was cut from six to five (*Rocky Mountain News* 1991). This necessary downsizing caused considerable consternation among DIA bond holders and in the bond markets. In March 1991, Standard & Poor dropped its DIA bond rating to BBB-minus, one grade above "junk" status (Flynn 1991).

The cargo airlines, such as Federal Express, were also dismayed with the airport because their proposed facilities were to be located on the north side of the airport, away from the main entrance and the major connecting highways. Accordingly, several decided to abandon DIA and base their operations at another regional airport. This was a serious blow to the financial viability of DIA because cargo shipments had been growing both consistently and dramatically and were expected to provide a significant share of DIA's revenues. Finally, after long and difficult negotiations involving the city, the carriers, and the FAA/DOT, the carriers agreed to use DIA, and the city moved the cargo facilities to the more-accessible south side of the airport. Once again a significant redesign was required.

Even the rental-car companies demanded changes. The original design called for the cars to be parked in the nearby parking garages. The companies complained about the high costs involved, and eventually the city agreed to a shuttle bus system that would carry the passengers to a remote parking area (Russell 1994).

Such developments, which added appreciably to the cost of the project, did little to assuage several local critics. The project did not, however, play a major role in the 1991 mayoral elections. Federico Peña had decided earlier not to run for reelection, and Wellington Webb was elected mayor. Peña would go on to become President Clinton's Secretary of Transportation. DIA was held out as a major reason for his selection. Though some consideration was given to slowing down or postponing the airport, the new administration decided to proceed with the original fast-track schedule, a decision that produced serious problems, primarily because of the problems created by the automated baggage system.

The baggage system, the cost of which had by now escalated to about $200 million, was to be the largest and most sophisticated in the world. It was designed to move 700 bags per minute to their specific load points in less than 10 minutes through a system of 4000 individual carts traveling at speeds up to 24 miles an hour over 17 miles of track suspended from basement ceilings. The system, which included 5000 electric motors, 2700 photocells, 59 bar-code reader stations, 311 radio frequency readers, and more than 150 computers, workstations, and communication servers, was expected to have all of its carts operating simultaneously during peak hours (GAO 1994; Rifkin 1994).

Unfortunately, this system, which involved state-of-the-art technology, proved to be a disaster. It encountered numerous mechanical and software problems so that, when tests were run on a small loop of the system, the result was misloaded bags, jammed carts, spilled luggage, and general chaos. The repeated failures of the baggage system led directly to several postponements of the airport's opening from October 28, 1993, to December 19, 1993; to March 9, 1994; and then to May 15, 1994. By May 1994, with negligible progress made on fixing the system, Mayor Webb announced that the airport would be delayed indefinitely, with no new opening date set, until the problems could be completely addressed. During the summer of 1994, intensive baggage system discussions involving United, the city, BAE (the system designer), and external consultants were held.

As a result of these discussions, it was decided in September 1994 to drastically simplify the automated system and to build a $50 million traditional system to serve as a backup. Work on completing the automated system would be focused on United's Concourse B at first and, once that part was operational, the system would be expanded to Concourses A and C. To serve these other concourses, an alternative system of conveyor belts, traditional tugs and carts, and manual operations would be implemented to enable the airport to open on February 28, 1995. This decision led to major protests by the other airlines, especially those located at Concourse C, the farthest from the terminal, which feared that United would gain a significant competitive advantage since other airlines' baggage would be delivered much more slowly. Nevertheless, Mayor Webb decided that this was the surest path to getting the airport opened as soon as possible.

Another major headache confronted the city in 1994 as Continental Airlines announced it was dismantling its Denver hub operations and shifting its fleets and crew to its new CALite service. Although claiming the decision was not airport-related, Continental's announcement in March 1994 came on the heels of yet another delay of the opening of the airport and contributed to a further downgrading of airport bonds to junk status by May 1994. DIA, originally predicated on the need to serve three hubbing airlines, was now down to just one, United, which was itself restructuring through an employee buyout, while other carriers in the financially strapped airline industry were showing very little interest in the Denver hub.

At this point, the success of DIA hinged largely on one hubbing airline, contradicting previous optimistic pronouncements by airport supporters that other carriers would savor the opportunity to hub in Denver. The FAA had earlier based its traffic projections on the likelihood that American Airlines would establish a major hub operation at DIA by 1995 (Peat Marwick Mitchell 1988). Later, while assessing the possibility of hubbing in Denver, American Airlines' CEO Robert Crandall wryly referred to DIA as a "field of dreams" and denounced the "build it and they will come" approach to airport planning exemplified in the Denver case.

Altogether, the problems encountered at DIA, especially those with the baggage system, caused four postponements of the airport's opening and resulted in substantial delays and cost overruns. The troubled automated baggage system's cost escalated to $360 million (from an original estimate of $193 million), the retrofitted backup system added another $50 million, and the four delays resulted in millions of dollars in interest payments due before the airport could generate any income. It has been estimated that at least $460 million in additional, unnecessary expenditures were required to open the airport, thus pushing the overall cost of the project, at the time, to well over $4 billion, compared to the original estimate in 1988 of $1.7 billion (Flynn 1995). The airport finally opened on February 28, 1995, but it has continued to make headlines as it has encountered various difficulties with its radar, people-moving, and baggage-handling systems.

A year after the airport opened, at least a dozen airport investigations were under way. The Denver District Attorney was investigating allegations of shoddy workmanship, minority contracting irregularities, and cronyism in the awarding of contracts. The Securities and Exchange Commission (SEC) was investigating whether the city disclosed sufficient details about DIA's delays and other problems to prospective bond purchasers. It also was investigating the relationship between bond underwriters' campaign contributions and their municipal contracts. The FAA was investigating $402,000 the city spent defending its affirmative action program for minority contractors. The U.S. Department of Transportation was investigating the city's management of the project (*Rocky Mountain News* 1995). Several civil lawsuits against the city were also pending; however, as discussed in Chap. 9, federal and local investigations were terminated after failing to produce any indictments.

Although the full story of how DIA was planned and implemented probably will never be told, our goal is not to write a detailed history of the project but to analyze the process on the basis of the best information available, in a scholarly manner. Furthermore, we seek to do so by placing the DIA story within a larger theoretical context. Only in this way can an adequate explanation be obtained. Understanding a single case may be of interest to many, but as social scientists we seek to derive generalizations of theoretical import. Only in this way can the lessons derived from the DIA experience be of wide relevance.

The megaproject syndrome

As we have seen, DIA grossly exceeded its original budget, opened more than a year late, and began operations at passenger and flight-frequency levels far lower than those anticipated. It encountered passionate opposition, and its implementation encountered serious difficulties. Such an outcome is not surprising—in fact, it is commonplace when dealing with megaprojects. These projects possess the following characteristics—they are very expensive, very large, and very complex. Such projects have historically encountered problems that lead to cost overruns, delayed openings, financial difficulties, and an inability to meet original objectives. They can be found in all countries and all sectors.[4]

Megaprojects can be divided into two categories. The first are those that comprise large projects that turn out to be either totally unfeasible or produce catastrophic consequences. Specific examples include the atomic airplane, the space shuttle *Challenger*, and the Chernobyl nuclear power plant disaster. A second category includes large projects that are often labeled "white elephants"—projects that encounter significant difficulties and whose environmental, financial, and other costs turn out to be far higher than necessary. Faulty planning leads to significant problems. But the problems can be overcome, leading some projects to become, in time, a praised icon and a valued success. Hirschman regarded this phenomenon as inevitable and as evidence of human creativity (Hirschman 1967). Examples include the San Francisco Bay Area Rapid Transit System (BART), the Sydney Opera House, the Aswan Dam, and perhaps one day, the Chunnel.

Neither category necessarily enhances human welfare. Of course, manifest disasters are to be avoided if at all possible. The "white elephant" may be less costly in terms of human suffering, but it imposes heavy, unnecessary costs on communities. Although the additional burdens may be affordable, at best, scarce resources are wasted; at worst, the project becomes a millstone that weighs heavily on the community for years.

Such projects do little to promote sustainable development. This concept, which has gained widespread acceptance, blends environmental, economic, and social concerns in an ethically based way. It

4. See Hall (1982), Hirschman (1967), Skamris and Flyvberg (1996), and Szyliowicz (1991).

was originally defined as "development that meets the needs of the present without compromising the ability of future generations to meet their own needs" (World Commission on Environment and Development 1987). Transportation is an important element in achieving sustainable development, and all its modes, including aviation and its infrastructure, have been increasingly subject to analyses from this perspective (Transport Canada 1996).

Megaprojects generally are problematic in terms of sustainable development. Being a megaproject by any definition—size, complexity, scale, and impacts—airport development must be viewed in a new way, since it is obvious that the narrow traditional approach has not created the kind of system, nationally and internationally, that best meets current and future needs.

Planning and decision-making

Many scholars have attempted to analyze the megaproject phenomenon and to explain why project outcomes are so seldom satisfactory. They have identified such factors as size, technological complexity, uncertainties, and the lack of appreciation of the local environment.[5] What remains unclear is why planners and decision-makers prove unable to design, manage, and implement large projects given these realities. To answer this question, it is necessary to consider the dominant approach to planning, especially in transportation.

The decision to build DIA and the planning process responsible for its realization are typical of the ways in which transportation projects generally and airports specifically are designed and implemented. The process is based on the rational comprehensive model, which has its roots in the more scientific approach to public administration developed in the early 20th Century (Wachs 1985). Later, this approach was institutionalized within the rapidly growing field of urban planning and became the accepted paradigm for planners in all sectors. The postwar trend toward greater quantification and precision in planning through data collection, statistical manipulation, and the emergence of new powerful computer applications contributed to the appeal of the rational comprehensive model. New methodologies reflected in more sophisticated forms of cost-benefit analysis, technology assessment, risk and forecasting analysis, program

5. See, in addition to the sources cited in footnote 4, Feldman (1985), Murphy (1983), Schulman (1980), and Steinberg (1985).

management techniques, and environmental impact statements were all based on the rational comprehensive approach.

Its impact upon planning and policy-making has been immense because it provides a systematic framework within which problems can be analyzed in a way that produces optimal results. Specific steps include the identification of the problem, value clarification and goal setting, generation of alternatives, analysis of the consequences of each alternative, and selection of the best alternative on the basis of objective criteria. Sometimes monitoring and evaluation are included. To implement such a process, it is necessary to gather all relevant information—the essence of the model is its comprehensiveness—and attempt to identify the optimal solution on the basis of exhaustive analyses of available options.

The rational comprehensive model has become the dominant paradigm among planners and engineers because of its clearly defined steps and its emphasis upon optimality. To a considerable degree, this has been the case for transportation planning generally (Wachs 1985) and for airport planning specifically. Although the discussion that follows focuses on the U.S. pattern, we must emphasize that this pattern has generally been followed throughout the world.

In the early years of aviation, airport planning in the United States was conducted predominantly by local and state governmental entities (Ashford and Wright 1992). In fact, the Air Commerce Act of 1926 prohibited the federal government from constructing or operating airfields and airports, though federal aid for airport building has been available since 1933 (Sampson et al. 1990). But as a result of the growth in importance of civil aviation to the nation in subsequent decades, the federal government became increasingly involved in establishing guidelines and supporting airport planning activities. By the time deregulation started in 1978, the federal government, in the form of agencies such as the FAA, and local governments (or regional authorities) had become the key participants in matters of airport planning.

Today, planning is guided by the FAA's specification of the airport master planning process. This process, clearly a "rational" model, dominates airport planning around the world because ICAO's recommendations closely mirror those of the FAA.[6] The FAA has for many years published specific prescriptions that mandate the application of a rational comprehensive approach. The master plan is expected ". . . to provide guidelines for future airport development which will satisfy aviation demand in a financially feasible manner, while at the same time resolving the aviation, environmental, and socioeconomic issues existing in the community" (Ashford and Wright 1992). These rather ambitious goals are to be achieved through the application of a rational planning process with the following dimensions:

1. Organization and preplanning
2. Inventory of existing conditions and issues
3. Aviation demand forecasts
4. Requirements analysis and concepts development
5. Airport site selection
6. Environmental procedures and analysis
7. Simulation
8. Airport plans
9. Plan implementation

These elements are combined in a sequential process as illustrated in Fig. 1-4 (FAA 1985).

Many elements of this model can be criticized. On a general theoretical level, it contains all the weaknesses inherent in the rational comprehensive model. These can be summarized as follows: (1) its informational requirements are unrealistic; it is simply impossible to expect planners and decision-makers to have access to all the necessary data; (2) seldom are all the value preferences known or specific goals and objectives agreed upon; (3) it ignores the role of power and other political variables that are present in every megaproject; and (4) it assumes the existence of a powerful unitary actor. On the applied level, it makes the assumption that it is possible to accurately forecast demand projections, an assumption which has been widely discredited (de Neufville 1976, 1991). We shall return to this point in Chapter 2 when we consider the forecasts that were made in the case of DIA. Equally troubling is the naive way in which critical socioeconomic and political variables are treated.

6. For an English perspective, see Lewis (1990).

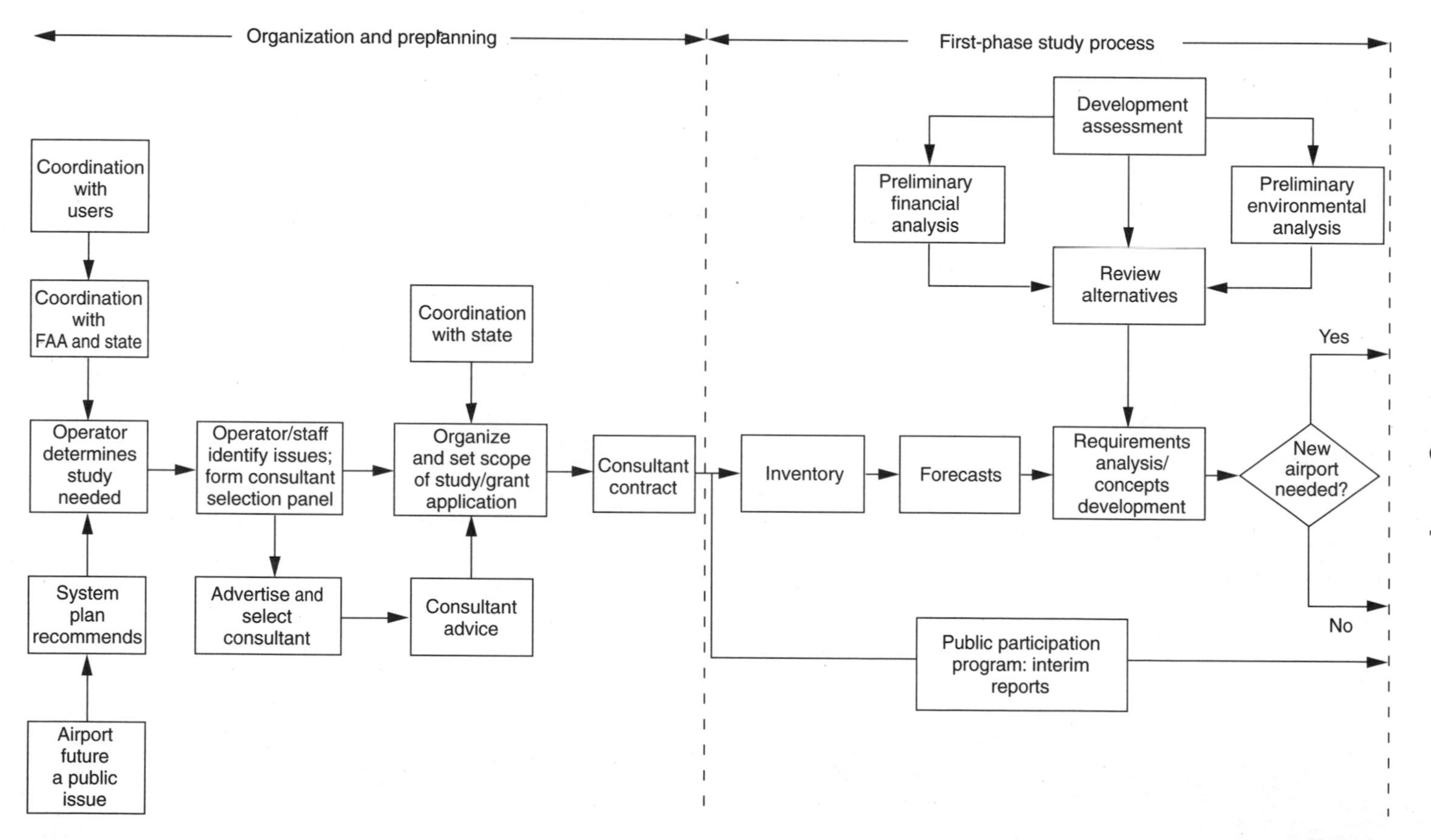
Organization and preplanning
First-phase study process
Coordination with users
Coordination with FAA and state
Operator determines study needed
System plan recommends
Airport future a public issue
Operator/staff identify issues; form consultant selection panel
Advertise and select consultant
Coordination with state
Organize and set scope of study/grant application
Consultant advice
Consultant contract
Inventory
Forecasts
Requirements analysis/ concepts development
Public participation program: interim reports
Development assessment
Preliminary financial analysis
Preliminary environmental analysis
Review alternatives
New airport needed?
Yes
No

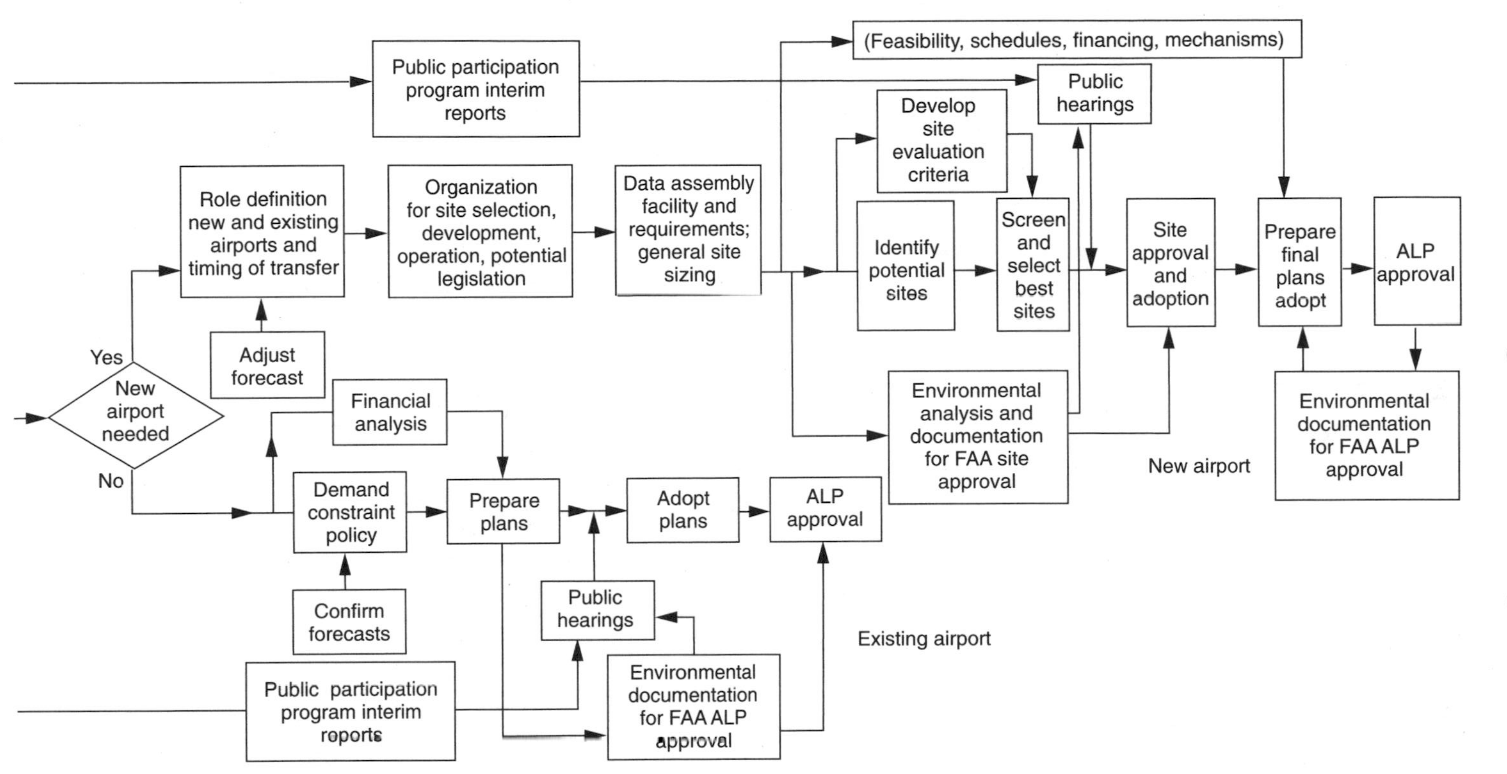

Fig. 1-4. *The Federal Aviation Administration's airport master planning process.*

These are key elements in decisions involving airport development, but they are treated as externalities by this approach. As a result, airport projects encounter numerous problems that might be avoided with an approach that explicitly considers these other variables. It is no exaggeration to suggest that the existing approach is "doomed to failure" (de Neufville 1991).

Some of DIA's planners were aware of the deficiencies in the FAA's approach and deviated from it in various ways—especially in terms of the environmental assessment and the pattern of public participation—but that approach still permeates current practice in airport planning. In the remainder of this book we shall consider how the process actually worked at DIA, the extent to which the criticisms mentioned are relevant, the available theories, and the ways, if any, in which it can be improved.

The related literature

Unfortunately, little help is available from the literature on airport planning. Although the literature dealing with the rational actor model and other planning and decision-making approaches is extremely rich and there is a valuable body of work on megaprojects, the topic of why and how airports are designed and implemented has received limited attention. With one significant exception, the available works are overwhelmingly technical in their orientation. Richard de Neufville has clearly identified the inherent difficulties of applying a rational approach to airport planning and has been an early advocate of the need to adopt a "strategic approach" (de Neufville 1976, 1991). No books, however, analyze the planning and implementation of a major airport. No studies analyze what happened, and why, at Dallas/Fort Worth International Airport (1974), Pittsburgh International Airport (reconfigured in 1992), Munich Flughafen (1992), or Osaka International Airport (1994).

The vast majority of works are textbooks and manuals that deal with general principles of airport planning, administration, design, regulation, security, and finance. Some works in this category do deal with specific airports (Allen 1979; Hillestad et al. 1993; Kaplan 1994; Okabe and Tomuro 1994; and Powell 1992). None, however, deal analytically with how the decisions to build the facility were made and implemented nor provide detailed information on the airport planning process.

A few scholars have been concerned with various political aspects of airport planning, primarily in the 1970s (Bromhead 1973; Budden and Eernst 1973; Nelkin 1974). Since then, however, only two scholars dealt with this topic, and their work deserves widespread recognition. Elliott J. Feldman and Jerome Milch have focused explicitly upon the relationship between technology and democracy, using airport development as a focus, both in Canada (1983) and in a major cross-national study of eight airports in five countries (Feldman and Milch 1982). Their work sets a high standard for comparative analyses. It demonstrates clearly the degree to which economic interests drove the push to solve the capacity problem by building new airports in all countries. We shall discuss the degree to which this finding applies to the DIA case in Chapter 3.

Our intention is to fill what appears to be a major gap in the literature by addressing both political and technical factors within the framework of a detailed case study which is sensitive to contemporary international developments. Accordingly, we shall, under the heading "Lessons learned" at the end of each chapter, draw out the generalizations that are relevant for airport planners and decisionmakers everywhere.

References

Albin, Robert. 1994. Building Denver International Airport was a 15-year obstacle course. *Denver Post.* February 26.

Allen, Roy. 1979. *Great Airports of the World.* New York: Charles Scribner's Sons.

Ashford, Norman and Paul H. Wright. 1992. *Airport Engineering,* 3d ed. New York: John Wiley & Sons, Inc.

Bromhead, Peter. 1973. *The Great White Elephant of Maplin Sands.* London: Elek.

Budden, Sandra and J. A. Eernst. 1973. *The Movable Airport: The Politics of Government Planning.* Toronto: Hakkert.

de Neufville, Richard. 1976. *Airport Systems Planning.* Cambridge, MA: MIT Press.

————. 1991. Strategic planning for airport capacity: An appreciation of Australia's process for Sydney. *Australian Planner* 29:4

Denver Post. 1983. DRCOG voted 28–11. June 20.

Federal Aviation Administration (FAA). 1974, 1994. Airport Activity Statistics.

————. 1985. Airport master plans. Advisory Circular AC 150/5070-6A.

Feldman, Elliot J. 1985. *Patterns of Failure in Government Megaprojects.* In Samuel Huntington and Joseph Nye, Jr. Boston: The Center for International Affairs and University Press of America.

Feldman, Eliott J. and Jerome Milch. 1982. *Technocracy Versus Democracy: The Comparative Politics of International Airports.* Boston: Auburn House Publishing Company.

————. 1983. *The Politics of Canadian Airport Development.* Boston: Auburn House Publishing Company.

Flynn, Kevin. 1991. Planners: It'll be cheaper to finish airport. *Rocky Mountain News.* March 17.

————. 1995. Who botched the airport baggage system? *Rocky Mountain News.* January 29.

U.S. General Accounting Office (GAO). 1994. New Denver airport: Impact of the delayed baggage system. Washington, DC: GAO.

Hall, Peter. 1982. *Great Planning Disasters.* Berkeley, CA: University of California Press.

Hill, Leonard. 1995. Nordic newcomer. *Air Transport World.* February 1:80.

Hillestad, Richard, John Owen, and Donald Blumenthal. 1993. *Airport Growth and Safety: The Schiphol Project.* Santa Monica, CA: Rand.

Hirschman, Albert O. 1967. *Development Projects Observed.* Washington, DC: Brookings Institution.

Kaplan, James. 1994. *The Airport: Terminal Nights and Runway Days at John F. Kennedy International Airport.* New York: William Morrow & Co.

Kowalski, Robert. 1994. Turbulence marks DIA history. *Denver Post.* March.

Lewis, Nick. 1990. Airport planning. *The Planner,* August 24:13-15.

Liden, John. 1995. Southern China sees an airport boom. *The International Herald Tribune.* June 12.

Mecham, Michael. 1994. Growth outpaces Asian airports. *Aviation Week & Space Technology.* August 29:54.

Meredith, John. 1995. Room to boom. *Airline Business.* January:38.

Murphy, Kathleen J. 1983. *MacroProject Development in the Third World.* Boulder, CO: Westview Press.

Neilan, Edward. 1994. Turbulence ahead for Asia's new designer airports. *Tokyo Business.* August:30.

Nelkin, Dorothy. 1974. *Jetport: The Boston Airport Controversy.* New Brunswick, NJ: Transaction Books.

Okabe, N. and T. Tomuro. 1994. *Kansai International Airport Passenger Terminal Building.* Tokyo: Process Architecture Co.

Peat Marwick Mitchell. 1988. Report on Denver International Airport.

Powell, Kenneth. 1992. *Stansted: Norman Foster and the Architecture of Flight.* London: Fourth Estate.

Rifkin, Glenn. August 29, 1994. What really happened at Denver's airport? *Forbes.*

Rocky Mountain News. 1991. Chronology. March 17.

————. 1995. An update on DIA inquiries. January 29.

Russell, James S. 1994. Is this any way to build an airport? *Architectural Record.* November.

Sampson, Roy J., Martin T. Farris, and David L. Schrock. 1990. *Domestic Transportation: Theory and Practice*, 6th Edition. Boston: Houghton Mifflin Company.

Schulman, Paul. 1980. *Large Scale Policy Making.* New York: Elsevier.

Skamris, M. K. and B. Flyvberg. 1996. Accuracy of traffic forecasts and cost estimates on large transportation projects. Paper presented at Transportation Research Board annual meeting. Washington, D.C. January.

Steinberg, Gerald. 1985. Comparing technological risks in large scale national projects. *Policy Sciences.*

Szyliowicz, Joseph S. 1991. *Politics, Technology, and Development.* St. Antony's College Series. London: McMillan Press Ltd.

———— and A. R. Goetz. 1995. Getting realistic about megaproject planning: The case of the new Denver International Airport. *Policy Sciences*, vol. 28. Norwell, MA: Kluwer Academic Publishers.

Transport Canada. 1996. *The Greening of Aviation.* Ottawa: Transport Canada.

Trautmann, Peter. 1993. The need for new airport infrastructure. Paper delivered before International Conference on Aviation and Airport Infrastructure, Denver, Colorado, December 5.

Vancouver International Airport Authority (VIAA). 1995. Expansion '96.

Wachs, M. 1985. Planning, organizations, and decision-making. *Transportation Research.* 19A:521–531.

Wiley, Karen B. and Steven L. Rhodes. 1987. The case of the Rocky Mountain Arsenal: Decontaminating federal facilities. *Environment.* April:16-20, 29-33.

World Commission on Environment and Development. 1987. *Our Common Future.* Oxford, England: Oxford University Press.

2

Three hubs to one: The folly of forecasting in the dynamic airline industry

"As repeated retrospective studies demonstrate, the long-term forecasts needed for master planning are 'always wrong.'"—RICHARD DE NEUFVILLE (1991) PROFESSOR AND CHAIR, TECHNOLOGY AND POLICY PROGRAM, MASSACHUSETTS INSTITUTE OF TECHNOLOGY

"[In promoting deregulation] I talked about the possibility that there might be really destructive competition, but I tended to dismiss it. And that certainly has been one of the unpleasant surprises of deregulation."—ALFRED KAHN FORMER CHAIR, CIVIL AERONAUTICS BOARD[1]

"If you build it, will they come?"—ANONYMOUS

The decision to build a new airport or to expand capacity is—or should be—driven by estimates of future demand. Forecasting is one of the key steps in a rational planning effort. Without some evidence that existing capacity will be severely constrained, no new airport can be built. But credible forecasts must be grounded in the realities of the contemporary airline industry, which has become especially volatile in today's deregulated environment.

The decision to build Denver International Airport was predicated on the assumption that it would serve as a hub facility for at least three major airlines and that passenger enplanements would continue to

1. Quoted in Velocci 1993.

grow exponentially. In the mid-1980s, this assumption seemed plausible to planners and consultants because United, Continental, and Frontier Airlines were all hubbing at Denver, and traffic at Stapleton had nearly doubled in the eight years from 1978 to 1986 to more than 17 million enplaned passengers.

But historically, the performance of planners and consultants in air-passenger forecasting has been notoriously poor. In the first two postwar decades, aviation forecasters consistently underpredicted actual passenger volumes, as tremendous growth in aviation during the 1950s and 1960s occurred as a result of unanticipated rapid technological advances. Since 1970, forecasters have generally overpredicted growth in traffic (Ashford and Wright 1992). Forecasting has become even more difficult in the wake of U.S. airline deregulation, international liberalization and globalization, and the increased volatility of passenger traffic that has resulted. In practical terms, forecasting has become more guesswork than science.

Still, politicians and airport planners cannot escape the need to estimate and prepare for future growth. The practice of forecasting itself must be improved for better airport planning. But more importantly, recognition and acknowledgment of the increased uncertainties in aviation forecasting must be built into the planning process so that the inevitable errors made in passenger forecasting will not result in disastrous consequences from poor or uninformed decision-making.

Aviation forecasting

Forecasting air travel demand is an extremely difficult task, but it is essential to make reasonable estimates of future capacity needs at airports. Not only do local governments and airport authorities need forecasts for airport master planning, but central governments, commercial airlines, and aircraft manufacturers rely on forecasts for their own planning purposes (Ashford and Wright 1992). Forecasters must compile and analyze a considerable amount of data, including historic trends in aircraft, passenger, and freight traffic volume for the region under study; population and economic growth characteristics of the region; trends in national and international traffic; geographic factors influencing transportation requirements; and airline industry dynamics, such as existence and degree of competition at an airport with respect to pricing, travel time, service frequency, and quality (Horonjeff and McKelvey 1994). Furthermore, it is no exaggeration to say that air traffic forecasting has become much more complex over time as recent deregulation and globalization trends in the airline industry have greatly exacerbated the volatility of air traffic patterns.

To meet these challenges, various forecasting techniques and methodologies have been developed. They include forecasting by judgment; trend projection and linear, exponential, and logistic curve extrapolation; market share models; and econometric models such as multiple regression or logit models for trip generation, trip distribution, and modal choice analysis (Horonjeff and McKelvey 1994). Although these have become increasingly sophisticated, aviation forecasting remains a very subjective process that can, and often does, result in widely differing forecasts depending on the assumptions and techniques that are used. Consultants, for instance, are hired by cities and local planning agencies to conduct specialized forecasts. These carefully crafted technical studies can produce persuasive justifications for building or not building projects based on decisions regarding which numbers are used and the type of assumptions made about the future. Obviously, ethical dilemmas permeate the entire consulting industry, and aviation forecasting is certainly not immune (Wachs 1990).

The U.S. Federal Aviation Administration (FAA) also conducts forecasting for the entire U.S. aviation system, involving national, hub, and terminal area level forecasts (Ashford and Wright 1992). The FAA applies a step-down process whereby aggregate national level enplanement forecasts, based on multiple regressions involving predicted levels of Gross Domestic Product (GDP), yields, and revenue passenger miles, are used to estimate hub and individual terminal area enplanements (Ashford and Wright 1992). The FAA acknowledges in its aviation forecasts for the fiscal years 1995 to 2006, however, that:

> *in general, [econometric] models and equations are simple portrayals of a complex system. They cannot account for a number of political, social, psychological, and economic variables and for all the interrelated actions and reactions that eventually lead to a particular set of results. It is particularly important, therefore, that the initial model results are reviewed, revised, and adjusted to reflect the analysts' best judgment of the impacts of the events occurring or expected to occur during the forecast period.*

Any forecast is therefore inherently subject to great uncertainty. Accordingly, observers have suggested that many different serious forecasts be generated and considered in any new or expanded airport planning exercise. Furthermore, because of such uncertainties, planning should be as flexible as possible to minimize risks associated with a highly volatile external environment. We discuss this topic in detail in Chapter 10.

Also problematic are the issues of capacity and delay. Capacity refers to the processing capability of a facility over a period of time (Horonjeff and McKelvey 1994). When capacities of airports or other facilities become strained by increasing demand, delays occur. Alternative concepts of capacity include "practical capacity," which corresponds to "reasonable" or "tolerable" levels of delay, and "ultimate" or "saturation" capacity, which is the maximum number of aircraft that can be handled given constant demand (Ashford and Wright 1992). Capacities for airports are typically calculated in units of operations (flight arrivals and departures) per hour. Factors that determine capacity include the number, layout, and design of the runway system; air traffic control procedures; characteristics of demand; and environmental conditions in the airport area. Since ultimate capacity depends on so many diverse factors, it is not always certain whether physical expansion of airport capacity is always the best solution to accommodating increasing demand. Add the problem of accurately forecasting demand, and costly capacity increases become ever more difficult. As we shall see, this problem is an international phenomenon—it is unfair to single out the FAA or any one organization for criticism.

Nevertheless, there are clear cases in which ultimate capacities of airports are being reached, and expansions or new airports are necessary given current and projected levels of demand. This situation is especially true in the Asia-Pacific region, which represents the world's fastest growing air transport market.[2] As the International Air Transport Association's (IATA) John Meredith notes, "Without drastic improvements, the forecast growth will soon overwhelm the capacity of Asia-Pacific's aviation infrastructure" (Meredith 1995).

Over the next two decades, the world air transport market is projected to grow between 5 and 6 percent a year, while the North

2. Demand for air transportation in the Asia-Pacific region grew 12.1 percent per annum between 1985 and 1990 (IATA 1994).

American market is anticipated to grow at only about 4 percent a year (Dempsey 1995a). That growth, however, will not be evenly distributed across cities or regions.

The Orient Airline Association predicts 7.5 percent annual growth for the Asia-Pacific region through the year 2000. Such estimates are corroborated by the findings of other groups. The IATA predicts between 7 and 8.6 percent growth through 2010. The Organization for Economic Cooperation and Development (OECD) predicts inter-Asian traffic growth of between 8 and 9 percent over the next two decades. McDonnell-Douglas predicts 9.7 percent through the year 2010. And the People's Republic of China (PRC) is anticipated to enjoy traffic growth in the range of 13.6 to 14.7 percent.

In 1991, China's passenger and cargo volume grew by 28 percent; in 1992, 33 percent; and in 1993, 20 percent. This growth has placed enormous strains on the capital requirements of the commercial aviation sector and has caused serious safety and operational problems. The PRC has concluded that its airlines and airports need capital and operational expertise and has recently opened both to foreign investment. The CAAC (the Chinese air transport ministry) will designate two of its airlines for foreign investment/operations, allowing foreign investment up to 35 percent and foreign voting rights up to 25 percent (*China Daily* 1994). Among the most intriguing opportunities that appear to be on the table is the possibility of setting up a joint venture to build an airport in China. Construction costs on mainland China are a fraction of what they are almost anywhere else in the world. By the year 2000, eastern China will have 22 new airports, while 10 more will have been expanded and upgraded (Trautmann 1993). It has been estimated that China will need 400 new airports over the next decade (Schoof 1995).

Overall, the Asia-Pacific region's share of total global scheduled passenger traffic will grow from 25 percent in 1985 to 51 percent in 2010, thereby displacing North America as the world's busiest commercial aviation market (IATA 1993). One source has summarized the explosive growth in air traffic predicted for this region during the next several decades:

> *IATA has forecast that the total scheduled traffic to, from and within Asia-Pacific (and this does not include the expanding charter market) will grow from a 1990 figure of 87.3 million passengers to 132 million by 1995 and to 189 million by the*

end of the century. In other words, it will more than double within the normal gestation period of a new airport, which is about 10 to 12 years. Which in turn means that either airport capacity has to double in that time, generating more departure slots, parking bays, and airport terminal space, or traffic growth will be severely stunted, thousands of dollars are going to be wasted in holding patterns and passengers are going to find travel in the region a very frustrating experience (Hardeman 1993).

International airport forecasts

As we have noted, major airport initiatives have been launched in this region and elsewhere. This section summarizes the recent experiences of these international airports with regard to the issues of forecasting and capacity.

Munich's Franz Josef Strauss Airport

Munich's Franz Josef Strauss Airport handled 13.5 million passengers in 1994, 26 percent more than Munich's Riem Airport in its final year (O'Driscoll 1994). The new airport can handle 17 million passengers a year. The terminal can be expanded to accommodate 30 million passengers annually (Trautmann 1993). Like the airport it is replacing, the new airport will have noise curfews, restricting takeoffs and landings to between 6 A.M. and 10 P.M., but allowing 28 additional operations per day for Stage 3 aircraft between 5 and 6 A.M. and 10 and 12 P.M.

Munich's capacity will take some of the load off other German airports. Willi Hermsen, CEO of Munich's airport authority, notes, "Munich came just in time to take part of the overflow from Frankfurt." Already Munich boasts that its airport is Germany's most punctual, with an 85 percent on-time rate for international flights and 92 percent for domestic flights. It guarantees 35-minute connections and 8.5 minutes for luggage to reach baggage claim (O'Driscoll 1994).

Osaka's Kansai Airport

Although Osaka's Kansai International Airport has but a single 3500-meter (11,400-foot) runway, it is ultimately expected to be able to handle up to 454 arrivals and departures a day, accommodating

68,000 passengers and 3000 tons of cargo (*Kyodo News International* 1994). Ultimate capacity is 160,000 takeoffs and landings, transporting 30.7 million passengers and 1.4 million tons of cargo per year (*Aviation Week & Space Technology* 1994b). Because aircraft approaches are over the ocean, Kansai is Japan's first 24-hour-a-day airport (Lassiter 1994). This will substantially enhance cargo utilization, for cargo landings can be cleared at night and transported through Japan when highway traffic is light, thereby saving warehousing expenses (Black 1994).

Original projections of 600 flights per week, made when the Japanese economy was robust, have fallen. The airport opened at about two-thirds of its original expectations (*Aviation Week & Space Technology* 1994a), with 337 international weekly flights—compared with 630 when the airport achieves full operation (Black 1994)—and 469 domestic weekly flights (*Aviation Week & Space Technology* 1994c). That was still 40 percent above the capacity of the airport it replaced, Itami, which had serious capacity limitations because of noise slot controls. Itami will be left open for domestic operations (Lassiter 1994), although flights will be reduced from 340 a day to 121 (*Aviation Week & Space Technology* 1994a). Nevertheless, airport officials have concluded they need to begin airport expansion almost immediately.

Macau International Airport

The new Macau International Airport (MIA) is anticipated to handle 2.7 million passengers during its first year of operation, 4 million by the year 2000, and 6 million by the year 2010 (*Travel Weekly* 1995). But delays in construction of Hong Kong's new Chek Lap Kok Airport have suggested that these numbers should be revised upward (*Phillips Business Information* 1994). CAM (Macau Airport Company) anticipates that MIA will gain up to 15 percent of Hong Kong's passenger traffic (*World Airport Week* 1995a). Initial annual capacity of MIA is 4.5 million passengers and 120,000 tons of cargo. International flights are anticipated to account for 72 percent of operations (Donoghue 1995).

Hong Kong's Chek Lap Kok Airport

Hong Kong's Chek Lap Kok Airport (CLKA) will open with only one runway, although its location suggests potential 24-hour utilization. Kai Tak's maximum slot-controlled utilization rate was 28 flights an hour, although it sometimes reached 36. First year volume at CLKA is anticipated to be 28 million passengers and 1.4 million tons of cargo (Mok 1993). Chek Lap Kok is anticipated to take as many as 43 movements an hour. The new airport is expected to handle 35 million passengers a year, demand that is expected to materialize by the year 2002 (*Aviation Week & Space Technology* 1994a). The airport will open with one runway, with a second to be commissioned two years later, in 1999 (Mok 1993). CLKA's ultimate capacity is anticipated to be 87 million passengers and 9 million tons of cargo annually (Darmody 1993).

Kuala Lumpur Sepang International Airport

Phase 1 of the New Kuala Lumpur International Airport (NKLIA) includes two parallel runways capable of handling simultaneous wide-bodied aircraft takeoffs and landings, main and satellite terminals, with a capacity of up to 25 million passengers a year. With 25,000 acres, the airport has sufficient room for expansion to five runways so as to raise capacity to 45 million passengers annually (Hill 1993).

New Seoul International Airport

The site for the New Seoul International Airport (NSIA) is predominantly landfill of an area of Inchon Harbor between the islands of Young-jong and Yong-jong, to create nearly 50 square miles of space (Fentress 1995). The new airport's original master plan included a single 3750-meter runway equipped with CAT-IIIA navaid equipment. It will be capable of handling 170,000 aircraft movements, 1.7 million tons of cargo, and 27 million passengers annually (*World Airport Week* 1995b). Two additional runways have since been added to the plan to handle the anticipated 10 percent annual growth in traffic. Four runways may be built, with ultimate completion in the year 2020 (*World Airport Week* 1995d). Airlines have urged that the new airport be opened with two runways, instead of one, as is designated in the airport master plan (*World Airport Week* 1995c).

The cargo terminal will be 175,000 square meters in size, with expansion ultimately to 806,000 square meters, and will be capable of handling 7 million tons of freight (*World Airport Week* 1995c). When ultimately built out (in about the year 2020), the airport will have four runways with a capacity of 530,000 aircraft movements, 100 million passengers, and 7 million tons of cargo (Shin 1993).

Bangkok Nong Kgu Hao International Airport

The first phase of the new Bangkok airport will accommodate 30 million passengers. The airport is projected to handle 38 million passengers in 2010 and 100 million at full build out in 2023 (*World Airport Week* 1995d). This compares with the 12-million passenger capacity at Don Muang, which is also being expanded, raising capacity to 25 million by 1997, three years before Second Bangkok International Airport (SBIA) opens (*Travel Trade Gazette Europa* 1993). Such additional capacity at Don Muang led Airports Authority of Thailand to call for limiting SBIA to only its first phase.

This overview suggests that in many cases, the forecasts are quite optimistic. How and why forecasts are subject to uncertainty and so often prove dubious is vividly demonstrated by the Denver case.

The history of forecasting in Denver

The following sections explore the history of aviation capacity and forecasting in Denver. Accordingly, we begin with the development and expansion of Stapleton Airport.

The early limits to growth at Stapleton Airport

Ever since the old Denver Municipal (later renamed Stapleton International Airport) was opened in October 1929, it had to expand periodically to meet the evolving demands of aviation growth in Denver and the nation in subsequent decades. In fact, it seemed that the airport was always in a state of construction or reconstruction. As soon as one expansion plan was implemented, politicians and planners were already crafting new master plans to accommodate the next round of anticipated traffic (Miller 1983).[3] Starting from an initial plot covering 640 acres, Stapleton expanded to more than 10 times its original size. From a humble beginning with only 40 employees, four small gravel and dirt runways, three airlines, and one small terminal building, by 1983 it had grown into a 10,000-employee facility serving 16 trunk and regional airlines, 10 commuter

3. Much of the following discussion about Stapleton Airport is based on Miller 1983.

carriers, and numerous general-aviation aircraft with four long concrete runways, a four-story terminal building, and four concourses. Before DIA was completed, a fifth concourse was built at Stapleton.

Some analysts postulated as early as the 1940s that the long-term aviation needs of the Denver area would require additional airport capacity outside of Stapleton. A 1948 joint study by the Bureau of Business and Social Research at the University of Denver and the Bureau of Business Research at the University of Colorado recommended significant expansion of runway and terminal facilities at Stapleton and the construction of 20 new airfields to handle general-aviation activity. By the 1950s, observers anticipated that Denver's commercial airport facilities would have to be relocated because Stapleton lacked the necessary expansion land that was required to prepare for the coming "jet age." Stapleton had significantly expanded its land area since 1929, but urban growth moving outward from the city center was circumscribing the airport. Residential, business, or county property foreclosed expansion to the west, south, and east of Stapleton, while to the north lay the Rocky Mountain Arsenal, an Army facility that had been manufacturing nerve gas since 1943 and was sited in the middle of an 18,000-acre expanse of mostly open land. Although expansions in the 1940s and 1950s continued to rely on existing Stapleton land, Denver mayors and airport officials realized that additional land was needed to keep pace with accelerating technological advancements in aviation.

The only realistic expansion alternative for Stapleton was onto the Rocky Mountain Arsenal. As early as 1954, the city began asking the Army about acquiring Arsenal land for future runway expansion. Originally requesting only 252 acres, the city soon realized it would need much more to plan for a jet-age airport and increased its request to 3000 acres a few years later. In 1958, the city proposed two airport expansion plans, both relying on land from the Arsenal. One was a shorter-range plan that called for a new 12,000-foot north-south runway extending onto the southern end of the Arsenal, while the other was a long-range plan involving three 12,000-foot runways and a large terminal to be built on the Arsenal. The only (somewhat significant) hitch to either plan was getting the Army to agree to cede the land.

From the 1950s until the mid-1980s when the decision to build the new Denver International Airport was finally made, the city of Denver had to contend with the myriad institutional, political, and environmental problems of acquiring Arsenal land for airport expansion purposes. In the late 1950s, the Army was not at all inclined to allow Denver to expand aviation activities onto the Arsenal even as chemical weapon production was being curtailed, and it forestalled the requests as much as it could. The Army actually ceded 2000 acres of Arsenal land to the U.S. Air Force in 1958 rather than giving it to Denver. Nevertheless, through an arrangement involving the Army, the Air Force, and the new FAA, which had been recently created to handle air traffic and safety issues as part of the 1958 Federal Aviation Act, Denver was eventually allowed to have 679 acres of Arsenal land in 1959 to build a new north-south runway.

Although the more ambitious Arsenal expansion plan was put aside, it was not forgotten. Soon thereafter, in 1962, Denver Mayor Richard Batterton requested 1355 acres of Arsenal land from the U.S. General Services Administration (GSA), since the Army had recently declared the acreage to be surplus land. The intention was to provide additional land buffers for the new runway and for another proposed north-south runway to be built 8000 feet to the east of the one currently under construction. After more prolonged discussions, the GSA put the land up for sale in 1964, but Denver was only able to purchase 805 acres on the west side of the new runway. Denver intended to build the additional runway, however, and felt that the land would still be available in the future when it needed it. This assumption was a mistake.

Technological developments and accelerating growth in aviation during the 1950s and 1960s affected Denver in a profound way. Regular jet service began in 1959, the "jumbo jet" (Boeing 747) first landed at Stapleton in 1970, and the advent of supersonic commercial aircraft was anticipated shortly. In 1961, Stapleton ranked third nationally in total aircraft operations and accommodated more than 2 million passengers; by 1970, it was handling more than 7 million passengers (Leonard and Noel 1992). In recognition of this growth, the city unveiled a new master plan in 1967 that called for a new terminal, new concourses, strengthening and lengthening of current runways, and a new 13,500-foot north-south runway to be built northeast of the current north-south runway. The new runway would have to be built on Arsenal land, so the plan necessitated a request for 6915 acres. This time around, the land was far more difficult to acquire.

Shortly after the announcement of Denver's 1967 master plan, officials from neighboring Adams County went on record to "vehemently oppose" any airport expansion onto the Arsenal. The Arsenal land had once belonged to Adams County, and the county wished to reacquire it should it become available. Adams County was also concerned about residential growth near the Arsenal and did not want to see that growth foreclosed by impacts from a vastly expanded Stapleton. In a preview of things to come, Adams County's opposition to Stapleton expansion was so strong that, in the summer of 1968, Adams County told Denver to move its airport entirely to a new location, even suggesting that 25,000 acres of Adams County land east of Aurora might be made available for this purpose. In May 1968, Denver Mayor William McNichols stated that if Arsenal land could not be acquired for expansion, Denver would be forced to begin airport construction on a brand new site. In August 1968, the *Denver Post* concurred that if a long-range airport expansion onto the Arsenal (like that first proposed in 1958) was not possible, then a new airport location would be necessary.

The wrangling over Arsenal land in the late 1960s and early 1970s had additional effects that would later become instrumental in the justification to move the airport to another site. Due in part to Adams County's strenuous opposition, the FAA initially denied Denver's bid to acquire Arsenal land. Together with continued reluctance from the Army, Denver eventually had to settle for only 622 acres of Arsenal land in 1971 for its new north-south runway, rather than the 6915 acres requested in 1967. In January 1971, the Denver city council agreed with a proposal from the city of Aurora (located to the south of Stapleton) that the new runway should be built entirely to the north of Interstate 70, negating the necessity of additional tunnels under the runway (and, incidentally, negating noise impacts on Aurora), but resulting in more encroachment on Arsenal land. Because of Stapleton and Arsenal land constraints, Denver planned to build its new parallel runway only 1600 feet east of the existing north-south runway. In inclement weather when instrument flight rules (IFR) were in effect, the FAA required a parallel runway separation of at least 3500 feet during simultaneous landing and takeoff operation for safety reasons. Having runways only 1600 feet apart meant that one runway would have to be shut down when IFR conditions were encountered.

The Denver Regional Council of Governments (DRCOG), the Colorado State Planning Commission, and the U.S. Department of Transportation all approved the runway expansion plan despite the limited spacing between runways. The Professional Air Traffic Controllers Organization (PATCO) was the only group that strongly suggested that the runways be placed at least 3500 feet apart.

Denver went ahead with the 1600-foot runway separation plan. Airport director Don Martin defended the decision by arguing that there simply was not enough land to increase the separation. If the new runway could have been constructed with the appropriate separation, one of the most often repeated arguments for the necessity of building a new airport would have disappeared.

Aviation growth in Denver: Past, present, and future

The most compelling argument for a new airport was the phenomenal growth in air traffic that Denver had experienced since 1929. The city's political and business leaders promoted Denver as one of the nation's centers for aviation activities. Their efforts and Denver's geographical position as the regional center of the inland West resulted in Denver consistently ranking among the nation's top 10 airports based on numbers of passengers and aircraft operations (Table 2-1). This was especially noteworthy given that Denver usually ranked only in the top 25 or 30 cities in the country based on population.

One of the first long-range forecasts of aviation growth at Denver was estimated by transportation consultant James C. Buckley, who conducted a study on growth at Stapleton for the City of Denver in the early 1950s. Buckley argued for short- and long-term expansion plans based on his forecast of 2,046,387 total passengers by 1980. As was typical of early aviation forecasts, he vastly underestimated future growth. The actual 1980 totals were 10 times larger than his forecast. Other early studies in the 1950s and 1960s also tended to underestimate actual growth. The sharp rise in worldwide passenger enplanements during this time was due to many factors, including rapid population growth, increasing industrialization and urbanization in developing countries, changes in the industrial structure of developed countries, and, most importantly, rapid changes in technology (Ashford and Wright 1992). It was also during this time that many structural changes in the industry, such as the transition from

Table 2-1. Growth in air traffic at Denver Stapleton Airport, 1955–1995

Year	Total passengers	Enplaned passengers	Total flights	Total aircraft operations
1955	1,157,125	55,087	235,000	
1956	1,260,507	55,599	276,000	
1957	1,444,942	59,803	275,000	
1958	1,524,575	61,320	272,000	
1959	1,814,195	789,319	67,505	297,000
1960	2,052,544	906,929	68,389	308,194
1961	2,236,497	1,083,039	80,791	267,386
1962	2,384,765	1,158,126	83,716	246,913
1963	2,684,401	1,310,571	85,065	228,731
1964	2,898,242	1,426,464	102,282	312,758
1965	3,330,639	1,674,778	116,793	328,871
1966	3,977,187	2,014,976	132,143	420,925
1967	5,078,502	2,489,573	155,515	444,802
1968	5,980,845	2,963,031	195,125	395,120
1969	6,998,340	3,347,514	212,510	365,135
1970	7,429,150	3,662,777	209,575	357,849
1971	7,827,192	3,851,877	203,581	349,189
1972	9,426,161	4,756,268	218,582	350,721
1973	10,560,158	5,261,262	226,844	362,503
1974	11,482,532	5,751,319	226,922	401,250
1975	12,026,415	6,024,951	232,164	386,456
1976	13,698,742	6,874,340	264,603	418,393
1977	15,281,842	7,666,672	309,997	465,936
1978	18,934,054	9,481,431	338,933	466,645
1979	20,542,682	10,344,114	358,085	486,300
1980	20,848,864	10,506,552	371,861	480,578
1981	22,601,877	11,345,486	379,265	479,766
1982	24,553,248	12,357,544	380,025	466,889
1983	25,247,105	12,774,142		458,088
1984	28,806,349	14,535,197		512,489
1985	29,897,644	15,053,438		495,286

Year	Total passengers	Enplaned passengers	Total flights	Total aircraft operations
1986	34,706,474	17,435,727		524,821
1987	34,069,189	16,996,846		520,836
1988	31,797,747	15,829,021		503,095
1989	27,568,033	13,732,399		463,797
1990	27,432,989	12,773,147		484,040
1991	28,285,189	14,173,874		488,254
1992	30,877,180	15,437,711		506,706
1993	32,626,956	16,320,472		552,422
1994	33,133,428	16,589,168		530,839
1995	31,036,622	15,518,311		475,932

Sources: U.S. Civil Aeronautics Board and U.S. Federal Aviation Administration, Airport Activity Statistics, 1970-1994; Stapleton International Airport, Summary of Activity, 1955-1994.

propeller to jet aircraft, were occurring. These technological advances dramatically changed the nature of aviation by increasing speed, efficiency, and productivity, while reducing real prices for consumers. Thus, the number of airline passengers increased dramatically in response to the improvements in air travel, a development that simply could not have been foreseen accurately.

In 1971, the DRCOG began a major long-range study, the Denver Regional Airport System Plan. With the help of consultants Peat Marwick Mitchell, DRCOG completed the study in 1973 and published it in 1974. The purpose of this study was to develop a plan that would address the Denver region's aviation needs through the year 2000 (DRCOG 1974, vol. 1). Commercial takeoffs and landings, which amounted to 212,891 in 1970, were forecasted to be more than 600,000 by the year 2000. Stapleton capacity was reported to be 450,000 (DRCOG 1974, vol. 3). From 3,663,000 enplaned passengers in 1970, the study forecasted 9 million by 1980, 20 million by 1990, and 35 million enplaned (not total) passengers by the year 2000 (DRCOG 1974, vol. 3).[4] While the forecast for 1980 was reasonably accurate (actual number was 10.5 million compared to 9 million forecasted), the 1990 forecast was overestimated by 56 percent (ac-

4. Enplaned passengers are only those who get on a plane at an airport, including both originating passengers and connecting passengers. Total passengers refers to all enplaning and deplaning passengers, including origination, destination, and incoming and outgoing connecting passengers.

tual enplanements were 12.8 million compared to 20 million forecasted). In contrast to later forecasts, however, these were actually more conservative. Obviously, the passenger declines in the late 1980s wreaked havoc with this and subsequent forecasts. Peat Marwick Mitchell, which conducted the forecasting, hedged on its own estimates by stating that the chances were greater that the actual number of passenger enplanements and operations would be less than forecast than greater than forecast (DRCOG 1974, vol. 1).

Also in the Regional Airport System Plan, DRCOG analyzed seven alternative sets, including the "do-nothing" alternative, expansion within Stapleton, expansion onto the Arsenal, construction of a new airport on the Arsenal, and construction of a new airport elsewhere in the metropolitan area. With Peat Marwick's cautionary caveat in mind, DRCOG recommended alternative set 2: expansion within Stapleton. Other reasons cited for this choice were

1. The uncertain availability of Arsenal land
2. The high costs of relocating the airport
3. The preferences of air travelers, the air carriers, and those concerned with environmental impacts
4. The effect that the looming energy crisis might have on air transportation.

Opposing the expansion of Stapleton were residents in the nearby neighborhoods, especially Denver's Park Hill, who were increasingly bothered by louder jet aircraft noise from a busier-than-ever airport. These nearby residents wanted the airport moved, preferably (in true NIMBY fashion) as far away as possible. Interestingly, the FAA also preferred a new airport location, mainly for efficiency reasons. Perhaps because of these sentiments, DRCOG did not foreclose the possibility of building a brand new airport. In fact, it believed that a move of air carrier operations in Denver to an entirely new site by the year 2000 was probable (DRCOG 1974, vol. 1).

In conjunction with DRCOG's Stapleton expansion recommendation, the City of Denver began another master plan in 1976, and again hired Peat Marwick as consultants. The plan, officially announced on October 28, 1978, called for increasing the number of gates from 68 to 103 by 1990 and to 133 by 2000, and the parking capacity from 6000 to 7600 by 1990, and 11,200 by 2000 (Miller 1983). But by the time this plan was announced, increasing dissatisfaction with the Stapleton expansion alternative was already being voiced. Denver city council member William Roberts argued in May 1978 that Stapleton's ultimate capacity was being approached and that a new airport would be needed in another 12 to 15 years at maximum, so spending money on Stapleton expansion made no sense. Captain Tom Lindermann of the Air Line Pilots Association concurred by stating that Stapleton was already at capacity and that planning for a new airport should begin immediately so that it would be ready in 10 to 12 years. Denver Mayor William McNichols, believing that a replacement airport might be necessary in 25 to 30 years, commissioned a new long-range site-selection study. DRCOG (and its Peat Marwick consultants, again) agreed in August 1978 to undertake the study, which became Phase 1 of the Metro Airport Study (Miller 1983; DRCOG 1983). At the same time, momentous changes were occurring in the U.S. airline industry that would significantly impact airport planning in Denver.

The Airline Deregulation Act of 1978

By the late 1970s, the nature of aviation had changed dramatically. Technological developments and rapid growth had contributed to massive expansion in the airline industry, as numbers of passengers and aircraft operations continued to soar.

Much of this growth had occurred since 1938, when the Civil Aeronautics Act was promulgated, creating the Civil Aviation Authority (CAA)—later renamed the Civil Aeronautics Board (CAB)—for the purpose of regulating and promoting the development of civil aviation. The CAA was given authority to regulate airline service (including safety), entry, pricing, mergers, and intercarrier agreements. In 1958, the regulatory responsibility over safety was transferred to the newly created FAA, while economic regulation remained with the CAB.

Before an airline could serve a given route, it had to obtain a "certificate of public convenience and necessity" from the CAB, which controlled entry and exit. Many of the service patterns established by the carriers in the wake of the 1938 legislation continued into the 1970s, as each trunk or local service airline became affiliated with particular routes, cities, and regions. For example, in 1938, United Airlines was authorized to serve a major east-west linear route from New York to San Francisco through Cleveland, Chicago, and Denver. United became a major airline in each of these cities and developed substantial facilities at each of these airports. Likewise, Continental later was awarded routes serving Los Angeles, Denver, and Houston. Local service airlines, such as Allegheny (later to become USAir), Frontier, Ozark, and Piedmont adopted corresponding regional emphases.

But by the 1950s and 1960s, economists were beginning to question the efficacy of continued regulation in the growing and maturing airline industry. They argued that regulation was inherently inefficient and that it might no longer be needed as a safeguard against the threat of monopolies and destructive competition, because the airline industry was expanding and behaving more like a "perfectly competitive" market, or so they believed. By the 1970s, additional economists and a growing number of policy-makers began to concur that regulation in the airline industry was inefficient and was denying airlines the ability to offer consumers more price and service options. Congressional hearings chaired by Senator Edward Kennedy, with substantial assistance from Stephen Breyer (later a U.S. Supreme Court justice), investigated the theory and practice of regulation in the airline industry and concluded that a major change of government policy in this area was appropriate. A consensus emerged across the political spectrum that government regulation was contributing to price inflation and restricting growth in the airline industry. The political battles were led in Congress by Senators Edward Kennedy and Robert Packwood, with President Jimmy Carter a strong deregulation proponent.

Consequently, Congress promulgated the 1978 Airline Deregulation Act, which was signed into law by President Carter in October 1978. In essence, airline deregulation meant that the Civil Aeronautics Board would no longer have authority over entry, exit, and fares. Deregulation was supposed to be phased in over a period of a few years, so that after 1982, any and all carriers that were "fit, willing, and able" could serve any domestic route at any price they wished. The CAB would at least nominally continue to serve through 1984, at which time it would be "sunsetted," and any remaining responsibilities would be transferred to the Department of Transportation. The FAA would continue to be responsible for all safety and traffic control matters. Needless to say, this transformation in government policy would have a major impact on the nation's airlines, and those serving Denver were certainly not immune.

Airlines in Denver prior to deregulation

Ever since the first regular air routes were established, certain air carriers had become associated with service to Denver, continuing these linkages into the era of deregulation. Among the first to serve Denver was Western Air Express (which later became Western Airlines) when it took over an airmail contract to serve Denver, Cheyenne, and Pueblo in 1927, two years before Denver Municipal Airport opened (Miller 1983). Except for a hiatus from 1934 to 1944, after the federal government canceled all existing airmail contracts, Western was a fixture in the Denver market for many years. Under deregulation, Western moved its hub to Salt Lake City. Western was merged into Delta Airlines in 1986 (Miller 1983; Dempsey and Goetz 1992).[5]

United Air Lines (whose name was officially changed in 1943 to United Airlines, Inc.) was formed in 1931 as a management corporation for four small air companies: Varney Air Lines, Pacific Air Transport, Boeing Air Transport, and National Air Transport (Miller 1983). In 1934, W. A. "Pat" Patterson became president of United. He was largely responsible for turning United into one of the world's largest and best airlines during his 30 years as president and, later, CEO, of United. After amalgamation, United inherited and retained the main transcontinental airmail route from New York through Chicago to San Francisco. Originally, this airmail route stopped at Cheyenne, Wyoming, instead of Denver for the same reason that the transcontinental railroad had earlier bypassed Denver: the 14,000-foot Colorado Rocky Mountains (see Chapter 4). It was felt that early aircraft would have difficulty with quickly attaining the altitude necessary to fly over the nearby mountains (Miller 1983). Thus, Denver was faced with a repeat of its situation in the late 1860s, and it had to rely on a spur air line to Cheyenne to link it to the main route.

United began its service to Denver from Cheyenne in 1935. But only two years later, United officially received authority to serve Denver directly along the main line. United soon began shifting its activities to Denver. In 1943, United moved its pilot training facility from Cheyenne, and in 1947 United announced its major operations base would be moved to Denver (Miller 1983). On the eve of deregulation, United had become one of Denver's leading trunk

5. Much of the following discussion about the airline industry and deregulation is based on Dempsey and Goetz 1992.

carriers, accounting for approximately 32 percent of the Denver market (FAA 1978).

The other major trunk carrier serving Denver was Continental Airlines, founded by Robert F. Six in 1937. Six had been general manager of El Paso-based Varney Speed Lines (different from United's Varney Air Lines) which had operated an airmail service between El Paso, Texas, and Pueblo, Colorado (Miller 1983). At the same time United acquired the Cheyenne-Denver mail route from Wyoming Air Service, Six was able to acquire the Pueblo-Denver route (Miller 1983). Six then moved the company headquarters to Denver and changed its name to Continental Air Lines, Inc.

By the 1950s, Continental had grown significantly and expanded its route network through interchange agreements with other airlines to include service to Los Angeles, St. Louis, and Houston. In 1959, Continental became the first all-fan jet airline in the world. In 1963, Six moved Continental's headquarters from Denver to Los Angeles, officially stating that Los Angeles had become the airline's geographic and economic focal point. Unofficial reasons for the move included speculation that Denver's attempts to increase landing fees at Stapleton to pay for airport expansion drove Continental away and that Denver was unwilling to give Continental a tax break to keep its headquarters in Denver. (Similar reasons were again heard when Continental announced in 1994 that it would be dismantling its hub operations in Denver.) One other unofficial reason was that Six's third wife, actress Audrey Meadows, preferred to live in Los Angeles rather than Denver (Miller 1983).

During the 1960s, Continental enjoyed record profitability as it expanded into international transpacific service through Hawaii to Guam, Japan, Taiwan, Korea, the Philippines, South Vietnam, and Thailand. Despite its corporate move to Los Angeles, Continental continued to add service to Denver from Portland, Seattle, San Antonio, Houston, and New Orleans (Miller 1983). When deregulation started, Continental, like United, was firmly entrenched in Denver, serving approximately 19.5 percent of the market (FAA 1978).

Frontier Airlines was the other large carrier to be headquartered in Denver and to have a significant presence for many years. In 1950, Denver-based Monarch Airlines, which had been formed as a local service carrier in 1946 to serve small communities in the Rocky Mountain region, merged with two other smaller companies, Challenger Airlines and Arizona Airways, to form Frontier Airlines. Frontier grew into a major regional carrier during the 1950s and 1960s, expanding service throughout the Rocky Mountain, Southwest, and Great Plains regions. In 1967, after it merged with Fort Worth-based Central Airlines, Frontier enjoyed the fourth largest route system in the nation (after United, Eastern, and Delta). Just prior to the onset of deregulation, Frontier was already operating a hub-and-spoke system out of Denver, serving 89 cities in 20 states as well as Canada and Mexico. It was a leading carrier in Denver at that time with about the same share of the market as Continental (FAA 1978).

The impact of airline deregulation: I

After President Carter signed the Airline Deregulation Act of 1978 into law in October, all the nation's airlines were faced with the realization that the industry's ground rules had changed, and they were in the middle of a totally new ball game. No longer would service entry and exit or fares be controlled by an independent federal regulatory agency. Instead, the airlines themselves, ostensibly disciplined by the "invisible hand" of the free market, would determine patterns of air service and fares.

The most significant operational change brought about under deregulation was the airlines' near-universal adoption of hub-and-spoke route structures. Hub-and-spoke networks were deemed to be more efficient than the previous linear route structures because airlines could accommodate larger volumes of passengers from many different origins by funneling them into a hub and then flying them out again to their desired destinations, thereby offering substantially more city-pair service opportunities. Hubs also gave airlines more control at key airports in their systems, allowing them to establish more market power opportunities, raising fares more than 20 percent above competitive levels for origin-and-destination (O&D) passengers. The marketing and yield advantages of hubbing apparently outweigh their high costs as reflected in relatively less efficient equipment, gate, fuel, and labor utilization (Dempsey 1995a).

In the years following deregulation, airlines quickly began to increase their share of passengers at the hubs to the point that many airports became dominated by single carriers. For example, from 1977 to 1987, Northwest's market share at Minneapolis grew from 45.9 to 81.6 percent, USAir's share in Pittsburgh grew from 43.7 to

82.8 percent, and TWA in St. Louis went from 39.1 to 82.3 percent. Interestingly, in Denver, it took longer for single-carrier domination to finally occur. As late as 1985, United, Continental, and Frontier were evenly dividing the Denver market, and each had major hub operations in full force.

Another major effect of the airlines' shift to hub-and-spoke networks was the dramatic increase in aircraft and passenger traffic at the hub airports. For example, in the seven years from 1980 to 1987, aircraft departures increased by more than 100 percent at newly established hubs in Charlotte, Cincinnati, Newark, and Raleigh/Durham. For Chicago, Dallas/Fort Worth, and Washington airports, departures increased at each city by more than 60,000 flights in the same seven years. Passenger enplanements also exploded in these and other hub cities. Since most of these airports were unprepared for the sudden increases in traffic, major congestion and delays became endemic throughout the system by the mid-1980s. With all of this increased traffic, safety concerns were raised as airports and air traffic controllers were stretched beyond capacity to cope with the extra demands. Even though the number of aviation accidents fortunately remained low, the margin of safety was being narrowed. (*Newsweek* magazine proclaimed 1987 as the "Year of the Near Miss"; Morganthau 1987, pp. 20, 24). In both 1986 and 1987, consumer complaints regarding the airlines' poor service reached unprecedented levels. To the Department of Transportation, the solution to these problems was increasing airport capacity to meet the hub airport demands. For a nation that had not constructed a new airport since 1974, many believed that the time had come for "more steel and concrete" solutions to the capacity problems.

Planning and forecasting in Denver during the early years of deregulation

The FAA regularly conducts air traffic forecasts for the nation as a whole and separately for the major airports. In 1979, the FAA conducted a series of airport forecasts, including one for Denver. It projected that Denver would enplane 14.6 million passengers by 1985, and 17.9 million by 1990 (FAA 1979, p. 38). These forecasts lent additional fuel to the debate over whether a new airport was needed. Indeed, the Department of Transportation had already placed Denver on a list of U.S. cities that "might" need a new airport (Miller 1983).

Meanwhile, on the local front, DRCOG was engaged in its Metro Airport Study. In Phase 1, completed in 1980, six potential sites for a new airport were identified and initially evaluated, and three were selected for further consideration. After the potential sites were announced, the City of Denver found that expanding Stapleton directly onto the Arsenal was not one of the three DRCOG alternatives. Thus, the city of Denver hired Peat Marwick Mitchell to conduct a separate plan based on Stapleton expansion onto the Arsenal (Denver 1982). Also as a part of DRCOG's Phase 1, the ever-busy Peat Marwick prepared another airline passenger and aircraft operations forecast. From an actual 1976 base year figure of 6.4 million enplaned passengers, DRCOG forecast 13.1 million by 1985, 17.2 million by 1990 (overestimated actual enplanements by 35 percent), 28 million by 2000, and 50 million by 2020. Aircraft operations were projected to rise from 221,081 in 1976 to 322,270 in 1985; 369,650 in 1990; 493,240 in 2000; and 750,000 in 2020 (DRCOG 1983). See Table 2-1 for actual figures through 1994.

Incorporating these projections in Phase 2 of the study, DRCOG evaluated the suitability of the three alternative sites, plus the city's Stapleton/Arsenal expansion, and narrowed the selection to two alternatives: a site 40 miles east of the metro area between the towns of Watkins and Bennett and the Stapleton/Arsenal expansion site. Finally at the end of Phase 3 of its Metro Airport Study in 1983, DRCOG's Aviation Technical Advisory Committee and its board of directors recommended "that DRCOG select the Rocky Mountain Arsenal alternative as the solution for the Denver region's long-term airport development needs and that the City and County of Denver be asked to take all necessary steps to implement the Board's decision, and that close cooperation with Adams County would be part of the working arrangement for expansion onto the Arsenal" (DRCOG 1983).

Events occurring during the 1983 to 1986 period, however, meant that DRCOG's recommendation would never be implemented. Continued opposition from residential neighborhoods near Stapleton and from Adams County over expansion onto the Arsenal, in addition to newly elected Denver Mayor Federico Peña's willingness to reconsider the issue, resulted in the January 1985 Memorandum of Understanding between Denver and Adams County and paved the way for the new airport. But what drove the process forward, lending an air of urgency to the air-capacity dilemma, was the unbelievably sharp increase in passenger traffic during these three years, along with the ensuing congestion and delays. Enplaned passengers grew by nearly 5 million from 12,774,142 in 1983 to 17,435,727 in 1986, a 36 percent increase. Most of this rise was attributable to the increase in connecting passengers as United, Continental, and Frontier vied with each other to establish and expand hub operations at Denver. Congestion and delays increased accordingly, and Stapleton became a major bottleneck in the nation's air traffic system. It was especially difficult when inclement weather limited airport operations to only one north-south and one east-west runway as IFRs went into effect.

The earlier decisions to build new runways only 900 to 1600 feet away from the old ones, because of land constraints, came back to haunt the city as air traffic congestion and delays proliferated. By the mid-1980s, Stapleton had become one of the most congested airports in the country, with annual air traffic delays of more than 38,000 hours; the FAA considers an airport congested if it experiences at least 20,000 hours of delay (GAO 1991). Stapleton had become a black hole among the nation's airports, giving Denver's national reputation a black eye as a result. These circumstances and their consequences laid the groundwork in providing the technical justification and the political impetus for building the new airport.

In 1985, the FAA conducted Terminal Area Forecasts for the period from 1985 to 2000 for all large hub airports. From a 1984 base of more than 14 million enplaned passengers, the FAA forecasted that Denver would have 21.7 million by 1990 (overestimating actual enplanements by 69 percent) and 36.3 million by 2000. It also projected that by the year 2000, Denver would surpass Atlanta, Los Angeles, Dallas/Fort Worth, and New York's Kennedy airports in passenger enplanements, and would be ranked second (only Chicago would be higher) in the nation in this measure (FAA 1984).

In 1986, a year after the Memorandum of Understanding was signed with Adams County, the City of Denver began to mobilize for a new airport by hiring eight separate consulting firms to begin work on a new master plan. Once again, Peat Marwick Mitchell was hired for forecasting, in addition to work programming and financial analyses. From a 1984 base of 14.5 million, Peat Marwick forecasted 25.8 million enplaned passengers by 1995 (a 72 percent overestimate, as actual enplanements in 1995 were about 15.5 million), 30.7 million by 2000, 42.96 million by 2010, and 55 million by 2020 (Denver 1986).

The last official forecast before the two airport referenda were held and groundbreaking began was conducted by the FAA in its 1987 Terminal Area Forecast for Denver. The FAA projected 28 million passenger enplanements by 1995 (overestimating actual enplanements by 87 percent) and 34 million by the year 2000 (FAA 1987).

Both the Peat Marwick and FAA forecasts, and all other forecasts that projected robust growth, were based on the following set of factors:

1. Continuing long-term growth in the Denver region. Denver had experienced a major population boom in the 1970s and early 1980s, and the forecasters were heavily influenced by that recent growth. Yet by the mid-1980s, population growth in Denver had slowed considerably, following its traditional boom-bust cyclicality. The long-term population growth prospects for Denver do remain strong, although the population forecasts used to predict future air travel were probably too high.
2. Continuing strong long-term economic growth in the Denver region. Similar to the population trends, Denver's economy went through a boom in the 1970s and early 1980s when the data used in the forecasts were being generated. Although there were (and are) good reasons to believe that Denver's economy will remain strong, the rate of economic growth used in the air traffic forecasts was probably too optimistic.
3. Denver's geographical location. A strong argument in favor of continued large increases in air passengers through Denver is its geographical location in the center of the relatively isolated western United States. No other large cities (more than 2 million in population) are within 500 miles of Denver. This relative isolation and its location midway between the Midwest and the West Coast make Denver an attractive place for an airline hub operation. Denver's strategic role as a hub was a very strong factor in the high passenger forecasts, so much so that the city felt that if one of the hubbing airlines were to leave Denver, other airlines would flock in to establish their own hub operations. The FAA had actually based its optimistic traffic projections on the likelihood that American Airlines would establish a major hub in Denver by 1995 (Peat Marwick 1988).
4. Structural orientation of the Denver regional economy. Denver is a major regional service center, specializing in

telecommunications, computing and information processing, banking, and other producer services. These functional specializations rely on and use air travel more frequently than, say, a heavy manufacturing orientation. It was thus expected that Denver would have more air passengers per capital than many other cities because of its economic orientation.

5. High educational attainment of Denver's population. Denver has a higher-than-average percentage of college graduates and those with graduate degrees. Since more highly educated people tend to fly more often, it was believed that Denver would exhibit higher passenger numbers than other cities of comparable size.

While recognizing the underlying validity of these factors, the Peat Marwick and FAA forecasts tended to exaggerate their importance. Other inaccuracies can also be discovered. For example, the Peat Marwick report of 1986 states: "Evidencing a long-term commitment to Denver, a number of airlines, including United, Continental, and Frontier, have entered into an agreement to move to a new air carrier airport, which is estimated to be operational by 1995." This statement is both surprising and misleading since the airlines had actually opposed the new airport starting in 1984 through 1990, when Continental became the first airline to sign a gate lease commitment at DIA. And, of course, the FAA believed American would be hubbing at Denver by 1995, even though American Airlines CEO Bob Crandall denounced the airport as a "field of dreams."

Not everyone had as much faith in these projections as the City of Denver and the FAA. Two other aviation forecasts were made in the late 1980s that were not as optimistic. In 1988, Landrum and Brown conducted a study under contract with United Airlines, which projected no traffic growth in 1989 and 1990 and only modest 1.9 to 3.3 percent increases through 1995. The other forecast was made by independent consultant Michael Boyd of Aviation Systems Research Corporation, who projected increases of only 1.1 percent through 1995 (Abas 1991). Even these more conservative forecasts turned out to overestimate actual traffic (Fig. 2-1).

But the last two official forecasts, conducted by Peat Marwick for the City of Denver in 1986 and by the FAA in 1987, were massively overestimated. Certainly, the rapid increases in traffic at Denver from 1983 to 1986 had contributed to these upward adjustments. But it must also be remembered that the City of Denver and Adams County had already agreed to the concept of a new airport by this time and that the FAA was already on record as supporting a new airport. Anticipating a battle over political support and funding, there was undoubtedly extra incentive to use and publicize those forecasts that were based on the most optimistic assumptions. This is not an unprecedented practice; the history of planning is full of examples in which forecasts have been made based on either optimistic or pessimistic assumptions, depending on external (especially political)

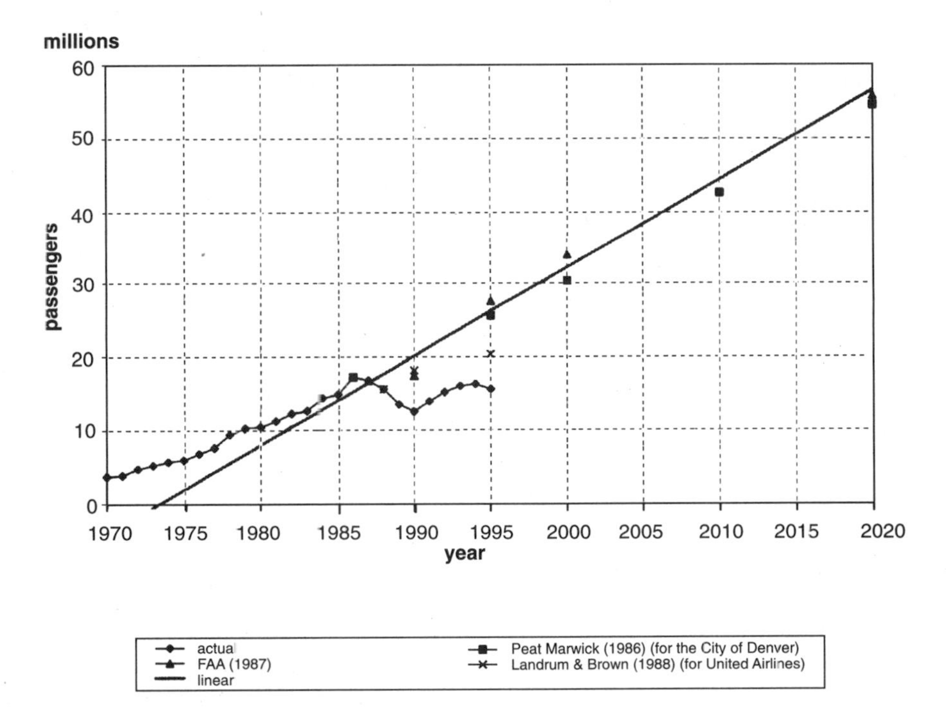

Fig. 2-1. *In 1989, the FAA, the City and County of Denver, and United Airlines were offering significantly different projections of traffic demand at Denver. United's were discounted by the city but proved to be the most accurate of the three.*

factors (Feldman and Milch 1982). Wide swings in forecasts can occur depending on changes in only a few assumptions. In this case, powerful political and economic interests had already begun to swing in favor of the new airport, and the optimistic forecasts certainly did nothing to slow the momentum. Table 2-2 compares the various forecasts.

Yet even aside from the optimism that pervaded the official forecasts, no consultants predicted that air passenger enplanements at Denver would actually decline each year from 1986 to 1990. In order to understand why this decline occurred, we need to revisit what was happening in the airline industry overall and how the unfolding of the deregulation drama was playing cruel tricks on the fate of Denver's aviation future.

The impact of airline deregulation: II

In the first few years after deregulation, many airline industry observers and much of the traveling public felt that deregulation was a success (Goetz and Sutton 1997).[6] Numbers of flights and passengers had increased dramatically as airlines offered a range of new, low, discount fares. New airlines, such as People Express, America West, and Midway, were formed, and together with former local service, intrastate, and charter airlines, they began to challenge the dominance of the 10 major trunk airlines—United, American, Delta, Eastern, TWA, Western, Pan Am, Continental, Braniff, and Northwest (Meyer and Oster 1984). In the face of this competition, the 10 former trunks saw their market share of domestic revenue passenger miles slip from 87 to 75 percent between 1978 and 1983 (Bailey et al. 1985).

In the next five years (1983 to 1988), the airline industry experienced a massive wave of bankruptcies, mergers, and acquisitions. More than 100 carriers folded or were absorbed, industry-wide concentration increased, and the majors reestablished their dominance. The largest carriers were eventually able to realize significant economies of scale, scope, and network density through their frequent flyer programs, airport gate space, and powerful computer reservations systems. By 1990, the combined market share of the nine largest airlines (American, United, Delta, Northwest, Continental, USAir, TWA, Pan Am, and Eastern) had surpassed earlier levels to reach 92

6. Much of the following discussion relies on Goetz and Sutton (1997).

Table 2-2. Comparison of enplaned passenger forecasts for Denver (millions of passengers)

Forecast	1980	1985	1990	1995	2000	2010	2020
DRCOG/PMM&C (1974)[1]	9.0	—	20.0	—	35.0	—	—
FAA (1979)[2]	11.4	14.7	17.9	—	—	—	—
DRCOG/PMM&C (1983)[3]	—	—	13.1	17.2	28.0	—	50.0
FAA (1984)[4]	—	—	21.7	—	36.3	—	—
EIS (1986)[5]	—	—	20.0	—	33.7	—	—
Denver/PMM&C (1986)[6]	—	—	—	25.8	30.7	43.0	55.0
FAA (1987)[7]	—	—	17.7	27.8	34.3	—	56.3
Landrum & Brown (1988)			18.4	20.6			
FAA (1990)			27.4	33.2			
FAA (1991)				17.7	22.0		
FAA (1992)				17.3	21.5		
KPMG Peat Marwick—two hubs (1992)				16.5	19.6		
KPMG Peat Marwick—one hub (1992)				13.1	14.4		
FAA (1993)				17.4	22.5		
Actual	10.5	15.1	12.8	15–16	—	—	—

1. Denver Regional Council of Governments (DRCOG) and Peat Marwick Mitchell 1974. Regional Airport Systems Plan.
2. FAA. 1979. FAA Aviation Forecast.
3. DRCOG and Peat Marwick Mitchell 1983. Metro Airport Study.
4. FAA. 1984. FAA Terminal Area Forecasts FY 1985–2000.
5. FAA. 1986. Final Environmental Impact Statement for Expansion of Stapleton Airport.
6. City and County of Denver (Peat Marwick Mitchell). 1986. New Airport Master Plan Study, Briefing Paper Number 2: Forecasts and Airport Requirements.
7. FAA. 1987. FAA Terminal Area Forecasts: Denver.

Source: *Denver Post*. 1993.

percent of domestic revenue passenger miles (Williams 1993). Concentration levels increased further after 1990 as Pan Am, Eastern, Braniff, and Midway[7] ceased operating and as other carriers experienced severe financial instability.

In Denver, upheaval in the airline industry started to take its toll. After vigorously supporting the 1978 Airline Deregulation Act, United Airlines encountered difficulties in its transition to deregulation, slipping in national market share from 21.5 to 18.7 percent of revenue passenger-miles by 1983 (Bailey et al. 1985). United reasserted itself after a mid-course correction in which it shifted its operational strategy to a multiple-hub, feeder route system. In 1983, it began to focus on strengthening its primary hubs at Chicago, Denver, and San Francisco by adding spokes, frequencies, and one-stop connections through the hubs. Its strategy at Denver was simple—to become the dominant carrier, driving off Frontier, Continental, or any other competitor that stood in its way. Frontier and Continental found themselves locked in a cage with a 500-lb gorilla who was determined to occupy the entire cage.

Continental Airlines, another of Denver's stalwart airlines, began the era of deregulation with the fourth-strongest balance sheet among trunk carriers. Within five years it was filing for bankruptcy protection. In 1978, Continental CEO Robert Six retired, and Alvin L. Feldman, who formerly headed Frontier, was handed the reigns of power. Feldman faced many challenges in those early years, as Continental struggled to develop a dual-hub system based in Denver and Houston with a large fleet of wide-bodied aircraft ill-suited to the demands of a hub-and-spoke route structure. The biggest challenge, however, came from Texas International CEO Frank A. Lorenzo, who, after acquiring 48.5 percent of Continental's stock, made a tender offer to take over the airline in 1981. Despite warnings from Feldman that his plan would burden Continental with massive long-term debt, Lorenzo forged ahead and took control. In 1983, during a labor dispute with machinists, Lorenzo placed Continental into bankruptcy, reorganized the company, and, in the process, shed himself of preexisting labor contracts and turned Continental into a low-cost, cut-rate airline. These tactics earned Lorenzo the reputation as a union-buster and resulted in long-lasting enmity between Lorenzo and airline labor organizations.

7. Other companies embracing the same name have since restarted operations.

During the same period, Frontier Airlines, another of Denver's leading carriers, found itself in a Darwinian struggle for survival. As late as 1982, Frontier was operating a solid hub in Denver, offering 38 nonstop flights compared with United's 24 and Continental's 16, and had been consistently profitable during the previous 10 years. But in 1982, United began a move to dominate Stapleton Airport by increasing flights out of Denver by a third and directly challenged Frontier in many of its markets. Also, seven other large carriers had started to serve Denver since deregulation began, including American, Eastern, Northwest, Piedmont, and Southwest. By 1983, Denver was the most overserved market in the nation. United (which had added more than 100,000 seats per week since 1978) had increased to 174 departures per day, 27 gates, and 40.8 percent of Stapleton's traffic, Frontier had 138 daily departures, 51 gates, and 24.3 percent of the market, while Continental had 18.5 percent. Frontier was being squeezed by a determined United on one side and by a new low-cost Continental on the other. Average fares in Denver were only $127 in 1984—among the lowest in the country—and then fell 8.3 percent in 1985 and another 4.6 percent in 1986 because of the ferocious three-hub carrier competition, and both the numbers of flights and passengers exploded. Frontier, however, could not withstand the onslaught. It began to record significant losses from 1983 to 1985. It asked for wage concessions from employees, and it was forced to begin liquidating assets. It sold 25 of its 51 Boeing 737s to United in May 1985. Finally, Frontier was sold to Donald Burr's People Express airline in November 1985 (Dempsey and Goetz 1992, pp. 70–71).

But People Express itself was in no shape to wage a competitive battle with United and Continental in Denver. By 1986, Donald Burr's dream of a worldwide enterprise was fading rapidly, because People had tried to become too big, too fast. People's "low-fare, no-frills" approach worked at the beginning when it first started as a small hub operation out of Newark and chose routes that did not directly challenge the majors. But when faced with competition from People Express, the majors matched the low fares on a capacity-controlled basis and gained passengers who appreciated not having to pay extra for lunch and for checked bags. When People Express bought Frontier and attempted to institute the "no-frills" approach in Denver, loyal Frontier customers reacted negatively and chose to fly on other airlines. In desperation, People Express tried to reinstitute full service, but it was too late.

In July 1986, People Express agreed to sell Frontier to United and began to sell off Frontier's assets to United while the deal was pending. United was able to acquire $43.2 million worth of assets from Frontier, including two hangars and six gates at Denver. By 1986, United had grown to control nearly 40 percent of the Denver market, followed by Continental at 28 percent and People Express–owned Frontier at 18 percent. Once it had acquired some of Frontier's prized assets, United began to take a hard-line stance in its negotiations over the acquisition of the rest of Frontier.

In the meantime, People Express continued to lose money and was forced to shut down Frontier on August 24, 1986. Only a month later, Frank Lorenzo offered to purchase both People Express and Frontier, and by February 1987, they were both subsumed into Continental. After a settlement with United over the transfer of assets from Frontier, Continental came away with most of Frontier's aircraft, in addition to three hangars and two concourses (C and D) at Denver's Stapleton Airport. With the acquisition of Frontier, Continental surpassed United with 236 daily departures compared to United's 218 and briefly held the dominant position at Denver from June 1987 to May 1988 (Fig. 2-2). The era of three hubbing airlines in Denver was over.

The impact of airline deregulation: III

With Frontier and People Express gone, the Denver market became a duopoly for Continental and United. In some respects, it resembled more of a "shared monopoly," as each carrier attempted to recoup some of the losses incurred in the battle with Frontier. Accordingly, airline ticket prices at Denver rose by 17.6 percent in 1987 and by 39.2 percent in 1988. Without Frontier, passenger enplanements at Denver declined sharply, exacerbated by the fare increases. Moreover, United began to peel aircraft off its Denver hub to establish a hub at Washington Dulles Airport, while Continental did the same to establish a hub at Cleveland.

A number of reasons have been offered to explain the unprecedented four straight years of decline in total and enplaned passengers at Denver in the late 1980s. Denver and Colorado had already entered into another of its periodic economic "busts," this time brought about by the withdrawal of energy company exploration activities in the wake of oil price declines in 1983. A significant portion of Denver's real estate market, which had expanded dramatically during the late 1970s and early 1980s, lay dormant in the wake of the collapse. Office vacancy rates in downtown Denver reached an alarming 40 percent. In addition, the skiing industry in Colorado had a number of poor years in the late 1980s, contributing to the reduced travel demand to Denver. Also by the late 1980s, as planning for DIA had become more serious, both United and Continental had gone on record against building a new airport in Denver.

Since most of the passenger decline was attributable to connecting passengers, it has been alleged that the two hubbing carriers rerouted their traffic away from Denver to make it appear that Denver did not have enough passenger traffic to require a new airport. Both United and Continental had fought a bloody battle to oust Frontier from the Denver market and establish dual control at Stapleton. Neither carrier was eager to see a new airport, with its increased gate space and capacity, attract other carriers possibly to establish hub operations and pose competitive threats. Even after groundbreaking on the new airport began in September 1989, neither carrier was willing to concede anything to the City of Denver while new airport gate lease negotiations were pending. Finally, in the summer of 1990, Continental broke the ice by signing a lease agreement with the city and committing to the new airport; United followed suit in the summer of 1991, and interestingly, passenger enplanements at Denver began to increase again.

The passenger revival at Denver in the early 1990s was, however, a slow one due to a string of unfortunate events that resulted in a financial nightmare for the airline industry. The onset of the Persian Gulf Crisis in August 1990 brought about increases in jet fuel prices and declines in air travel, especially for international destinations. The nation also entered an economic recession, which further depressed demand. Making matters worse, the airline industry itself was woefully unprepared for a series of bad years. In the wake of massive fleet expansions, mergers, acquisitions, and corporate "raids," the industry had become bloated with debt and excess capacity. From 1990 to 1993, the U.S. airline industry lost nearly $13 billion, more than all the profits made since the Wright brothers' inaugural flight at Kitty Hawk. Historic airlines, such as Pan Am, Eastern, and Braniff, were forced to cease operations. Continental, TWA, and America West entered Chapter 11 bankruptcy (the second time for Continental), while Northwest and USAir threatened bankruptcy.

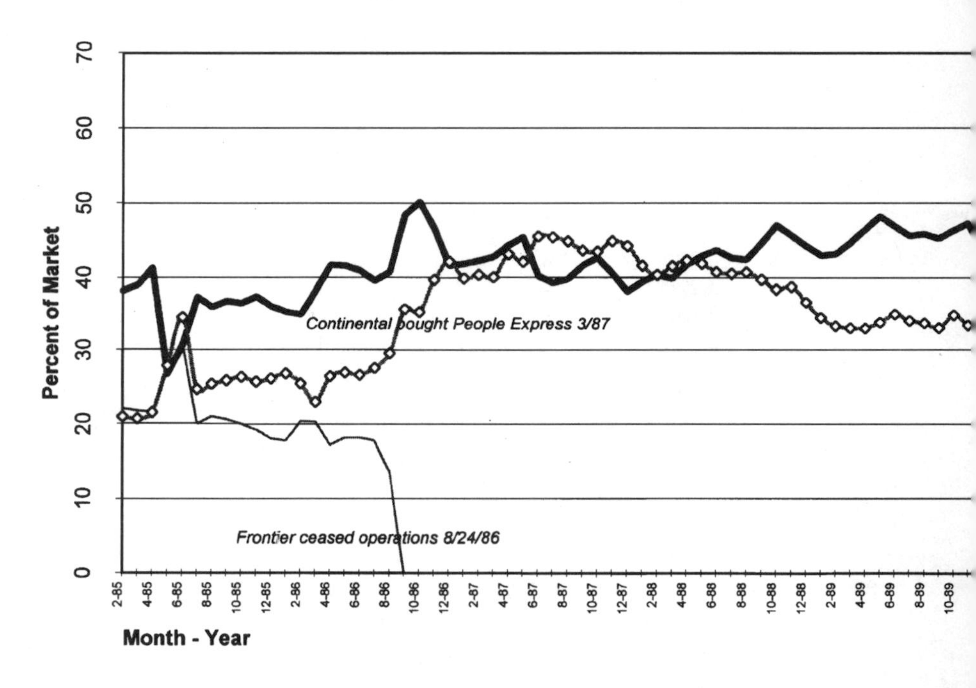

Fig. 2-2. *United, Continental, and Frontier's domestic market shares at*

United, American, Delta, and every other major airline, with the lone exception of Southwest, lost money during this period.

Prior to this bleak time, United and Continental had resumed their battle for supremacy over the skies in Denver. Shortly after Frontier and People Express went out of business, Continental became the leading carrier in Denver. The status was short-lived, however, as United quickly regained the lead in May 1988, a position it never again relinquished. In the late 1980s, Continental peeled aircraft out of Denver to establish a hub at Cleveland. At the same time that CEO Frank Lorenzo was folding Frontier and People Express into Continental, he also absorbed New York Air (a double-breasted nonunion Texas Air subsidiary created to pressure its labor subsidiaries to concede wage and work rule rollbacks) and was working on consuming Eastern.

To many, Lorenzo epitomized what had gone wrong in the airline industry since deregulation began. Ever since he first took Continental into bankruptcy in 1983, invalidating all existing labor contracts and

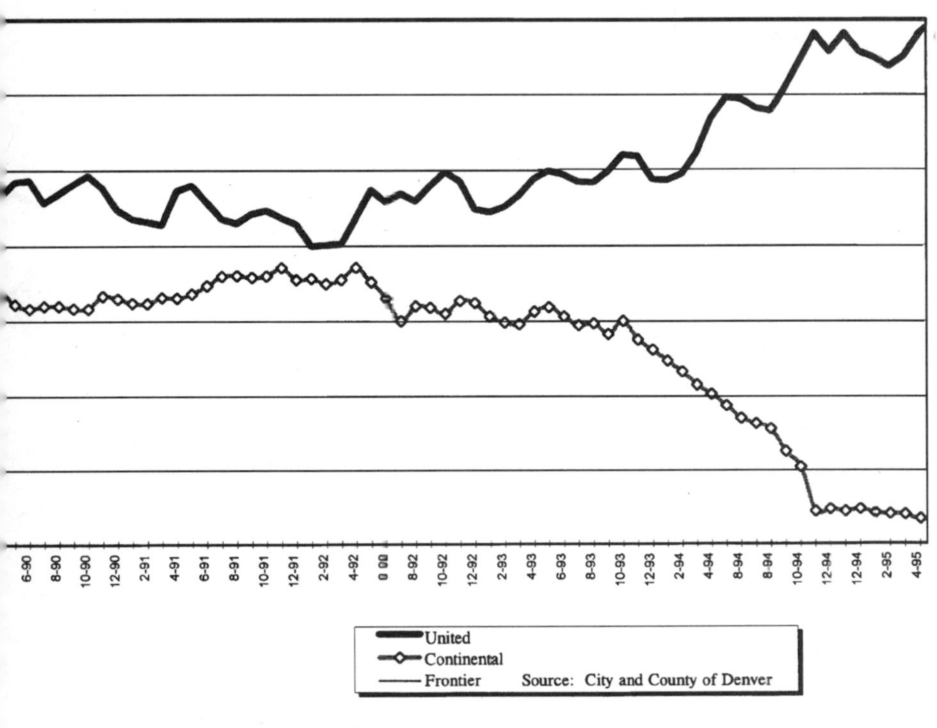

pleton International Airport from 1985 to 1994.

firing all of Continental's 12,000 employees, Lorenzo has been anathema to airline workers. He recklessly grabbed as many airlines as he could through highly leveraged financing schemes and stripped acquired carriers of their assets as his holdings drowned in debt. The following editorial from *Airline Business* sums up what many felt was Lorenzo's "contribution" to the industry:

> *The trouble with Lorenzo is that his only genuine successes have been in creating an empire of misfits which has accumulated debts of over $5 billion, in attracting undiluted hatred from his workforce, in bringing on an unprecedented investigation by the DOT into his fitness to manage an airline, and in his blatant efforts in asset-stripping.* (Airline Business 1988)

With Lorenzo at the helm, Continental Airlines became notorious for poor-quality service, especially for overbooking and canceling flights, having a poor on-time record, and losing luggage. In a head-to-head matchup with United and most other carriers, Continental found itself on the losing end. Toward the end of the bitter Eastern episode in which a labor-management impasse drove one of the industry's pioneer airlines into the ground, and facing increasing scrutiny regarding his fitness to run an airline, Lorenzo sold his shares in Continental Airline Holdings to Scandinavian Airline System for $30 million in 1990 and agreed to leave the airline industry for a period of at least seven years. A year later, Continental filed for Chapter 11 bankruptcy for the second time in eight years.

Among the last things Lorenzo did as CEO of Continental was to sign a lease with the city of Denver in the summer of 1990 for 20 gates at the new Denver International Airport. Because Continental became DIA's first hub carrier, it was able to reserve the closest concourse (A) to the main terminal building, and the city agreed to build a pedestrian bridge linking the terminal directly to the concourse.

The advantages Continental gained from signing the first lease at DIA were never fully realized because Continental decided in March 1994 to abandon its hub operations and relinquish the Denver market to United. Continental had already downsized its presence in Denver from a high of 258 flights a day (including Continental Express) in February 1992 to only 107 flights in March 1994. As part of its restructuring plan to emerge from Chapter 11 bankruptcy, Continental started a new low-cost spin-off airline called Continental Lite, or CALite, to be centered in the southeastern United States, and it needed the planes from Denver for this service. CALite was based on the model of Southwest Airlines, the only major airline to turn a profit in the early 1990s. Continental was never able, however, to shake the effects of the Lorenzo era in Denver or the myriad of predatory practices aimed at it by United. In early 1992, Continental reached its last high-water mark of 38 percent of the Denver market, close but not enough to surpass United's 40 percent at that time. United then increased its capacity by 30 percent over the next several years in what appeared to be a deliberate move to oust Continental once and for all (Dempsey 1995b). United pulled away and never looked back as it steadily increased market share leading to Continental's decision to pull out. Toward the end, Continental was losing $10 million a month at its Denver hub and could not face the prospect of continuing hub operations at the more expensive new airport.

Although Continental began operations a half century earlier as an airline flying between Pueblo, Colorado, and El Paso, Texas, and was for many years a premiere long-haul and Rocky Mountain airline, Lorenzo had so alienated its workers with his 1981 leveraged buyout and 1983 bankruptcy and unilateral breach of all union contracts that the angered labor force had retaliated by causing service to decline sharply. High-yield business traffic began to fly United, Frontier, or other higher-service carriers. Throughout the 1980s, Continental tried to repair its public relations by being a model corporate citizen—for example, by giving away free Continental travel at charity functions. By far, Continental was a better corporate citizen in Denver than United. But even with Lorenzo's departure in the early 1990s, Continental had little goodwill left with the public, and its traffic base was predominantly fare-driven. Thus, a departure from Denver would cost Continental little in terms of lost goodwill, would arrest losses alleged to be $10 million a month, and would enable it to attempt to replicate the operations of linear-route Southwest Airlines, the only consistently profitable carrier in the industry.

Another factor that may have contributed to Continental's departure was the City of Denver's refusal to allow Continental to continue to fly several planes per week into its maintenance facility at Stapleton Airport once DIA opened. The agreement that the city reached with the Park Hill neighborhood and Adams County stipulated that once the new airport was open, Stapleton would be closed to any aircraft landings and takeoffs. The city strictly enforced this agreement, so Continental was forced to close its facility. Continental had lobbied hard during the summer of 1993 to keep the maintenance base open. But in the end, the city would not consent, despite the loss of 1500 jobs to the local economy. The city did offer to help finance a new maintenance facility for Continental at DIA, but the increased costs of this option dissuaded the financially beleaguered airline.

Continental sought to keep its existing Stapleton maintenance base open with an assurance that it would need only one north-south runway open for a maximum of three flights a day. The benefit to the city would be that it would keep 1500 good-paying jobs and at least 30 flights a day by Continental at DIA to discipline United. But the Park Hill community, to the west of Stapleton, feared the noise would never go away. Some residents even perceived that it would be letting the "camel's nose under the tent"—they believed Continental would try to deplane passengers at the closed airport. In the end, the politicians did nothing to save the Stapleton maintenance base, the jobs that went with it, and even the presence of a second hub carrier that might not have been lost.

Stung by criticism that he did nothing, Governor Roy Romer subsequently insisted that Continental was in such disarray at the time (in the midst of Chapter 11 bankruptcy) that it was difficult to help the carrier because it couldn't decide what it wanted. True, a power struggle had erupted inside Continental between Bob Ferguson, the CEO, who wanted out of Denver, and Louis Jordan, the president, who insisted that Denver was crucial to Continental. When the maintenance base went down, Jordan left Continental, and Ferguson took Continental out of Denver and began CALite, a financial disaster that subsequently led to his ouster. Had Jordan won the power struggle, the Denver hub might not have been eliminated. (Jordan subsequently took a senior position at ValuJet, which became the second largest carrier at Atlanta.)

DIA director Jim DeLong later insisted that the Stapleton maintenance base would not have kept Continental in Denver, pointing to the fact that when Continental's maintenance base was closed, it transferred maintenance to Los Angeles for only a few months, when it decided to outsource maintenance. DeLong insists that Continental would have outsourced maintenance even if the Stapleton base had remained open.

One thing is certain. There is no certainty in predicting what an airline in financial distress might have done, even with the benefit of hindsight. And under deregulation, nearly all major airlines were in financial distress. U.S. airline net profit margins were poor before deregulation, averaging only 2.4 percent from 1960 through 1977. They were catastrophically worse in the first 15 years of deregulation, when they fell to –0.4 percent (Figs. 2-3 and 2-4). As previously mentioned, the airline industry lost more than $13 billion in the first four years of the 1990s alone. An erroneous perception that deregulation resulted in tens of billions of dollars in consumer savings coupled with a strong ideological antigovernment laissez-faire movement shackled Congress from addressing the economic collapse of this important infrastructure industry or its cause—deregulation. For the record, real yields in the airline industry fell 2.5 percent annually from 1950 to 1978, then fell at a significantly slower rate, only 1.7 percent, after deregulation (Dempsey 1995a). Thus, air fares were falling at a faster rate before deregulation than after it.

Continental announced that it was scaling back its Denver operations in March 1994, and the withdrawal proceeded throughout the following year (Table 2-3). Continental dropped 30 routes through

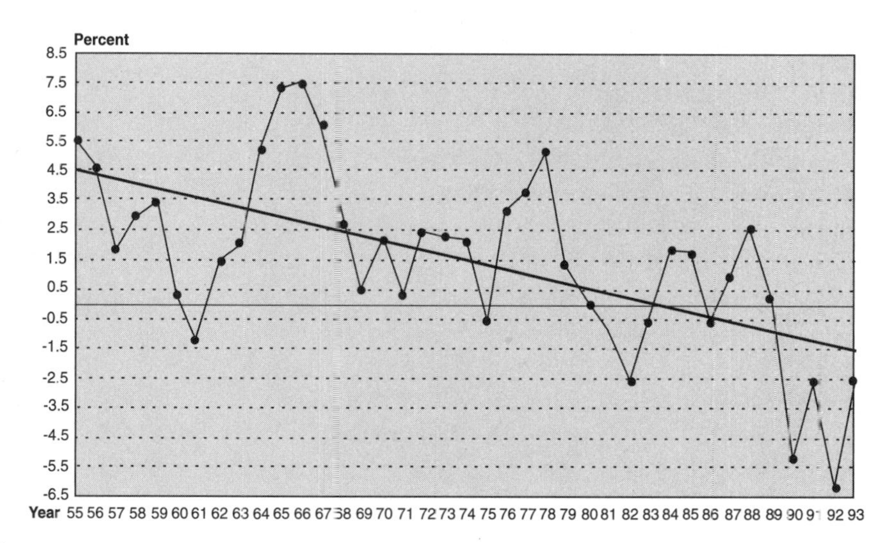

Fig. 2-3. *U.S. airline industry net profit margin from 1955 to 1993.*

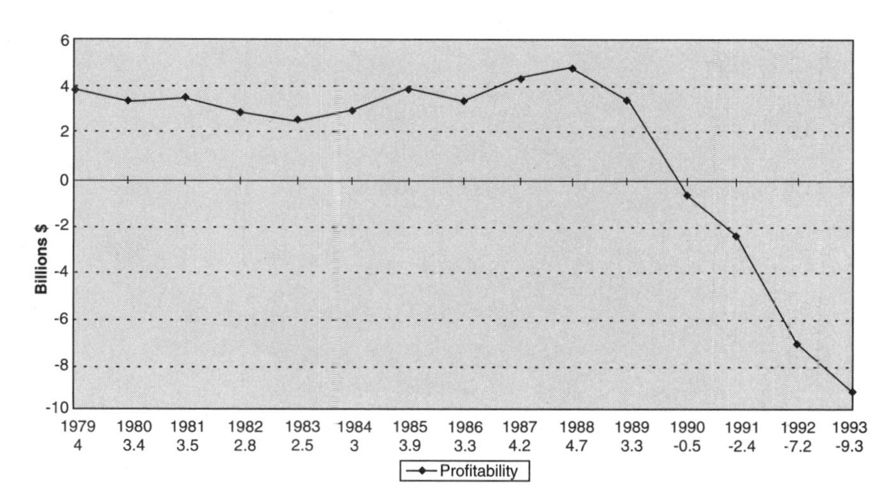

Fig. 2-4. *Accumulated airline profitability in billions of dollars from 1979 to 1993. The erosion of airline profitability after deregulation caused the industry to lose all the accumulated profit it had earned since the dawn of commercial aviation in the 1920s. Actually, these data have not been adjusted for inflation, so they tend to overstate the financial crisis. Nonetheless, by the early 1990s, the debt of nearly all major airlines had been downgraded to junk status.*

Table 2-3. Recent service cutbacks at Denver

Routes to	Airline	Date service discontinued
Milwaukee	Continental	9-7-93
Fargo, ND	United	9-8-93
Great Falls, MT	United	9-8-93
Sioux City, IA	United	9-8-93
Bismarck, ND	Continental	10-1-93
Detroit	Continental	10-1-93
Minot, ND	Continental	10-1-93
Spokane	Continental	10-1-93
Honolulu	Continental	11-1-93
Sacramento	Continental	12-16-93
San Jose	Continental	12-16-93
Houston/Hobby	United	1-15-94
New Orleans	Continental	2-1-94
Lubbock, TX	Continental	2-1-94
Wichita, KS	Continental	2-1-94
Bozeman, MT	Continental	3-9-94
Missoula, MT	Continental	3-9-94
Ontario, CA	Continental	3-9-94
San Antonio	Continental	3-9-94
Tucson	Continental	3-9-94
Tulsa	Continental	3-9-94
Tampa	Continental	5-1-94
Billings, MT	Continental	6-1-94
Portland, OR	Continental	6-1-94
Reno	Continental	6-1-94
Colorado Springs	Continental	6-1-94
Oklahoma City	Continental	6-23-94
Orange County	Continental	6-23-94
Grand Junction, CO	Continental	7-1-94
Jackson Hole, WY	Continental	8-1-94

Denver route suspensions (9-93 to 8-94): 30

Suspended by Continental on 10-31-94: 23

Albuquerque
Boston
Chicago O'Hare
Dallas/Fort Worth
El Paso
Indianapolis
Jacksonville
Kansas City
Las Vegas
Los Angeles
Milwaukee
Minneapolis
New York LaGuardia
Omaha
Orlando
Philadelphia
Phoenix
Salt Lake City
San Diego
San Francisco
Seattle
St. Louis
Washington Dulles

Additional suspensions by United: Denver to Moline, IL; Peoria, IL; and Springfield, MO on 2-8-95

Additional suspensions by Continental: Denver to Chicago Midway on 4-2-95; and London, England, on 4-17-95

August 1994 and an additional 23 on October 31, 1994. As Continental pulled out, United stepped up its frequencies to take advantage of the void. United's goal of becoming the dominant carrier in Denver has finally been realized—by mid-1996 it controlled more than 70 percent of the Denver market, and with its United Express code-sharing affiliates, nearly 80 percent of the DIA market. Continental has only 13 flights a day at Denver.

Passenger traffic, which had been increasing since 1990, began to decline after Continental's pullout. In 1995, only 30 million total passengers flew into or out of DIA, down from 33.1 million in 1994 (Fig. 2-5). A contributing factor in this decline is the fare increase strategy that United enacted as it realized more monopoly opportunities. Average "Y" class air fares in Denver rose 46 percent from 1994 to 1995, largely due to United's pricing (Dempsey 1995b). In many cases, fares have been so high that Denver travelers have chosen to fly out of the Colorado Springs airport, located 75 miles south of downtown Denver. A new low-cost airline, Western Pacific, started hub operations in Colorado Springs in 1995 and has attracted a sizable number of Denver travelers who prefer driving the extra distance to avoid the high fares in Denver. In the summer of 1995, regular bus and van shuttle service to Colorado Springs from Denver was inaugurated to tap into the growing exodus.

Several new small, low-cost, low-fare airlines emerged to serve Denver—MarkAir (which moved its operations in Chapter 11 to Denver from Alaska, then collapsed entirely in late-1995), Vanguard (headquarted in Kansas City), and Frontier (a new airline with many of the same officers as the original airline by that name) [Caulk 1995]. All

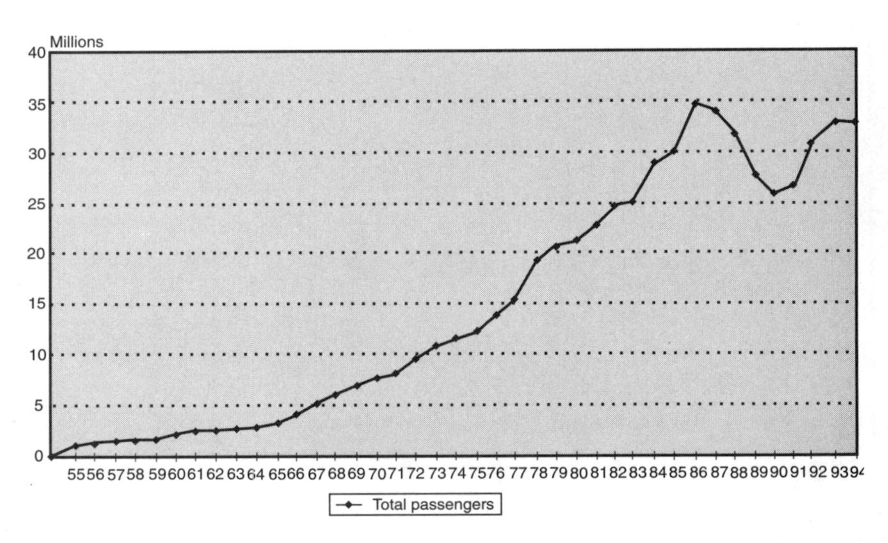

Fig. 2-5. *Total passengers at Stapleton International Airport from 1955 to 1994. Passenger demand peaked in 1986, when the city was served by three hub carriers, then fell sharply as the city began to pursue a new airport. By 1994, traffic still had not returned to 1986 levels.*

find United engaging in various types of predatory conduct—flooding new entrants' routes with excess capacity, matching discount fares only on flights in close proximity, paying travel agents commission overrides to steer business toward United, engaging in computer reservations system (CRS) bias against flights that connected with the low-cost entrants, and for the most part, refusing to code-share or offer joint fares with carriers other than those flying turboprop aircraft.

For most Colorado communities (and many small communities throughout the Rocky Mountains and Great Plains region), the result was they would be served from Denver only by a United Express affiliate flying turboprop aircraft and charging sky-high air fares, even in markets that had sufficient traffic to sustain jet service. The remedy was obvious. Federico Peña's Department of Transportation need only declare such conduct to be an unfair and deceptive practice under the Federal Aviation Act. But Peña had embraced code-sharing in the international arena and was slow to grasp the impact a discriminatory code-sharing arrangement might have domestically.

In October 1994, senators from Colorado, Montana, and North Dakota asked Secretary Peña to examine the problem (Baucus et al. 1994). In August 1995, North Dakota's Senator Dorgan insisted that Secretary Peña either eliminate all code-sharing or require that all airlines offer ticketing and baggage, joint-fare and code-sharing relationships with any airline that seeks it on a nondiscriminatory basis. Efforts were made to enlist the support of Governor Roy Romer in encouraging Secretary Peña to deal with code-sharing in a way that would enhance competitive air service for Colorado's western slope airports, but inexplicably (as in the effort to retain the Continental Airlines maintenance base at Stapleton Airport), Romer didn't want to get involved.

To informed observers, Peña's reluctance to act was also dismaying. The Federal Aviation Act gives the transportation secretary the power to prohibit "unfair and deceptive practices." As Denver's mayor, Peña conceived and built DIA, while Romer sold it. As Secretary of Transportation, Peña was the only person in the nation in a position to ensure that Denver would have competitive air service to discipline United. Did United possess such enormous political power that neither Romer nor Peña would stand up to it? As we shall see, Wellington Webb, Peña's successor as Denver's mayor, recognized UAL's power and granted it major concessions that profoundly influenced the implementation of DIA and its subsequent character.

The 1995 passenger numbers in Denver are much lower than what the forecasters from Peat Marwick or the FAA predicted during the 1970s and 1980s. Even forecasts made in the early 1990s are proving to be considerably off the mark. These regrettable forecasting results underscore the necessity of recognizing the inherent unpredictability of aviation forecasting and the need to adopt flexible planning strategies that can adapt to changing conditions.

Conclusion: Forecasting and planning

The aviation industry has always been highly sensitive to fluctuations in national and regional economies. Even throughout the regulated period, passenger traffic was closely correlated to cycles in the growth of the Gross Domestic Product. Under deregulation, however, volatility in airport traffic has increased substantially. Airlines can choose to serve or abandon markets as they please. The adoption of hub-and-spoke routing structures contributed to the volatility because an airport that served as a hub would experience greatly expanded traffic, while the same one serving only as a spoke would not. The airlines thus could determine which airports would experience massive traffic growth and which would languish. As strategies changed or airlines went out of business, airports experienced wild swings in traffic. Airports and their cities became more susceptible to the vagaries of the airline industry and, in a sense, more dependent on the success of particular airlines.

When Eastern was grounded in 1991, substantial impacts were felt in Miami and Atlanta, where many workers no longer had jobs. Kansas City, which had become Eastern's mid-continent hub, felt the effects of hub withdrawal for the second time in less than 10 years (TWA had pulled out in 1983). Similarly, Raleigh/Durham, North Carolina, Nashville, Tennessee, and San Jose, California, built major infrastructure for American Airlines to hub, only to have American determine after a few years that those hubs were not viable. Dayton, Ohio, built infrastructure for Piedmont, whose hub was closed down there after Piedmont was acquired by USAir. In Denver, the departure of the original Frontier Airlines, and later Continental, has meant that jobs, service, and passengers have declined, and fares have risen sharply. Denver has a brand-new, expensive airport built originally for three hubbing airlines, but only one is left to use it.

Lessons learned

The implications of these trends for airport planning can be summarized as follows:

- *Do not rely exclusively on aviation forecasts; they will nearly always be wrong*. The most obvious problem in airport planning specifically, and in planning generally, is created by the issue of forecasting. Projections of future demand are notorious for the large margins of error that are commonly involved. In this case the only forecasts that were accepted were the overly optimistic ones conducted by the FAA and the city's consultant, Peat, Marwick, Mitchell & Co. The more conservative ones were dismissed. The reasons for this phenomenon are many and complex. One cannot ignore the cognitive, political, and ethical dimensions involved in the client-consultant relationship. Perhaps more important in this case, however, are the inherent difficulties in forecasting passenger enplanements in the highly volatile aviation industry.

  Although any forecast can prove erroneous and estimates of passenger demand have traditionally been subject to significant errors, the new deregulated environment has greatly exacerbated this tendency, creating unprecedented uncertainty and volatility. In the DIA experience, we saw an airport planned and built for three hub airlines. One disappeared, and a second found itself unable to fulfill its agreements, twice made, to occupy a full concourse at DIA. Deregulation has unleashed an environment of unprecedented economic anemia for the airline industry. As a consequence, nearly all major U.S. airlines had their debt downgraded to "junk" status (Dempsey 1995a).

  The critical nature of aviation forecasts for airport planning and design and of demand forecasts for any type of project requires that new ways of generating more credible forecasts should continue to be developed and utilized. But more importantly, the planning process itself must be structured in a manner that permits change in the face of significant errors. Since forecasting is likely to remain more of an art than a science, the need for such a planning process is paramount.

- *As many credible forecasts as possible should be generated and seriously considered by planners and decision-makers.* No one has an exclusive franchise in predicting the future. Therefore, all forecasts need to be evaluated and assessed in making important decisions such as those concerning new airports.
- *Do not overlook alternative methods of resolving capacity problems.* Infrastructure expansion is extremely expensive and often difficult to achieve. Better utilization of existing airport infrastructure may be enhanced, for example, through a peak-period pricing structure to flatten the demand curve. Other innovative, but heretofore "taboo" strategies may become necessary.
- *The role of airlines must be fully considered in any future forecasting exercises.* Forecasters should not separate the state of the airline industry, and that of particular airlines, from airport planning and demand forecasting. Too many variables associated with airline decisions regarding hub operations, levels of service, and pricing can, and do, have dramatic effects on passenger enplanements at airport. Future forecasting methodologies must build in the role and viability of airlines at specific airports.

References

Abas, Bryan. 1991. Flying blind. *Westword.* March 6–12.

Airline Business. 1988. Gratitude or vitriol? August.

Ashford, Norman and Paul H. Wright. 1992. *Airport Engineering*, 3rd Edition. New York: John Wiley & Sons, Inc.

Aviation Week & Space Technology. 1994a. Growth outpaces Asian airports. August 29.

————. 1994b. Japanese carriers launch new service. September 5.

————. 1994c. Kansai International. September 12.

Bailey, Elizabeth E., David R. Graham, and David P. Kaplan. 1985. *Deregulating the Airlines.* Cambridge: MIT Press.

Baucus, Brown, Campell, and Dorgan (Senators). 1994. Letter to DOT Secretary Federico Peña. October 24.

Black, Alexandra. 1994. Trade and tourist boom from new airport. *Inter Press Service Global Information Network.* September 9.

Caulk, Steve. 1995. Frontier steps up expansion. *Rocky Mountain News.* August 3.

China Daily. 1994. Airlines clear path for foreign investment. May 27.

Darmody, Thomas. 1993. The design and development of world class airports. Paper presented to the IBC International Conference on Airport Development & Expansion, Hong Kong, October 28.

de Neufville, Richard. 1991. Strategic Planning for Airport Capacity: An Appreciation of Australia's Process for Sydney. *Australian Planner* 29:4.

Dempsey, Paul. 1995a. Airlines in turbulence: Strategies for survival. *Transportation Law Journal* 15:35-3.

——— 1995b. Rip United Airline's hold from DIA. *Denver Business Journal.* August 25–31.

———. and Andrew R. Goetz. 1992. *Airline Deregulation and Laissez-Faire Mythology.* Westport, CT: Quorum Books.

(Denver) City and County of Denver. 1982. Potential expansion of Stapleton International Airport into the Rocky Mountain Arsenal: Final report. Prepared by Peat, Marwick, Mitchell & Co., San Francisco, CA.

———. 1986. New airport master plan study. Briefing paper number 2: Forecasts and airport requirements.

Denver Post. 1996. GP Express files. Chap 11. January 12.

Denver Regional Council of Governments (DRCOG). 1974. Denver Regional Airport System Plan. Vol. 1: Recommendations.

———. 1974. Denver Regional Airport System Plan. Vol. 3: Forecasts.

———. 1983. Metro Airport Study: Final Report.

Donoghue, J. A. 1995. The Pearl-Y gateways. *Air Transport World.* February 1.

Dorgan, Senator Byron. 1995. Letter to DOT Secretary Peña. August 4.

Federal Aviation Administration (FAA). 1978. Airport Activity Statistics of Certificated Route Air Carriers. Washington, DC: Federal Aviation Administration.

———. 1979. FAA Aviation Forecasts: Denver.

———. 1984. Terminal Area Forecasts Fiscal Year 1985–2000.

———. 1987. Terminal Area Forecasts: Denver.

———. 1995. Aviation Forecasts—Fiscal Years 1995–2006.

Feldman, Eliott J. and Jerome Milch. 1982. *Technocracy Versus Democracy: The Comparative Politics of International Airports.* Boston: Auburn House Publishing Company.

Fentress, Curtis. 1995. Revitalizing the excitement of travel. *Passenger Terminal '95.*

General Accounting Office (GAO). 1991. New Denver Airport: Safety, Construction, Capacity, and Financing Considerations. GAO/RCED-91-240.

Goetz, Andrew R. and Sutton, Christopher J. 1997. The geography of deregulation in the U.S. domestic airline industry. *Annals of the Association of American Geographers.*

Hardeman, Terence. 1993. The necessity for new airports in Asia-Pacific. Paper presented to the IBC International Conference on Airport Development & Expansion, Hong Kong, October 28.

Hill, Leonard. 1993. Asia's newest "dragon." *Air Transport World.* September 1.

Horonjeff, Robert and Francis X. McKelvey. 1994. *Planning and Design of Airports,* 4th Edition. New York: McGraw-Hill.

International Air Transport Association (IATA). 1993. *Asia/Pacific Air Traffic Growth & Constraints.*

————. 1994. *Asia/Pacific Air Traffic Growth & Constraints.*

Kyodo News International. 1994. Kansai International Airport inaugurated. September 5.

Lassiter, Eric. 1994. Japan to open much-delayed Kansai Airport. *Travel Weekly.* August 29.

Leonard, Stephen and Thomas Noel. 1992. *Denver: Mining Camp to Metropolis.* Niwot, CO: University of Colorado Press.

Meredith, John. 1995. Room to boom. *Airline Business.* January 1995.

Meyer, J. R. and C. V. Oster. 1984. *Deregulation and the New Airline Entrepreneurs.* Cambridge: MIT Press.

Miller, Jeff. 1983. *Stapleton International Airport: The First Fifty Years.* Boulder, CO: Pruett Publishing Company.

Mok, John. 1993. The development of Hong Kong's new international airport. Paper delivered at International Conference on Aviation & Airport Infrastructure, Denver, Colorado, December 8.

Morganthau, Tom. 1987. Year of the near miss. *Newsweek.* July 27. pp. 20, 24.

O'Driscoll, Patrick. 1994. Munich's facility reflects future of DIA. *Denver Post.* July 31, 1994.

Peat Marwick Mitchell. 1988. Review of Aviation Activity Forecasts.

Phillips Business Information. 1994. New Macau airport set to open July 1995. January 28.

Schoof, Renee. 1995. China booming with new airport development. *Seattle Times.* April 8.

Shin, Jong-Heui. 1993. Airport developments in Korea. Paper delivered at International Conference on Aviation & Airport Infrastructure, Denver, Colorado, December 8, 1993.

Trautmann, Peter. 1993. The need for new airport infrastructure. Paper delivered before International Conference on Aviation & Airport Infrastructure, Denver, Colorado, December 5.

Travel Trade Gazette Europa. 1993. Bangkok to add gateway. April 8.

Travel Weekly. 1995. Far East facility slates December opening. July 3.

Velocci, Jr., Anthony. 1993. Kahn tells airlines: Sit tight, cut costs. *Aviation Week & Space Technology.* August 16.

Wachs, Martin. 1990. Ethics and advocacy in forecasting for public policy. *Business and Professional Ethics Journal.* 9:1,2.

Williams, G. 1993. The Airline Industry and the Impact of Deregulation. Brookfield: Ashgate.

World Airport Week. 1995a. Macau Air officials see bulk of first year traffic coming from Kai Tak. May 23.

————. 1995b. Air traffic growth forces expansion of Seoul Airport construction. May 30.

————. 1995c. NSIA privatization opportunities. June 20.

————. 1995d. NSIA considers opening with extra runway. July 1.

3

The politics of DIA's development

"Politics, n. A strife of interests masquerading as a contest of principles."—AMBROSE BIERCE

Despite widespread evidence to the contrary, many persons still cling to the belief that decisions involving large technological projects are made on the basis of plans that have been drawn up in an analytically neutral way. The idea that politics is involved in such a process is often anathema to planners and others who seek to promote the concept of a rational approach to decision-making. Technical rationality has an important role to play in any project, but it is not the only rationality that is involved. Indeed, in every public works project, political rationality permeates the process. Such projects involve elected officials who always consider such issues as opposition and support. Moreover, one cannot easily distinguish between the "political" decision to embark on a project and its subsequent "technical" implementation.

The decision by the City and County of Denver to build a new international airport rather than expand the existing facility illustrates this point well. The decision process can be understood only if the nature of the power structure in Denver and in Colorado is taken into account. Although the Federal Aviation Administration (FAA) is expected to impose a technical rationality upon airport planning, every major element of the decision-making process was the outcome of intensive bargaining and negotiations between a large number of actors with varying interests, values, and resources. These ranged from leading political figures in Denver and Washington, D.C. (who unanimously supported the project), to individual citizens who formed organizations to oppose the project, to businesspeople, developers, contractors, the airlines, and governmental agencies such as the Environmental Protection Agency (EPA) and the FAA. Not only was much bargaining involved in determining whether Denver could and would build such a facility, but the decision itself became a controversial one that involved widespread public debate and the holding of two elections.

Comparative perspectives

That power and ideology play a key role in airport development in many (if not all) countries has been highlighted recently by a comparative study of eight airports in five countries—the United States, Canada, France, Italy, and Great Britain. In each case, the initial decision to build a new facility was driven by the same ideology (Feldman and Milch 1982):

> *The common approach in all five countries derived from a business mentality characterized by a belief in the value of economic growth and development. . . . Elites everywhere justified the development of civil aviation infrastructure . . . as a public good.*

This common ideology, however, did not lead to similar outcomes. In some cases the airports were built, in others, they were delayed for varying periods of time or even abandoned. The key variable determining the outcome was the nature of the political system and the role of various actors therein. The degree to which and manner in which the public participated, for example, differed greatly. In all cases, however, such participation had only a marginal impact; elite actors essentially controlled the process (Feldman and Milch 1982).

To what extent do these generalizations hold for DIA? The DIA decision permits us to shed additional light on this question by analyzing how this project was actually devised and implemented. We shall consider also the implications for planning and decision-making. The key questions that emerge, therefore, include the following:

1. How are airport decisions actually made?
2. Who are the key actors?
3. What is and should be the role of the public?
4. What is the role of the media?

To answer such questions, it is essential to begin with a discussion of the structure of power in the community.

Power and the city

In any city, a complex relationship exists among the public officials, the citizens, and the business community. The mayor and other leaders are concerned with the well-being of the community as a whole, because they serve at the pleasure of the citizenry. The quality of life that the city provides to its inhabitants, however, is largely dependent on the level of economic activity. Businesses generate jobs and provide a large share of the tax revenues. Hence the community is dependent on owners of private capital, who possess considerable autonomy. Elected officials can only strive to influence investment and other decisions made by the private sector. Inevitably, therefore, businesspeople, business elites, and their associations wield tremendous power. Their power is enhanced by another consideration: Because the city's political leaders have to win the support of the citizens to gain office, and that is usually a costly undertaking, candidates are beholden to the business community for financial and other support.

Businesspeople have always played an influential role in shaping the development of cities, but their power has increased in recent decades as cities have lost many of their middle and upper class residents to the suburbs and as competition between cities for new businesses has intensified. Growth is the universal holy grail, and every community wishes to promote it by stimulating the local economy and attracting job-creating enterprises. Decision-making in cities can therefore best be conceptualized as linking the inhabitants, the officials, and private capital in a symbiotic relationship committed to growth (Elkin 1987, p. 18).

Land use becomes the dominant concern because any development strategy inexorably involves changes in existing land-use patterns. This is the case whether one is dealing with urban renewal or with major infrastructure projects such as highways and airports. Thus, whenever a coalition committed to growth emerges, the businesspeople involved will have a strong interest in the ways in which changes in land-use patterns will affect existing investments and create new opportunities (Elkin 1987, p. 37). The labor unions—especially those representing the building and construction trades—will often join such a coalition, since major projects mean jobs for their members.

Although nearly everyone benefits from being part of a healthy, thriving community, the question of how the benefits of growth are distributed remains an important question. There is little doubt that business leaders will be concerned with the ways in which growth promotes their interests. For this reason some scholars view the city as a "growth machine" steered by the business community. In their words: "Cities can be viewed as 'growth machines' dominated by local business elites seeking to engineer development policy in a manner consistent with their economic interests" (Molotch 1976). Of course, the degree to which they are able to do so varies, depending on particular political configurations. But they will always be a powerful force due to the structure of the system within which the decision-making process takes place.

Therefore, it is natural that a business-government coalition be formed, that businesspeople be influential members of the coalition, and that it strive to design and implement a specific growth strategy for the city. This is, and has always been, the Denver pattern. Indeed, as is demonstrated in Chapter 4, this structure has existed since the earliest days of Denver's history and has consistently driven its growth, especially in transportation. Indeed, Denver's role as the regional center was made possible when its business leaders joined with its officials to make certain that the city would be linked to the transcontinental railroad.

To understand the political process by which the airport came about, it is necessary to consider the principal actors:

1. Elected officials, especially Mayor Federico Peña, Governor Romer, and the Colorado congressional delegation
2. Local businesspeople and their associations—chambers of commerce, hotel owners, Front Range Airport, landowners, investment bankers
3. The citizenry and various interests groups, such as the Park Hill Neighborhood Association, and the various organizations that were formed to fight the airport.

However, the process of decision-making within a city is not determined solely by officials, businesspeople, and citizens. It is also influenced by external actors. Much of the capital that city officials wish to attract is controlled by individuals and organizations who are not members of the community. The federal government has always played a role, as have those who extend credit to the city. And, often, similar coalitions in other cities are competing actively to achieve their own growth strategy.

This was clearly the case for the new airport. Denver did not have the ability to build the airport without annexing land (which it was constitutionally prohibited from doing without a vote of the adjacent jurisdiction's electorate) and without significant external financing. And other cities were competing for federal airport funding. Accordingly, we must also consider key external actors such as Adams County, which could provide the site on which Denver could build its airport; Congress, which provided the funds to initiate and legitimate the project; financial firms, which issued large amounts of bonds to pay for the project (they are discussed in Chapter 5); various federal agencies, such as the DOT and FAA; and the airlines, especially United and Continental.

These actors were involved for different reasons and to different degrees. To understand the decision to build the airport, we must consider the goals, resources, and policies that each followed, as well as analyze the way in which the political process operated, especially in regards to the role of the citizenry.[1]

The internal actors

Elected officials

The mayor Even though the airport was a significant regional asset and the Denver Regional Council of Governments (DRCOG), on which Denver was represented by the mayor and one elected representative, had discussed the issue of airport capacity since the early 1970s, every Denver official viewed the airport decision as one that was crucial to Denver's future. There was agreement that the surrounding counties and the state as a whole would benefit from an efficient, well-run airport. But Denver had to be the major decision-maker for the airport that replaced Union (rail) Station as its critical link to the nation and the world and hence to economic prosperity.

Denver's mayor has always been a major participant in any decision involving the airport. He actually played a dual role. On the one hand the mayor was the city's representative to the regional body that dealt with such issues as transportation (DRCOG) and participated in its studies and deliberations. On the other hand, the mayor could and did act independently, since Stapleton International Airport lay fully within his jurisdiction.

1. In this chapter, we have drawn on research carried out by several students in the University of Denver Graduate School of International Studies' Technology Policy Planning Seminar, especially Daniel Osazuwa and Jerry Young.

The history of the past two decades illustrates this point well. DRCOG had begun dealing with the issue of airport capacity in 1972, but Adams County, Aurora, and Commerce City objected strongly to any attempt to expand Stapleton. Accordingly, in 1978, DRCOG initiated a major study to explore the possibility of building an entirely new airport. Denver, however, in 1981, initiated its own study on the expansion of Stapleton and continued to fight for that alternative within DRCOG.

Federico Peña's election marked a turning point in the debate over the future of Stapleton and the City of Denver. Peña was elected in June 1983. Previous mayor Bill McNichols had failed to keep city streets clear after a freak May blizzard hopelessly clogged local traffic, thus calling into question McNichols' transportation management ability. A few years later, Peña's transportation management ability was also called into question. A major snowstorm struck Denver in late December 1987, but Mayor Peña decided not to plow the city's side streets, fearing the operation would be too costly. Instead, Peña decided to run garbage trucks up and down the streets to pack down the snow. In the subsequent two weeks, a deep freeze struck the city, the snow turned to ice, and much of Denver remained gridlocked because the garbage trucks had created deep ruts in the side streets that made them virtually impassable (Coates 1988). Thus Peña had to relearn the importance of not allowing weather difficulties to get in the way of the flow of transportation, a lesson he seemed to understand much better in the case of Stapleton Airport.

Like his predecessors, Mayor Peña was originally supportive of expanding Stapleton, but after taking office he initiated a review of the situation. He concluded that expansion was neither feasible nor desirable and initiated negotiations with Adams County to explore the possibility of achieving a cooperative arrangement that would permit a new airport to be built. He was apparently influenced by the actions of the Adams County commissioners, who vigorously opposed any attempt by Denver to implement the decision reached by DRCOG that expansion of Stapleton onto the Rocky Mountain Arsenal was preferable to building a new airport near Bennett, Colorado. Furthermore, it has been argued that any further expansion of Stapleton would result in an inefficient facility, especially in poor weather conditions. These negotiations culminated in the Memorandum of Understanding that allowed Denver to annex a parcel of land in Adams County on which it could build a new airport in return for various economic concessions.

This decision has been criticized for two reasons. The first challenged the idea that Stapleton could not be expanded. New investments were being made to increase capacity at Stapleton in the mid-1980s. A new concourse and a new commuter terminal were under construction, and the FAA, having completed an Environmental Impact Statement in 1986, had approved the building of two new runways. The first, a concept suggested by Stapleton Airport director George Doughty as an interim option and possibly even as an alternative to an Arsenal E-W runway, was a short N-S runway, which would divert smaller aircraft from the longer runways. The second runway, the E-W runway, would have been built on the southern edge of the Rocky Mountain Arsenal, an area that the Army had found to be not significantly polluted. These facilities would have permitted Stapleton to reduce delays and to handle the demand projected by the FAA until the year 2000 (Kowalski 1989a). Furthermore, many have argued that the short N-S runway could have been lengthened even though various buildings blocked easy expansion. In any case, shortly after his reelection in 1987, Mayor Peña canceled the east-west runway project, arguing that Denver should focus on building a new facility.

The second criticism focused on the argument that Stapleton was a major bottleneck in the national airway system and had to be replaced. Even without the addition of new runways, it was not clear to what extent Stapleton created delays. According to the FAA's own figures, the number of operations (per 1000) delayed 15 minutes or more was 27 at Stapleton, 25 at Atlanta, 76 at JFK, 88 at Chicago, 89 at Newark, and 115 at La Guardia (Fumento 1993a). It must be noted, however, that this figure can be very misleading as a measure of the capacity problem. If activity at a hub is reduced infrequently but dramatically, the impact is exponentially greater than if flights are routinely delayed for short periods.

Whatever the reasons for Mayor Peña's decision, many obstacles had to be overcome before the airport became a reality. The decision had to be ratified by the voters of Adams County and, despite the Mayor's (and Governor Romer's) initial opposition, by the Denver voters as well. In the first election, the mayor was soon relegated to the sidelines because the historical animosity between Denver and Adams County meant that many Adams County voters viewed Mayor

Peña with great suspicion. Governor Roy Romer had to fill the political vacuum and did so energetically, organizing a powerful effort and personally campaigning long and hard. We discuss this pivotal election in more detail later.

Mayor Peña played a somewhat more significant role in the Denver election. In the campaign, the mayor was supported by all the elected city officials—the 13 council members were unanimous in their support. His official role, however, was advisory only, since it is illegal for the city to use public funds for a political campaign. Accordingly, the time and resources spent by the mayor and some of the staff were private, and consisted essentially of providing information to the campaign staff on strategy and tactics and on the substance of the project.

Whatever his role in the political campaigns, however, it is Mayor Peña who made the critical decision to build a new airport, and it is Mayor Peña who saw to it that the decision would be implemented, despite numerous obstacles (Fig. 3-1). Most of the implementation fell to Mayor Peña's successor, Wellington Webb (Fig. 3-2). Webb's role is discussed in Chapters 5 and 9.

Fig. 3-1. *As Denver's mayor, Federico Peña was the key political figure in pushing for a new airport, but as President Clinton's Secretary of Transportation, he kept a relatively low profile on Denver's airport issues.*

Fig. 3-2. *Mayor Wellington Webb, standing in front of the Elrey Jeppesen statue, makes the dedication remarks on the opening day of Denver International Airport in February 1995. To his right are Mike Musgrave, Denver's director of public works, and Jim DeLong, Denver's director of aviation.*

Governor Romer Roy Romer, the popular governor of Colorado, was a very important figure in the struggle to gain the necessary support, especially, as we shall see, in the Adams County election. He campaigned vigorously to sway the voters, arguing that the project would benefit the entire region, while Mayor Peña stayed discreetly in the background. Wearing his trademark leather jacket, the governor made numerous early-morning visits to restaurants throughout Adams County in what was affectionately referred to as the "oatmeal circuit," and was able to diffuse the anti-Denver feelings. His efforts were probably the determining factor in the election outcome, though Romer was careful to distance himself from the airport during his 1994 re-election campaign after the project began to encounter serious implementation difficulties. By the time DIA finally opened, Romer lamented his support of the project, saying he would never have campaigned for it had he known it would cost so much.

The City Council On the whole, the Denver City Council played a passive role in the process of airport development, supporting Mayors Peña and Webb almost consistently. More vigorous oversight might have reigned in costs as construction ran overbudget, but the council instead tended to advance the interest of contractor-constituents. Al Knight of the *Denver Post* summarized the work of the Council's airport committee:

> *If it weren't for one thing, it might be possible to summon up genuine feelings of sympathy for the Denver City Council members who serve on the city airport committee. . . .*
>
> *The one thing that has prevented the buildup of sympathy is that the committee handles the job of airport oversight so poorly. It has more often than not had its attention focused on the small disputes and especially on the tawdry politics of minority- and women-owned businesses. In the process the city council has, again more often than not, been a rubber stamp for policies that have run up the costs. . . . (Knight 1994)*

Without the pork barrel that Denver's mayors and City Council members were able to direct toward favored constituents, essential political support for the airport might not have been forthcoming. Nonetheless, as discussed in Chapter 9, pork barrel politics drove up the cost of DIA and takes a bite out of the wallets of every passenger who uses it.

The Greater Denver Chamber of Commerce and the Business Community

During the economic boom that Denver enjoyed between 1978 and 1984, the Denver Chamber of Commerce more than doubled its membership, from 1500 to 3500. It took the initiative in developing an economic development program for the city which included a number of specific projects—airport expansion, a new convention center, light rail for the downtown, small business development, and air quality (Denver Chamber of Commerce 1985). Its priorities were indicated by the establishment of an Airport Committee in 1978 to specifically address the issue of the need for a new airport in coming decades.

By the following year, the chamber had endorsed the creation of a new regional airport to expand capacity at Denver. Over the years, the airport issue retained a high priority with the chamber, and its leaders were involved to some extent in all decision-making involving the airport. Its members participated in the DRCOG study and challenged their earlier position by endorsing the expansion alternative, arguing that expanding Stapleton was the only viable choice and that other options were a waste of time and money (Albin 1983). At that time, this was indeed the only feasible choice, because many believed, on the basis of historical precedent (though limited in scope), that the Arsenal land could be made available and earlier thoughts of building a new airport had been abandoned as being no longer necessary. Furthermore, given the political configurations that existed—Mayor Bill McNichols was committed to airport expansion onto the Arsenal land—a new airport was not feasible. Political change was necessary before such a project could be contemplated; this fact indicates that the power of the business community is not absolute that a delicate interplay of political leaders, businesspeople, and citizens is always at work.

Whatever its role in shaping the decision away from expansion toward a new airport, there is little doubt that, through its Airport Committee, the chamber played an important role in the discussions. It did so in various ways. Most obviously, it was able to maintain the continuity of the issue through numerous political administrations. Since 1978, governors (2), mayors (3), presidents (4), as well as numerous federal, state, and city officials, came and went. Through the chamber's efforts the airport remained on the local, regional, and even national agenda. Now that the project could at last become a reality, the chamber threw all its energies to secure the realization of the project and was involved more than ever in all phases of the decision-making process giving support that was "broad and aggressive." The Airport Committee, headed by Robert Albin, used a variety of formal and informal techniques in dealing with the City Council—offering official testimony, working with its staff, and so forth—but most of the work was done through direct contacts with the mayor (Hong 1993). The Denver chamber also conducted discussions with Adams County's Metro North Chamber of Commerce, which paved the way for the 1985 Memorandum of Understanding (Albin 1995).

The chamber's membership was largely supportive, and the 30-member board of directors that actually ran the organization were unanimously in favor. The chamber went into high gear and engaged in an aggressive campaign to ensure that all the financial, legal, and political obstacles that stood in the way of its implementation be overcome. It launched a major public relations campaign to solidify the support of the business community and the general public and exerted considerable political pressure on critics of the project.

The chamber's publications stressed the enormous economic benefits that the community would derive from the project. It constantly reiterated the results of one of its studies that demonstrated the positive impacts of other airports on economic development—Dallas/Fort Worth, a gain of $5.5 billion into the regional economy; Atlanta, a $7 billion gain, 200,000 new jobs, and 124 new companies. Furthermore, it constantly reminded the business community that Stapleton had reached its limit and that the FAA was predicting that Denver would be handling 72 million passengers a year by the turn of the century (*Airport Update* 1987, 1988; Cramer 1987; McConahey 1987).

The chamber also raised and spent enormous amounts of money on behalf of the new airport and other projects it supported. In November 1986, the chamber formed an affiliated organization, the Greater Denver Corporation (GDC), to promote the Denver area. Quickly accumulating 587 company sponsors and a war chest of $5 million, the GDC was a formidable economic and political power (Hornby 1990). Essentially, the GDC acted as a channel for airport proponents to finance the expenditures that selling the idea of a new airport required. Altogether, it spent about $2.5 million, of which $685,000 (27 percent) was allocated to the airport project (Colorado Attorney General 1991, p. 15). The Internal Revenue Service (IRS) ruled that these contributions could count as "ordinary business expenses," thus providing the donors with a substantial tax benefit. The GDC resisted revealing its donors but was eventually forced to do so through legal action. Leading the list were the two local utilities, US West Communications, $400,000, and Public Service Company of Colorado, $300,000. Trailing was the notorious Silverado Savings and Loan, which donated $95,000. MDC, a major Denver construction company, donated more than $100,000; Bill Walters donated $50,000; and the Fulenwider and Van Schaack families, prominent members of the Denver establishment who owned significant tracts of real estate in and around the new airport site, gave $50,000 each (Colorado Attorney General 1991, pp. 33–57; Carnahan and Hubbard 1995a; Fumento 1993b).[2] As we shall see, the two local newspapers also contributed to the campaign.

Each of these donors stood to profit greatly from any growth that the new airport would stimulate. But they would not have to wait decades for returns on their investments. The Van Schaack and Fulenwider contributions would be made only if Denver purchased their land. Even the GDC participated in the financial bonanza, receiving a grant of $325,000 (Carnahan and Hubbard 1995a; Fumento 1993b). The Public Service Company would, in return for spending $16 to 20 million to remove high-voltage lines, earn as much as $8.5 million a year for 20 years by providing the airport with power. One city councilman called this a "sweetheart deal for the utilities" and argued, unsuccessfully, that the city should generate its own power.

USWest also won a $24 million contract to wire the airport. But the significance of that contract extends well beyond its size—it guaranteed a financial bonanza to USWest because the city effectively ceded control of the telecommunications system at DIA. The fiber-optic system for voice, data, and video contains a digital switch that links the system to the local network. Hence DIA cannot engage in an increasingly common practice whereby airports generate revenues by providing "shared tenant service."

In this practice, the airport essentially functions as an independent phone company that provides its tenants with communication services. It negotiates rate discounts with long-distance companies, an arrangement that produces savings for the tenants and profits for the airport, which can also generate additional income by providing numerous "enhanced services" and from owning the pay phones. This arrangement has become commonplace, and airports in Kansas City, Washington, D.C., Houston, Orlando, and Los Angeles either have the service in place or are planning to implement it. Although it is still possible for DIA to set up such a service, to do so would be expensive—an estimated $300,000—and by then, the airlines might have set up their own systems. In the meantime, Denver has sacrificed the opportunity to generate an estimated $3 to 8.7 million a year (Locke 1992).

2. The contributions are itemized in Report of the Colorado Attorney General, November 1991.

The airport also attracted the attention of speculators and others who flocked to Denver to take advantage of its many opportunities. The 1980s has been labeled the "get rich quick" decade, and Denver attracted more than its share of speculative activities. The airport proved to be a magnet, but such activities were not limited to the airport development. After all, this was the era of the savings and loans scandals and the kinds of land deals discussed later were commonplace throughout the country.

In Chapter 9 we discuss these dealings in more detail and assess whether and to what degree the city was fleeced. Here, our purpose is to illustrate the ways in which speculators who were active supporters of the project and who contributed generously to political campaigns to help it become a reality were seeking profit from it.

Silverado S&L was at the center of this universe. It was heavily involved in real estate dealings around the airport. One of its first acquisitions, in September 1984, was the Little Buckaroo Ranch, a 39-acre wheat farm located in what was then prairie land far away from Denver's growth path but which, four months later, became a very valuable property. Then MDC, the major Denver builder, announced that it owned 877 acres bordering the farm. Following Mayor Peña's announcement, both entities expanded their land purchases. MDC purchased another 1300 acres. Bill Walters, an important Silverado stockholder and developer, purchased 4000 acres with a $26 million loan from the S&L, which did not request any guarantees or appraisals. Subsequently he received another $3 million for a down payment on another 1300-acre plot and $8 million for a share in another 1200 acres. Overall, either directly or through a series of partnerships, Silverado invested more than $70 million.

These were not investments for the long term. And such activities were not limited to the airport development. In keeping with what was by now accepted practice, almost as soon as a deal was consummated, the parties involved were selling and buying the properties to drive up the value. The Little Buckaroo Ranch, for example, was sold three times in two years, rising in price each time so that its value climbed from $1.5 to 4.3 million. The first sale was by Silverado to a former MDC executive to whom Silverado loaned most of the necessary capital. MDC provided the rest of the funding in exchange for a share in the profits when the land was resold. About a year later, another former MDC associate purchased the land. In this and numerous similar transactions, everyone made money. But, if one of the participants ran into financial problems, the others would suffer too. That happened in December 1988 when Silverado collapsed, at a cost to taxpayers of more than $1 billion (Wilmsen 1991, pp. 135ff).

Silverado and its associates contributed generously to the Greater Denver Corporation which, in turn, made extensive donations to political races, especially those involving people who would have an impact on the airport decision. By 1988, Silverado and its allies had contributed about $1 million to local campaigns and still more to congressional and presidential races (Wilmsen 1991, p. 134). They tended to hedge their bets, supporting both sides in any given election, either directly or indirectly. Larry Mizel, the chairman of MDC, was a renowned Republican fundraiser, but on his board sat Norman Brownstein, a prominent lawyer and Democratic fundraiser. Altogether, MDC raised $303,230, which it distributed to 29 candidates and five political groups in the 1986 gubernatorial election, the 1987 Denver mayoral election, and the 1987 Aurora mayoral election. The major recipients were Michael Licht, $98,700; Don Bain, $42,950; Ted Strickland, $29,050; Roy Romer, $24,250; Ken Kramer, $15,485; and Federico Peña, $12,320 (Attorney General 1991).

The flood of money greatly increased the cost of running for office. Richard Young, a leader of the airport opposition, pointed out that when Dale Tooley first ran for mayor in 1979, for example, his total expenditures amounted to $176,000; four years later, when Federico Peña was elected, expenditures amounted to more than $1 million (Young 1991), and his opponent spent twice as much. Money flowed to many other campaigns. In 1986 Republican Ted Strickland was running against Democrat Roy Romer. At first the contributions went to the Republican, who was leading in the polls. But when Romer caught up and seemed likely to win, the contributors switched their allegiance, donating about $100,000 to his campaign (Wilmsen 1991, p. 139).

Then came the 1987 mayoral campaign in which Peña was running for reelection against Michael Licht, an auditor, and Don Bain, a lawyer. Licht ran on an anti-airport platform, but he encountered difficulties when he was accused of using his staff in the campaign. Larry Mizel, the MDC chairman, visited him and offered to help. Licht changed his position on the airport, later saying, "You have to be flexible." Although $80,000 flowed to his campaign, the two leading vote getters were Peña and Bain. At first Don Bain was in the lead, and he received the more generous contributions. As a former Peña campaign official explained the situation:

> *We were substantially behind, and we didn't have any money . . . . There were four weeks to go in the campaign. At that point you don't ask questions. They were outspending us . . . . We had to get a lot of money fast. The way it usually works is that you go to someone who's been good to you in the past and see what they can do for you. The thing is that those people get a lot of access to the candidate. It's usually pretty subtle. They would say, "Boy, you could really help us out on something." You'd be hard-pressed not to be a little accommodating (Wilmsen 1991).*

In an era of media-blitz high-cost political campaigns, and in the absence of meaningful campaign-finance legislation, politicians often find they must make compromises in order to get elected.

It has also been suggested that Mr. Mizel played an important role in the decision to place the project on a fast track (Kilzer et al. 1995), but this appears to be merely one of the many tall tales surrounding the airport. Tom Gougeon, Mayor Peña's former aide, who was intimately involved in this and other major decisions, strongly objected to this allegation. Gougeon insists that the decision to complete the project in four to five years was actually made on the basis of careful analyses by the management team. They carried out detailed studies that balanced the additional financial costs involved in a longer timetable with the risks of accelerating the schedule and these served as the basis for the adminstration's decision. In Gougeon's words, "good or bad, the judgments were based on detailed analyses" (Gougeon 1995b). We analyze the consequences of this decision in Chapter 9.

In the meantime, Silverado collapsed, MDC lost $20 million, and Walters defaulted on his loans (Wilmsen 1991, p. 194). These developments attracted the attention of the Colorado Attorney General's office, which carried out an extensive investigation of MDC's political contributions and its links to the new airport. After spending more than 5000 investigator hours and conducting more than 225 interviews, the final report concluded that "Probable cause exists to believe that MDC committed numerous criminal violations relating to election laws....(but) the statute of limitations precludes the filing of any criminal charges" (Attorney General 1991). As far as the land deals were concerned, it found:

> *The land-loan transactions between MDC and Silverado are extensive and have been well documented . . . . Although MDC gave substantial amounts of money to the candidates for the elected offices that had substantial influence on the new airport, this investigation was unable to establish any quid pro quo by the candidates . . . .*

The Greater Denver Chamber of Commerce also used its powerful economic resources to provide major funding to the pro-airport campaign during the two referenda. It contributed $400,000 to the Adams County campaign and was responsible for the formation of the "Vote Yes" organization in the Denver election and largely for its financing. This group is discussed in more detail later. It acted to organize support for the project and to neutralize the growing opposition, though in reality, the chamber worked closely with the mayor's office to supply "Vote Yes" with most of its financial resources, its strategic analyses, and technical expertise. Thus, the "Vote Yes" organization, rather than representing a group of concerned individuals, as did "Vote No," actually was the voice of the business community, which successfully mobilized a strong volunteer effort. That the voters in Adams County and Denver endorsed the project was due in no small part to the efforts of the Chamber (*Colorado Economic Review* 1989).

The chamber and its allies also lobbied in Washington to gain federal funding for the project. According to Gael Clapper, director of the Colorado Forum, an organization of Colorado's 50 leading executives, the state's business leaders for at least a decade "followed the entire process, talking to the FAA, talking to our Congressmen or to the key people . . ." (Barnes 1993).

Without the work of the chamber and its related organizations, the airport would never have been built. Its officers, especially Richard Fleming (its CEO) and Robert Albin (the long-time chair of the chamber's airport committee), played a critical role in coordinating and directing efforts to build a new airport. They mobilized the business community, raised the necessary resources, and maintained pressure on doubters and opponents.

Above all, the chamber played a key role in building the powerful coalition that was necessary to help overcome local, regional, and national barriers. Why it chose to support a new airport rather than the expansion of Stapleton has been widely debated, because the decision to build a new airport would inevitably have a profound impact on the pattern of Denver's growth. Denver already was a dispersed community, consisting of three major activity zones inadequately linked with public transportation—the downtown core, the Denver Technological Center to the southeast, and Stapleton International Airport. Despite major efforts at urban renewal (including the construction of a new convention center, the pedestrian 16th Street mall, and architecture designed by the renowned I.M. Pei), downtown Denver long had been a major source of concern to the city, which continued efforts to revitalize it. As in most cities, major retail and office centers had moved to the suburbs.

The construction of a new airport would do little to help the downtown and would push Denver's development further to the east. Metropolitan Denver already was plagued with massive suburban sprawl, with less than 2 million people spread over a land mass the size of the five boroughs of New York. As the dominant mode of surface transport, the automobile was already creating intolerable concentrations of carbon monoxide and particulates, which made the city's air inhospitable when temperature inversions trapped it in the Platte River valley. If the new airport were to become, as its supporters argued, a growth pole, that growth would probably come at the expense of growth elsewhere in the community and was likely to make Denver a sprawling conurbation, much like Los Angeles. Other important decentralizing forces were already at work, perhaps making such an outcome inevitable, but obviously the airport would greatly increase the probability. Why the chamber, which represented numerous businesses in the downtown and nearby areas, would opt for a project that would not benefit and might well injure many of its members, requires an explanation.

The answer to this question lies in the nature of the organization. Although it had a large membership, the chamber's leadership was no longer what it had traditionally been, drawn from the downtown business elite. It had been renamed the Greater Denver Chamber of Commerce, a title that reflected its metropolitan focus. Now it was centered precisely in those companies and individuals who, as noted earlier, stood to benefit greatly, both in the short and long term, from any project that would promote growth and those involved in real estate and finance who could profit from the opportunities for speculation that would be generated by the new facility.

The external actors

Adams County

It can be argued that the new airport was conceived in Adams County, for that county fought the expansion of Stapleton and made available to Denver the land on which to build DIA. Its opposition (and that of other neighboring communities, including Aurora, Thornton, Commerce City, Brighton, Northglenn, and Federal Heights) to the expansion of Stapleton dated back to at least 1968 (as noted in Chapter 2). Opposition was based on environmental and economic considerations since this area would be affected by noise pollution and other adverse impacts. Most significant was the pattern of growth that was taking place. The Denver metropolitan area was expanding rapidly but primarily to the southeast, southwest, and northwest. Growth in the northeast, where Adams County was located, was geographically blocked by Stapleton Airport and the adjacent Rocky Mountain Arsenal. Furthermore, Adams County was losing businesses and would gain little from the expansion of Stapleton. On the other hand, if a new airport were built in the relatively unpopulated northeast area, Adams County would benefit directly from tax revenues as well as from the ancillary growth. Nor could one overlook the new jobs that would be created in the short term (construction) and in the long term (services and operations).

In addition to such concrete elements, one must also consider the history of relations between Denver and the surrounding communities. For years Denver had been regarded as an arrogant city seeking to expand at the expense of its neighbors while refusing to take their interests, especially water, into account (Denver had refused to share its vast water resources during drought years, especially between 1954 and 1962). Furthermore, Adams County was a largely unplanned industrial area, home to such facilities as a petroleum refinery, a cement plant, a dog track, and grain elevators. Its residents felt that Denverites considered them their social inferiors. These resentments, which were shared by other neighboring communities, were codified by the Poundstone amendment (1974) to the Colorado Constitution, which essentially prevented Denver from making any further annexations. Given such a context, it is not surprising that the two sides were polarized on the airport issue and that their stands were rigid and inflexible (Moore 1990).

A word about the Poundstone Amendment is in order. In 1969, a local federal judge ordered limited school busing to desegregate Denver's schools, because schools in the Park Hill section of the city (ironically, the residential area closest to Stapleton Airport) had been intentionally segregated. In 1973, in a landmark decision authored by Justice Brennan, the U.S. Supreme Court ordered that schools be desegregated "root and branch" and that busing be carried out citywide, despite the fact that efforts to keep the schools segregated had existed only in Park Hill (Dwyer and Sutton 1994; Keyes 1973). It was an era of aggressive social engineering by the federal judiciary.

Shortly thereafter, a suburbanite, Freda Poundstone, began circulating petitions for a ballot initiative to amend the Colorado constitution, prohibiting annexation by the City and County of Denver of any additional real estate without a majority vote of all residents of the county, any part of which Denver sought to annex. The Poundstone Amendment passed easily, largely because of the suburbanites' desire to protect their schools, and Denver's boundaries were effectively frozen in place. Over the years, Denver's schools had become more segregated as middle-class white families moved to the suburbs in what has been described as "white flight," many to take advantage of community schools (Amole 1995). In 1972, two-thirds of Denver's students were white (about 63,000 students); by 1995, only 29 percent, or about 18,000, were white (Dwyer and Sutton 1994; *Rocky Mountain News* 1995b). As in most forced-busing cities, the city's tax base did not keep pace with the suburban communities, while its social welfare burden grew and its population declined.

Now, for the first time in two decades, Denver had a chance to significantly increase its tax base. The annexation for DIA was massive, comprising 53 square miles, larger by far than the core of the city. This would free up the 7.5 square mile Stapleton Airport site for redevelopment and enable Denver to take advantage of suburban growth on the northeastern perimeter of the metropolitan area. Certainly, the opportunity for Denver to grab suburban land was a powerful motivator in driving the decision to build a new airport.

In fighting Denver's plan to expand Stapleton, the Adams County commissioners found ready allies among their neighbors and formed the Airport Coordinating Committee. In September 1981, they flew to New York, where they met with Eliot Cutler, who had served as special counsel to President Carter. Following President Carter's defeat, Cutler had jointed a prestigious law firm that included John Lindsay, former mayor of New York. Cutler developed a 25-page plan that they would eventually use as he continued to advise them. The first step called for a study of contamination at the Arsenal; later, in 1983, Adams County published a report on the advantages of closing and redeveloping Stapleton. Cutler's total bill eventually amounted to $7 million (Carnahan and Hubbard 1995a).

The strategy that Cutler devised consisted of three major elements. The first involved designating the Arsenal, which lay within the boundaries of Adams County, "as an area of state interest." Under Colorado statutes, such a designation would entitle Adams County to exercise considerable jurisdiction over the area. Such a filing meant that the Colorado Land Use Commission (CLUC) would convene and set up a special subcommittee to oversee actions regarding the Arsenal. And, Adams County could always take legal action to force the courts to review the CLUC's decisions. In this way Adams County greatly enhanced its bargaining position. Its actions ensured that attempts by Denver to implement its expansion plan would become far more costly and time consuming, whereas building a new airport would encounter far less opposition.

In addition, the Adams County commissioners would challenge the environmental impact statement (EIS), which, since the project involved significant federal funds, had to be filed with the Environmental Protection Agency (EPA). Under the National Environmental Policy Act of 1969, an EIS must be prepared before any major federal action that might degrade the quality of the human or natural environment. As required by law, the draft had to be made available to the public and the relevant agencies 90 days before the project began, and a final statement dealing with all the comments and objections that were registered had to be made public 60 days later.

The process by which an environmental impact statement is prepared offers many opportunities to opponents of a project to delay and perhaps even halt it. There are two common ways of fighting an EIS. The first involves making strenuous objections to the content and findings of the initial draft and to raise issues that are not covered. Sometimes certain sections may be omitted; more often, trivial issues may be neglected. In either case, these topics must be studied and included before the final EIS can be submitted, thus delaying the project. The second method of attack is a legal challenge in court on the basis that the information contained in the EIS is erroneous. If the court agrees to hear such a case, the final release of the EIS can be delayed for years. This happened, for example, in the case of Two Forks Dam, a project that was an important element in Denver's growth strategy but was stopped through the EIS process (Moran 1986, p. 19).

Adams County planned to attack the EIS on several key points. First, it would raise the issue of noise pollution and its impact on the neighboring communities. The second concern would be air pollution, since expanding Stapleton would inevitably lead to more traffic and hence more air pollution in Denver. The third issue would be the chemicals in the soil of the Rocky Mountain Arsenal, whose toxicity was by now widely recognized. In order to expand Stapleton, new runways would have to be built, a process that would involve plowing and leveling, which would create a great deal of dust and potentially release chemicals into the atmosphere. Adams County also planned to try to force Stapleton to assume responsibility for any chemical release caused by the construction, a move which, if successful, would create enormous legal liabilities for the airport.

The final step in Adams County's strategy was to prevent the army from transferring the land to Denver. Since the Department of Defense controlled the land, Denver required the DOD's permission to build. But Adams County claimed that the land had originally been part of the county and should revert to it. Once again, the specter of lawyers battling for years loomed on the horizon as the county filed suit in 1982 to prevent Denver from building on the Arsenal.

Despite the obstacles that this strategy placed in the way of expansion, it is not at all clear that it would have worked. Two key issues were involved: environmental impact and the ownership of the land. As far as the former is concerned, it is important to recognize that of the 27 sections of land contained in the Arsenal, only nine were designated as contaminated. Where the expansion would take place, "Along the south boundary some areas are pretty clean," according to Art Whitney, public information officer for the Rocky Mountain Arsenal. Both the City of Denver and the airlines felt that the airport's expansion could take place in the safe areas (Glass 1982).

Furthermore, Denver felt that it could obtain legal access to the required land. The FAA had officially notified Mayor McNichols that it would "be pleased to assist Denver in any way we can" in regards to expansion onto the Arsenal. Such an assurance was extremely significant; under federal law, government agencies are to cede land for airport use if the FAA makes such a request, unless the agencies can demonstrate a more important need. In this case, the army was considering abandoning the Arsenal for another location and had officially notified Mayor McNichols that Denver could probably have access to the southern section by 1986. On the other hand, Adams County officials were informed that the army would not release the land until comprehensive surveys had been carried out and methods of cleaning up the contamination identified. This was not expected to occur until 1990. Still, in May 1982, army undersecretary Paul Johnson stated "if everything goes on schedule, our mission at the Arsenal will be completed by 1986 and we would see no reason not to give the arsenal to a municipality if they need it, including the City of Denver for an airport" (Glass 1982). The airlines (Frontier, Continental and United, the three hubbing carriers) also supported the expansion onto the Arsenal. The DRCOG endorsed that alternative as well. Since the final decision would be made in Washington, these were powerful considerations.

Hence, it is not clear that Adams County's strategy would have prevailed, although it certainly could have delayed the process. This was an important consideration, since Mayor Peña wanted to solve the capacity problem as rapidly as possible. Even so, Denver might have been able to negotiate an arrangement with Adams County that would have permitted Stapleton to expand onto the Rocky Mountain Arsenal expeditiously. Such an arrangement would have involved economic concessions of various types and, as late as October 1984, Denver and Adams County were discussing a possible deal that would involve building a new east-west runway on the southern edge of the Arsenal and, over the long term, additional runways and a new terminal on the northern end of the arsenal (*Denver Post* 1995).

These possibilities became moot when Adams County unveiled the other part of its strategy. A new group of leaders who wished to promote a new kind of development had come to power. Recognizing that a new airport built on the undeveloped side of the county could make an enormous contribution, in the early 1980s they tried to persuade Mayor McNichols to change his mind about the desirability of expanding Stapleton. According to Denver Democratic Party Chairman Richard Young, in a private meeting, Eliot Cutler threatened to tie up the expansion project for 10 years, but McNichols responded that such a delay was no problem since planning the project would require a decade anyway.

Adams County, however, continued to press its plan. In January 1984, while filing its claim for "an area of state interest" designation, its leaders again indicated their willingness to permit Denver to annex land for a new facility. When Federico Peña was elected mayor in 1983, he brought a new orientation to the issue. Whereas Mayor McNichols did not have a high regard for Adams County and could not forge an alliance with it, Peña had a different perspective. Having reviewed his options, he changed Denver's policy and began negotiations to explore the feasibility of the new airport. These culminated, in January 1985, in the Memorandum of Understanding (MOU), which states:

> *The parties agree that a new airport to replace Stapleton should be constructed by 1995 or earlier . . . Denver will annex sufficient land in Adams County to accommodate the new airport . . . Denver agrees to complete transition to the new airport as soon as possible but no later than the year 2000, and agrees that after transition has been completed, no aircraft will be permitted to operate at the existing Stapleton International Airport.*

The decision to enter into such a deal was not universally welcomed within Adams County, because the important issues of noise pollution and development remained unsettled in the minds of many, especially those living in the part of the county where the new airport would be located. To quell the opposition, negotiations continued between Denver and Adams County, and by April 1988 agreement had been reached on the following key issues:

1. Denver would help pay for selected infrastructure development around the airport
2. Adams County would share in the economic benefits (including the jobs) that would be realized through the construction of the new airport
3. Noise limitations would be imposed upon the new facility to protect the inhabitants.

Despite such steps, uneasiness with the agreement remained strong, fueled to a significant degree by the historic divide between the two communities.

The federal government

The EPA As noted, in 1969 the National Environmental Policy Act (NEPA) was enacted, which subjected all major federal projects to an environmental assessment. From then on all airport projects would have to undergo an extensive environmental impact review in which the public would participate and whose results would have to be approved by the EPA and the FAA. The new Denver airport became the first project of its kind to be fully subject to the NEPA requirements, because it is the first airport to be built in the United States since the Dallas/Fort Worth International airport. That project (completed in 1974) was already well underway when NEPA was enacted and thus was exempted from meeting all of the act's stringent requirements.

Although the environmental impact assessment was a dominant concern throughout the planning process, the EPA itself played a relatively minor role. The bulk of the work done in preparing the final environmental impact statement (FEIS) was carried out by the city with the continuous approval of the FAA.

On April 29, 1988, the FAA published a Notice of Intent "to prepare an EIS and conduct an EIS scoping meeting to be held on June 6, 1988," at which all interested agencies could identify the topics that the EIS should cover. The final EIS was approved by the FAA and the EPA about a year later (see Chapter 6 for more information on the environmental impact process).

Congress Federal funding was crucial to the project's success. Even though the FAA developed priority rankings for many projects it endorsed, Congress had the final say—it had created the structure within which the process took place and held the purse strings. Congress would make decisions in such areas as the total amount that would be made available from the Airport and Airway Trust Fund, the annual disbursements, and the physical allocation of funds, all in a manner that provided a steady flow of capital. The fund was created by the Airport and Airway Development Act of 1970, which imposed a tax on airline passengers that would be disbursed to projects designed to improve the safety and quality of the airway system. By the late 1980s, nearly $15 billion had accumulated in the Airport and Airway Trust Fund. A major grant from its assets was crucial if DIA were to be brought to fruition, not only because of financial needs but also because it would legitimize the entire project.

The process began in 1987 with Senate hearings conducted in Denver on the administration of the Airport and Airway Trust Fund and the need for a long-term authorization bill. Fortunately for Denver, the times were extremely propitious. The administration of the fund was being subjected to harsh criticism because its resources were not being spent for its original purpose. Senator Tim Wirth pointed out that "the administration . . . has proposed financing 85% of the FAA's operational budget with monies that are currently in the Trust Fund . . . the purpose of the Trust Fund is not to serve as a back door method for financing the FAA" (Hearings 1987, p. 216). Furthermore, although the fund had built up a massive surplus, the administration was reluctant to spend it on improvements in the airway system, preferring to use it to minimize the budget deficit. As Senator Wendell Ford noted: "Could it be that the Office of Management and Budget enjoys using the Trust Fund's huge surplus to keep the overall budget deficit from looking worse than it already is? Must we play these ridiculous little games?" (Hearings 1987, p. 3).

Furthermore, Denver was able to make a strong case for allocating some of the surplus to DIA, because a study demonstrated that although local passengers had contributed more than $1 billion to the fund since 1984, Denver had received the least discretionary funding—$600,000 of the $487 million that had been disbursed to the top 25 airports between 1982 and 1988. Although this lack of funding was due to the debate over the future of Stapleton that raged during this period, the new airport provided the administration with the opportunity to redress the imbalance (Denver 1990). At the same time (mid-1980s), serious congestion problems became commonplace. Many advocates of deregulation believed that the problems were not the fault of the policy but resulted from a lack of new capacity to keep pace with rising demand. Many of the deregulation-policy advocates in the Reagan and Bush administrations and in Congress pushed to provide funding for more airports, especially for a new hub airport at Denver.

Although this provided a positive context, Congress still had to approve the allocation of funds, and securing them was no simple matter. Other states wanted to obtain airport funding, and many representatives and senators wished to minimize expenditures to balance the budget.

Colorado's congressional delegation, which unanimously supported the project (although some members of the delegation—notably Congressman Joel Hefley of Colorado Springs and Senator Bill Armstrong—had doubts that they largely kept to themselves), was quite a powerful one, especially for a small state. Senators Tim Wirth and Armstrong were on the budget committee, where they were able to prevent the diversion of monies to other activities, and Congressman David Skaggs served on the Aviation Subcommittee of the House Transportation Committee, although he was a relatively junior member, ranking 15th out of 21. No member of the Colorado delegation, however, was on the important appropriations committees.

That weakness was made up through personal ties to Senate Majority Leader Robert Byrd and especially Senator Frank Lautenberg of New Jersey, the chairman of the critical Appropriations Subcommittee on Transportation, who had a vacation home in Vail, Colorado, and was a good friend of Senator Wirth. His support was essential, because he blocked various efforts to eliminate funding for the project. In 1989, the House eliminated $102 million in funding for DIA, but Lautenberg persuaded the Senate to approve a $50 million appropriation and provisions that would enable Denver to receive the remaining $52 million in the future. Finally, Congress approved $40 million and defeated efforts to impose tougher restrictions on Denver's future requests for funding (Barnes 1993).

Coloradans proved to be appropriately grateful, and a steady stream of contributions began to flow to Senator Lautenberg. Altogether, after 1987 he received more than $100,000. As an associate of one of the contributors remarked, "What it was is a payoff" (Barnes 1993). Donations came from Democrats and Republicans alike, most of whom stood to benefit from the new airport. They included the former chairman of USWest and its political action committee, or PAC (USWest provided much of DIA's communications infrastructure); the head of Public Service Company of Colorado; Norman Brownstein, a prominent lawyer and Democratic fundraiser, and his associates (his law firm acted as special counsel for the bond issues); Walter Imhoff, the head of a bond house; and various real estate developers, including H. C. Van Schaack III, Leland Alpert, and John Fuller, a Republican whose contribution represented his first to a Democrat in many years. One prominent Denver businessman summed up the prevailing attitude: "It's the way Washington works. He did us a favor. I personally (contributed) as a gesture of thanks. I hope he would continue his support." Lautenberg's position, as explained by his chief of staff, was that "The Senator considered the airport on its merits and was convinced that it was a good investment" (Barnes 1993).

The FAA The FAA became involved in the DIA project during the early studies by DRCOG. Its leaders had long believed that the United States needed additional airport capacity to handle the growing traffic. Denver clearly headed the list. Stapleton was viewed as a major bottleneck in the national air transportation system because its runway separation did not meet the criteria for simultaneous instrument operations. Hence, the FAA argued, flight delays were commonplace, and the effects rippled throughout the country, particularly when Denver served as hub for three airlines, causing delays at numerous other airports. As noted previously, the seriousness of this problem, or more precisely, the contribution that a new airport would make toward eliminating delays in the national airway system, remained debatable.

In any event, the FAA worked closely with Denver area officials to explore ways of remedying what were viewed as inadequate facilities at Stapleton. Indeed, a former FAA official recalled traveling around the area looking for suitable sites for a new airport in the early 1970s. Furthermore, land was obviously available, and Denver's political establishment strongly supported the idea of enhancing airport capacity. A former FAA official recalled receiving a call from H. R. Haldeman, President Nixon's chief of staff: "I thought he worked for the FAA. We talked about the airport and then he asked me if I knew who Joseph Coors [a scion of the Coors Brewing Company, which is a major force in Denver] was . . . it clicked . . . He said if there was going to be a new airport, it would be built in Denver by Denver" (Barnes 1993).

Those words proved prophetic. The FAA does not intervene directly or take control of airport projects; its official role is to approve or disapprove specific proposals. In the early 1980s, the FAA believed that Stapleton expansion was the best way to deal with the capacity problem, and it funded a new runway. When Denver decided to build DIA, it prepared an analysis (spring 1987) that did not consider five alternative sites in any detail but essentially focused on its preferred choice. When the FAA rejected Denver's study on the grounds that "The study documentation does not support the . . . position that the only reasonable new airport site is the one preferred by Denver," the city objected vehemently. After all, five months earlier, the FAA had declared "The FAA is committed to working closely with Denver to process federal approvals as expeditiously as possible so that the city can realize an accelerated schedule for construction of the new airport" (Hubbard 1995). Clearly there was a fundamental difference between the FAA field officials and their superiors in Washington, the latter wishing, above all, to see a new airport built, the former concerned with the ways in which the capacity problem could best be resolved.

Peter Melia was the FAA official overseeing Denver's new airport. On January 27, 1987, he wrote a position paper predicting "heavy political pressure, particularly from the mayor, since anything negative regarding the new airport could impact the mayor's re-election and the Adams County annexation vote." On February 25, 1987, Melia wrote a three-page letter insisting that Denver do a better job of justifying the site for the new airport. According to newspaper accounts, Peña went ballistic, complaining to FAA's leadership in Washington and to the Secretary of Transportation (Hubbard 1995). At the same time, Colorado Senator Bill Armstrong met with the FAA brass. After a meeting between the head of the FAA and Mayor Peña on March 11, the FAA changed its position. Melia, and indeed, the entire FAA office, were no longer included in the process; a new FAA office was established for the project (Hubbard 1995; Kilzer et al. 1995). Now the FAA accepted Denver's position, deciding not to ". . . undertake a de novo comprehensive analysis of an optimum site and configuration for a new airport since CCD (City and County of Denver), as the sponsor and future airport proprietor, has the fundamental role of deciding whether or not and when to build a new airport, and ultimately for planning, constructing and operating the airport" (FAA 1989).

The degree to which the FAA was willing to accommodate the city is also reflected in its decision not to endorse or publish a draft letter it had prepared concerning the noise pollution provisions of the Intergovernmental Agreement that Denver and Adams County had signed, which specified particular noise restrictions (see Chapter 6 for more on the noise agreement). Those provisions were viewed by the FAA as an infringement of its authority, and it stated in its letter that it would not implement those provisions. Given the public's sensitivity to noise impacts, the city's officials immediately recognized the damage that such a statement would wreak upon its efforts to win the local election and swung into action immediately. George Doughty, Denver's director of aviation, drafted a formal response and, together with the city's Washington lobbyist, successfully persuaded various FAA and DOT officials to block the letter (Kowalski 1989c).

As the new project took definite shape, the FAA provided technical guidance, advice, and its formal approval, which was required for many aspects of the project (such as air traffic control and management procedures and facilities). The FAA's role was critical because of the financial contribution that could be provided by the Airport and Airway Trust Fund. Without the FAA's support, efforts to generate significant funding would obviously be greatly complicated. Furthermore, the FAA would spend between $150 to $200 million of its own funds on its facilities at the airport.

Gaining FAA support was probably the smallest problem that the airport proponents faced. From the outset, the FAA's forecasts and studies had served to justify the project that the FAA would be asked to help implement. Furthermore, the political power that the project's proponents were able to generate ensured that whatever opposition might exist would be overcome easily. Not surprisingly, therefore, the FAA was one of the strongest supporters of the project. It made certain that the project met all applicable criteria at each step. When the final proposal was published, the FAA was fully prepared to defend its merits and see the project completed. The FAA's enthusiasm was officially based on:

1. Its belief, based on its traffic forecasts (described in the preceding chapter), that such a facility was needed
2. That a new airport would eliminate the chronic delays caused by Stapleton
3. That it would increase safety and efficiency because of the runway and taxiway configurations.

The *Rocky Mountain News* later summarized how the city managed to manipulate the FAA into supporting DIA:

- The city got criticism of the project quashed at the highest levels in Washington.
- FAA analysts skeptical of the city's plans were stripped of airport duties.
- The city enlisted the aid of Department of Transportation and FAA officials in Washington to fend off strident opposition from the airlines.
- Just before Adams County voters decided whether to allow Denver to annex land for the airport, the FAA kept quiet about serious concerns regarding noise-control promises made to county residents (Hubbard 1995).

The airlines

After the demise of Frontier Airlines, Continental Airlines and United Airlines were the two major hub carriers operating at Stapleton. Keen competitors, they dominated the traffic, each accounting for between 30 percent and 50 percent of the total traffic from 1986 until 1994. They were opposed to the construction of a new airport, fearing that their operating costs would increase greatly. They also objected to the fact that they were not active participants in the decision process, even though they would bear the financial burdens and be responsible for most of the revenues that would be used to finance the project.

Despite their opposition, the airlines signed agreements with Denver (September 1985) to pay a certain proportion of the costs of building the new facility. But over time, they became more and more opposed to the project. In the words of a UAL press release, which was issued about the time that major improvements to its facilities at Stapleton were completed:

> *United is not opposed to building new airports when needed and economically justified; however, we have consistently maintained that every effort must first be made to enhance the capacity of existing airports. . . . There must be a thorough and professional evaluation that clearly established the financial ability of the airport users to reasonably absorb the cost of a new airport through airport landing fees, rents and other charges.*

By July 1987, the airlines had stiffened their position. They were critical of the escalating projected costs of the project and their limited ability to participate. When Mayor Peña decided to push the project's completion date to 1992, their concerns were heightened, and they refused to participate in the project, although their technical involvement never ceased, even during periods of acute conflict. Accordingly, Mayor Peña halted scheduled improvements at Stapleton, increased the landing fees, and decided to continue construction of DIA, even with the airlines in opposition.

The airlines chose to fight behind the scenes, carrying out studies that challenged the passenger forecasts that had been made by the FAA and Denver, as well as the projected savings from the new facility, and presenting them to the FAA. But they did not enter the political struggle, choosing to remain neutral throughout the elections. They did not support either of the groups that were attempting to secure the defeat of the project and did not participate in the campaign in any way.

Even after the project was well underway, United might have killed it by refusing to sign an agreement to operate at the new airport. Apparently United gave serious consideration to this possibility because of the high costs of operating there. However, it chose not to refuse for two reasons. The first, which the city had correctly anticipated, was that it had no alternative—its Chicago O'Hare hub was congested, and other locations possessed serious disadvantages. But one should not minimize the second—UAL also decided that Continental's financial weakness gave it a prime opportunity to expel it from the market and establish a fortress hub at DIA. Also, then-Secretary of Transportation Sam Skinner pressured United vigorously to sign a lease at the new airport.

The political process

The Adams County annexation election

Early signs indicated that the proposal would win because a plurality of the voters supported it (37 to 25 percent, with 37 percent undecided in July 1987). But as the campaign progressed, the opponents gained steadily, and by early May 1988, 46 percent were in favor and 46 percent were opposed. That, however, was the high-water mark for the opposition, which could not overcome the financial and political power of the proponents who had organized Partners in Progress for a New Airport (PIP).

As the name suggests, the proponents portrayed themselves as favoring progress and their opponents as disgruntled critics who sacrificed economic benefits and feared the future. They emphasized the jobs-creation aspects of the project—the local unemployment rate stood at 9.3 percent—and its implications for the long-term development of the county. The ballot that voters were given, for example, read, in part, "Studies have shown that the significant economic activity resulting from the construction and operation of the new airport should generate substantial net increases in county, city and school district revenues" (Kilzer et al. 1995).

PIP's campaign was carefully crafted and carried out with considerable political skill. PIP was organized a full year prior to the election and was richly financed by the Greater Denver Corporation, an arm of the Greater Denver Chamber of Commerce. Altogether, PIP raised and spent more than $1 million on the campaign, about a third of which was devoted to the last weeks of the campaign. At least a quarter of the funding came from the corporation.

PIP pulled together an experienced political group for the campaign. Richard Fleming, head of the chamber, recruited John Frew, once a key aide to former senator Tim Wirth and manager of his 1986 campaign, to head the effort. He was aided by Mike Stratton, who had helped manage Colorado Senator Gary Hart's presidential campaign, as well as three other experienced political figures. They had to overcome concerns about noise levels, pollution, increased taxes, and continued hostility to Denver.

A skilled political strategist, Frew coordinated the daily activities with "military precision." More than 1500 volunteers canvassed the voters, aiming to reach every one. They distributed thousands of brochures, leaflets, and other publications and were supported by a sophisticated computerized database on which every voter was ranked and precise neighborhoods targeted. Leading proponents made innumerable appearances at civic organizations, clubs, and forums of all kinds; Governor Romer toured the "oatmeal circuit" for two months, making countless early-morning visits to diners, coffee shops, and truck stops. A political consultant described the campaign in this way: "If you respect professionalism, it would bring tears to your eyes" (Carnahan and Hubbard 1995a; Harkavy 1994).

On the other side was an opposition divided into three separate groups—The Citizens Airport Coalition, which later became the Committee for a Better Airport (CBA), Ordinary People for a Better Adams County, and the provocatively named Stop Denver's Rip-Off of Adams County. The CBA was organized by several opponents of the project who possessed extensive political experience, had been involved in the airport issue for years, and enjoyed considerable grass-roots support. The Ordinary People group represented the Van Aire community, a group of about 100 families who owned their own planes and a private landing strip, the continued use of which would be threatened by the new airport. Rip-Off was the brainchild of a prominent Republican lawyer and political activist. These groups, however, were divided in their goals so that, despite various attempts at cooperation, the opposition remained fragmented, each pursuing its own tactics and objectives.[3]

The opposition had to fight the election with scant resources. CBA's total budget for the campaign was a mere $9200, half of which came from individuals, most in the form of modest contributions. The other two groups were far better funded, although the sums involved were modest compared to PIP's resources. The Van Aire group raised over $50,000, but it spent these funds ineffectively. Rip-Off raised $34,255, $29,500 of which came from another organization, Committee for a New Stapleton Airport, whose funding was largely provided by land investment firms in Oregon and Arizona (Moore 1990). Thus, the proponents spent 10 times more on the campaign than the Adams County opposition groups. Even so, the new airport proponents might well have lost if they had not been able to expend such large resources, especially in the last stages of the campaign. Also contributing greatly to the victory was the inability of the opposition to unite, to promote any kind of positive program, and to prevent itself from being portrayed as totally negative. The vote was 34,070 for, 26,828 against.

Still, the achievements of PIP should not be underestimated. It made the airport possible despite "the distribution of tangible, irreversible, negative externalities, and the obscurity of the benefits of a large-scale municipal public works project, strong (even vehement) suburban county antagonism, the aggressiveness of its detractors, [and] the intergovernmental complexity of the project's constituent parts" (Moore 1990).

3. Much of the discussion of the political activity in Adams County, including the election, is drawn from Moore (1990).

Although the results of the election favored the annexation, it is ironic that two of the three commissioners who supported it were subsequently defeated at the polls. James Nelms, one of the leaders of the opposition, fought to unseat the incumbents and won election as a commissioner. This action may well reflect the ambivalence that the voters of Adams County felt about the annexation—supporting it in May, voting against its leaders six months later.

The Denver election

This election was neither foreseen nor desired by the city and its supporters, who were surprised by the opposition that the project continued to engender. Several leading citizens, including former mayor Bill McNichols, former Republican state senator Bill Chenoweth, Former Denver Democratic party chairman Richard Young, and former Denver safety manager J. D. MacFarlane, were calling for a public vote even though the groundbreaking was only six months away. Young and Chenoweth, two former political opponents who became the leaders of the "Vote No" group, wrote privately to Mayor Peña, urging him to reconsider. The mayor responded by releasing the letter and attacking them politically. Governor Romer held a news conference at which he made his famous threat "to roll over and crush" the opposition (Carnahan and Hubbard 1995b). Undeterred, they proposed that the city council hold public hearings, and, finally, after some hesitation, it decided to do so.

Over two days, more than 300 people showed up to voice their opinions. The mayor's office arranged the order of speakers, giving supporters priority even though many opponents had arrived before them. In this way, the important 10 P.M. news shows would show widespread support for the project (Grelen and Gavin 1989; Kowalski 1989b). Of course, interested citizens with the necessary stamina could watch the entire hearings on a cable channel. After some delay, during which the opponents vowed that they would initiate a petition drive to force an election, if necessary, the Denver City Council finally voted for an election. The entire process affected many, including several council members, Governor Roy Romer (who had originally argued that such a vote would damage the project as well as Colorado's reputation), Chamber president Richard Fleming, and eventually Mayor Peña (who had insisted that an election was not necessary even though public opinion polls showed that more than half of the voters supported the project and that two-thirds wanted the opportunity to vote). Thus these people were persuaded that a special election was a political necessity, even though its results would not be legally binding (Gavin 1988).

The mayor's opposition to a referendum was based on recent history. In 1972, Colorado had been designated as the site of the 1976 Winter Olympic Games, and in 1985, the city had planned to build a convention center near the railway station. But even though business and political leaders strongly supported both projects, they had been defeated when opponents forced a public vote. Those events (particularly the Olympics vote) reflected an important change in the ways in which decisions about major projects would be reached in the state. Whereas the public had theretofore participated only indirectly, through its elected representatives, now people felt increasingly entitled to vote directly on major projects and could place an issue on the ballot with relative ease (Fig. 3-3).

Although Governor Romer subsequently retracted his intemperate threat to "roll over and crush" the opposition, such terms accurately reflected the tone of the campaign. Many important figures had doubts about the project—including former governor Richard Lamm, who later explained that he did not voice his reservations because of his loyalty to Governor Romer (Carnahan and Hubbard 1995b), and former Denver mayor Bill McNichols, who criticized the project once. Few, however, participated in the campaign. Indeed, enormous pressure was brought to bear upon anyone who was critical of the project. One publicized example is that of Mike Bell, once a senior vice-president at Boettcher and Company, a local investment firm, who prepared a study that showed that the airport could economically be mothballed. Peña and airport officials summoned Boettcher's executives to demand that the report not be released and the computer files destroyed. The executives agreed (Kilzer et al. 1995).

Both Young and Chenoweth were harshly attacked for their efforts to defeat the project. Chenoweth received hate mail from "people I thought were my friends, prominent guys in Denver. They called me every name." Young, an attorney, lost clients. Chenoweth recalls that on at least two occasions, while he was in Young's office, he heard clients calling to fire him, saying, "We can't get involved with you. You're destroying Denver." Dick Lamm described the situation as follows: "[Dick Young] felt Denver was making a terrible mistake. He

Metropolitan Denver Public Works Expenditures (in millions) 1985–1995

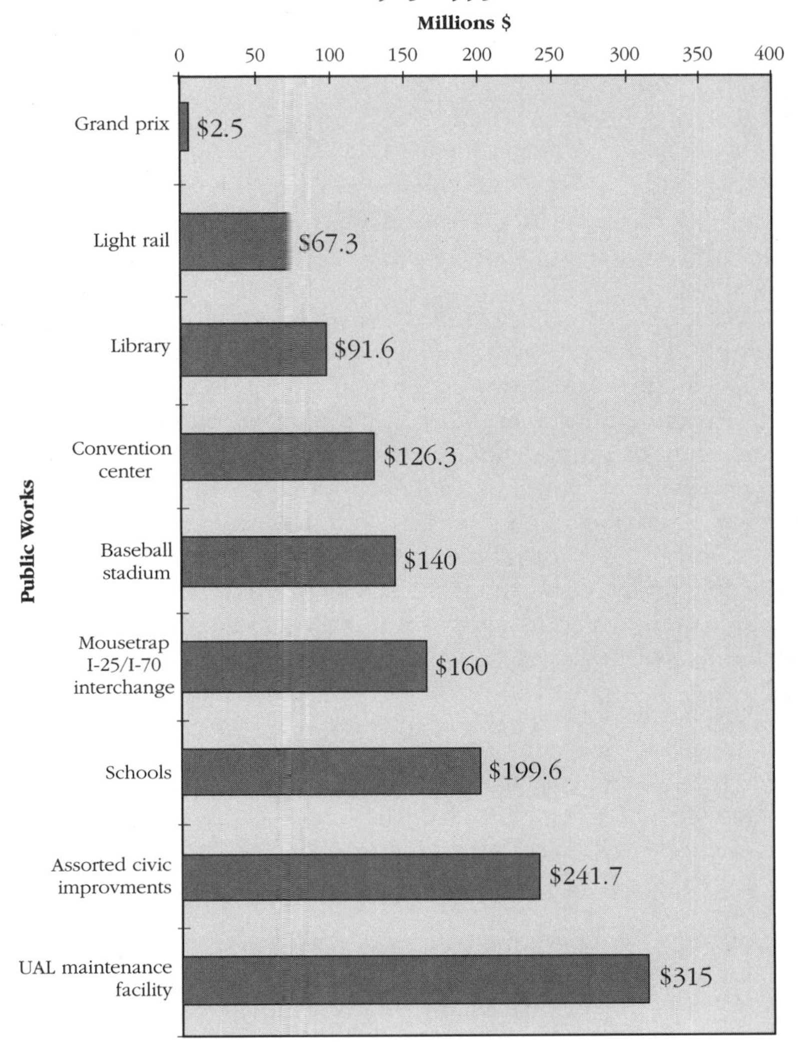

Fig. 3-3. *The new airport was only one of several major public works projects approved by the voters of metropolitan Denver to jump-start the stagnant Colorado economy of the 1980s. Although $315 million was approved for a United Airlines' maintenance facility at Denver, United unexpectedly walked away, instead building its major maintenance facility at Indianapolis.*

drove himself . . . and his law practice into the ground trying to prove it" (Carnahan and Hubbard 1995b). Young was a tragic figure, subsequently disbarred from the practice of law because of ethical financial lapses in his legal practice, perhaps caused by the exodus of clients precipitated by his high-profile opposition to DIA.

The opposition The leading opponents of the new airport, the Vote No group, was organized by Young and Chenoweth and enjoyed the support of a small band of citizens who feared that the proposed airport would be a disaster for Denver and its economy. Composed largely of grass-roots volunteers, it possessed only limited financial and technical resources.

Also fighting the project was a second organization, the Save Our Stapleton (SOS) Group. As the referendum approached, the owners of hotels near Stapleton Airport became increasingly concerned that the new airport would be approved by a majority of the voters. Accordingly, they organized themselves and raised sufficient funds to hire a well-known political consultant from New York, Roger Ailes, to prepare a series of television commercials. The move backfired because Denver's citizens proved highly resistant to the idea of having outside experts intervene in the election. Furthermore, SOS was effectively portrayed as a selfish organization concerned only with the well-being of the hotelkeepers.

This group was the mirror image of the Vote No organization. It possessed little grass-roots support, but was reasonably financed and organized. If these groups had combined their efforts, they would have achieved better results, but the differences in their philosophies and orientations prevented them from doing so—even though they shared a common goal.

The proponents The new airport proponents suffered neither from divisions nor lack of technical expertise or a shortage of financial resources, because they reassembled the team that had won the Adams County election, with John Frew once again in charge, and organized the Vote Yes group. It was run by a 15-member steering committee composed of representatives from the Chamber of Commerce, the mayor's office, and some business organizations, but John Frew was responsible for the overall campaign.

Its activities were closely coordinated with those of the city, especially the mayor's office and the Chamber of Commerce, which provided most of the funding for the campaign. Essentially, Vote Yes was a creature of the city and its supporters in the Chamber of

Commerce, who believed that the airport was an essential element in the future growth of Denver but who recognized the need for a "front" organization. Active public involvement by the chamber might have been a liability or a distraction. Accordingly, Vote Yes was established to assume responsibility for the advertising and media and for all other aspects of the election campaign, including mobilizing support and endorsements from numerous individuals and many neighborhood and other organizations.

The chamber chose to remain in the background to make Vote Yes appear as an independent organization that represented concerned individuals. The chamber sought to minimize its financial role. Not until 1990, when the Greater Denver Corporation was accused of failing to report two $50,000 contributions, was the chamber forced to provide a detailed accounting of its contributions to the Vote Yes campaign (Brimberg and Brown 1990; Brown 1990; Gavin 1990).

This time, too, the opponents were facing a formidable, well-heeled political organization that outspent its opponents by a four to one margin, raising $710,499 as compared to $187,823. This sum was raised mainly through large contributions—Central Bancorporation ($50,000), Public Service Company ($20,000), and the Greater Denver Corporation, closely affiliated with the chamber ($33,021). The SOS group also received its funds from a few large donors—Denver 2000, Inc., $100,000; Harsch Investment Corp., $30,000; Midwest Construction, $20,000; and another $3800 from two other sources (Carnahan and Hubbard 1995b).

The Vote No group, which raised about $30,000, did so largely through small donations from private citizens. The results were practically preordained because, as before, the opponents were divided, poorly funded, and hard-pressed to counter claims that the airport meant jobs, development, and the future. The issue was defined on the ballot as follows: "In order to create jobs, stimulate the local economy and meet future air transportation needs of Denver and Colorado, shall the City and County of Denver construct an airport . . . using no city taxes?" The vote was 70,122 for, 41,754 against.

The new Denver airport had become a symbol of the city Denver aspired to be. The airport would permit Denver to escape its "cow town" image, to become an international hub, to be viewed as a forward-looking community, and to escape the morass of the recession into which it had plunged. As Robert Scanlan wrote in a Greater

Denver Chamber of Commerce newsletter, "The day we open our new airport—one that could be the second busiest in the world—is the day that Denver becomes the Gateway to the West as well as the Gateway to the World." The airport's boosters were those who would profit from the project—businesspeople, lawyers, political leaders, real estate firms, developers, bankers, and union leaders.

The role of the media

While he was a newspaper publisher, Warren Harding succinctly identified his philosophy as "Boost, don't knock" (Hulteng 1985). In practice, this has meant that newspaper proprietors have watered down or slanted stories, not pursued them at all, or even not printed them because of possible "damage" to the community. This pattern exists in many cities and is not confined to the United States. Henrik Ibsen's play *An Enemy of the People* illustrates the universality of the problem. Many motives may be involved—a desire to please local advertisers or elected officials, a feeling of protectiveness, or a concern with growth (Hulteng 1985). Such feelings come naturally to the media elite because they are often active members of the community. One study revealed that 90 percent of publishers and station managers are involved with local business and professional associations even though 49 percent recognized the conflict of interest involved (Fink 1988). Not surprisingly, the media elite are usually active members of the "growth coalition."

Such a situation raises fundamental questions about the degree to which the media can perform its responsibility to protect the community by being a "watchdog." Today, most observers would agree that the media has a "social responsibility" to act in that way. It should provide accurate, balanced, complete, unbiased reporting in a manner that provides the audience with information that will enable it to make intelligent decisions on complex issues.

To what extent did the media perform this role in the airport discussion? Our focus is on the ways that the *Denver Post* and the *Rocky Mountain News*, the two leading newspapers, covered the story. The alternative avant-garde newspaper, *Westword*, carried several critical stories, based on investigative reporting, which provided much valuable information about the project. The TV stations took uncritical stances similar to those of the *Denver Post* and the *Rocky Mountain News* and often cooperated with them. For example, ABC-affiliate KUSA Channel 9 teamed up with the *Rocky Mountain News* for the month beginning April 16, 1989, that preceded the Adams County vote. Perhaps the worst offender was NBC-affiliate KCNC Channel 4, whose head was elected chair of the Greater Denver Chamber of Commerce. As one airport critic noted: "Never in recent memory have I seen so much shameless puffery disguised as news on television . . . The other channels have been doing it, too, but not nearly as much as Channel 4" (Amole 1994).

The *Denver Post* and the *Rocky Mountain News* acted as boosters for the project in various ways. Mayor Peña and Governor Romer were frequently quoted in articles and were featured in numerous photographs. As important political figures, both were naturally targets of the media's attention, and they used their position to advantage in their struggle to persuade Adams County voters to approve the annexation and Denver's voters to approve the project.

But the media did more than simply report the views of the mayor and the governor. Neither the papers nor the TV stations provided voters with an accurate or complete picture of the project. This was especially true during the key decision-making years when the stories that they ran tended overwhelmingly to support the project. On September 10, 1986, Chris Broderick wrote a story for the *Rocky Mountain News* titled "Airport Bonanza Forecast," which began, "A new Denver airport would be responsible for more than 230,000 direct and indirect jobs by 2010 to become the biggest economic catalyst in Colorado history . . . ." On April 24, 1988, about a month before the Adams County election, James Wright wrote an article titled "Jetport of the Future" for the *Rocky Mountain News*, which began "Airline passengers would be able to eat, sleep, shop for clothes and groceries, exercise and go to the movies without ever going outside Denver's new airport." The *Denver Post* was also enthusiastic about the project and its proponents, especially from an editorial standpoint. Chuck Green, an editor, wrote a piece, "Peña's Popularity Rating Defies His Record" (July 29, 1990), that was so flattering that Mayor Peña wrote a letter to the editor stating, "While I appreciate the generous comments Mr. Green made about my role in the efforts to strengthen our economy . . ." (August 22, 1990).

When opposition emerged and grew within Denver, the papers continued to put a positive spin on many stories. James Wright, writing for the *Rocky Mountain News* on the debate within Denver, covered it with a story titled "Rhetorical Battle Rages over Proposed Airport"

(January 29, 1989). Rather than merely describing this "rhetorical battle," however, he first supported and defended Mayor Peña. Then, when quoting Richard Young, the leader of the opposition, he attempted to negate his argument:

> *"If you look at other communities and chart their development, it would show that the airport is very, very irrelevant," said Richard E. Young. . . . "Plenty of other cities have developed without a new airport and plenty of cities that have built airports have not seen growth." But Denver airport developers can point to precedent. The 1977 expansion of Atlanta's Hartsfield International Airport has done more than improve the aviation climate. . . . Hartsfield has a $7 billion economic impact on Atlanta and caused the creation of 54,000 jobs since improvements were finished in 1980.*

The papers' editorials continued to support the project and challenge the claims of the opponents. The *Denver Post*, for example, charged that the opponents were "manufacturing myths by the planeload" (March 19, 1988). Another, titled "There Young Goes Again" (February 4, 1988) began, "Opponents of the Denver airport, lacking any real ammunition against the project, are firing blanks in hopes of creating enough smoke and noise to confuse the voters." In retrospect, what it called Richard Young's "increasingly wild and irresponsible charges" turned out to be sensible and accurate forecasts. On the other hand, each claim in the paper's attempted rebuttal—that the Greater Denver Corporation "eagerly publicized" its contributors, that DIA would not have fewer gates than Stapleton, and that its cost would be $1.7 billion, not $3 billion—has been falsified by history.

The press consistently ran many more positive than negative stories and editorials about the airport. One student, who for a seminar project studied the *Denver Post* from January 5, 1988, through May 16, 1989 (the date of the Denver election), found that the paper ran 683 pieces that were related to the airport, 463 of which dealt directly with the project. Of these, 239 were deemed "neutral," 72 percent (162) of the remainder were "clearly supportive," and 28 percent (62) were "clearly against." Still, the *Denver Post* was perceived by the Peña administration as being much tougher than the *Rocky Mountain News*.

Two columnists, Gene Amole in the *Rocky Mountain News* and Tom Gavin in the *Denver Post*, who always referred to DIA as "Dumbbell International Airport," consistently opposed the project. However, their comments were often offset by other columnists and editorial writers. Ironically, the media did fully cover demands for a public vote on the airport and this coverage of the demands for a public vote made such a vote inevitable. At first the project's leaders, especially Mayor Peña and Governor Romer, were opposed to a referendum, but media poll results that showed growing public sentiment for a vote, even by supporters, forced them to change their position.

Beginning in 1990, both the *Rocky Mountain News* and the *Denver Post* changed direction; they began publishing articles that raised serious questions about the project. This type of reporting reached its peak in the month prior to the airport's opening in February 1995. The Denver papers engaged in some measure of self-flagellation, admitting their role in selling the city's and Chamber of Commerce's spin on DIA to the public with overly optimistic projections of demand and overly conservative projections of cost. They had failed to engage in hard investigative journalism early on, when the critical decisions were made, and seemed to feel some responsibility for a project clearly out of control. In banner front-page headlines, the *Denver Post* described the project as "DIA: Dream to Disappointment" (Kilzer et al. 1995), while the *Rocky Mountain News* published a multipart series, "DIA: What Went Wrong?", which uncovered "new details about a monument to miscalculation" (*Rocky Mountain News* 1995a).

Whatever their motives, whether the reporters were uninformed or the press (as Richard Young has charged) deliberately kept the real costs from the public (Young 1991), or whether the publishers genuinely believed that the airport would benefit the community or they were simply loyal members of the "growth coalition," the papers, as these articles revealed, supported the airport not only through their coverage but financially as well. The *Denver Post* donated $60,000 to the GDC's campaign, and its then-publisher cautioned the editors and reporters about printing negative stories. The *Post*'s former Sunday managing editor, Al Knight, talked of the "growing tension" and of struggles to get critical stories in print. One such story dealing with commuting times caused George Doughty, the director of aviation, to go "ballistic," and, in a meeting with the editors, complained of a "pattern of negative coverage," even though its editorials had been and remained staunchly pro-airport (Kilzer et al. 1995).

Financially, the *Rocky Mountain News* contributed much less to the campaign ($12,000), but it was equally hostile to any criticism of the project. That the *Rocky Mountain News*—and the *Denver Post*, for that matter—did not adequately meet its journalistic responsibilities to the public is vividly attested to by the touching mea culpa of a former editor. In discussing his paper's coverage of the airport, he said:

> *It was a mistake, I know that now. I just bleed for the city. I think it's just tragic. I was wrong and so were most of the other people who had anything to do with it . . . This thing—it's just horrible (Carnahan and Hubbard 1995b).*

Columnist Gene Amole, DIA's most persistent critic, lamented:

> *If DIA fails, and I hope it doesn't, Denver media will have to share the blame. Had the airport been investigated aggressively from the start, the whole fiasco might have been avoided. But the media, including TV, radio and both daily newspapers, became boosters of what was really a monster public-works project (Amole 1995).*

In Denver, the press has a background of boosterism. Its performance in regards to the airport was consonant to this tradition. The media leaders were actively involved in the Greater Denver Chamber of Commerce, and, as mentioned, the newspapers both contributed to the campaign for the new airport. Essentially, the media largely abandoned its "watchdog" function and played an important role in the coalition promoting the new airport. Joanne Ditmer insightfully summarized the role that the media should have played as follows (Ditmer 1995):

> *The local media, individually and collectively, did not play their essential role of watchdogging for the public. That doesn't mean we had to be against the proposal, but there should have been a lot more coverage on the options and consequences before the vote. The media have the intellect, contacts and time to truly examine a question far more intensively than most individuals, and we almost unilaterally ignored that responsibility. Everyone was so busy being a booster boy that the elemental aspects of critically studying*

what the proposal meant, and what its repercussions could be, seem to have been ignored.

The role of the public

At first glance it would appear that the public played an important role in the decision to build the new airport and that its right to participate, a right that has come to be widely accepted, particularly where large technological projects are concerned, was respected by the business and political elites.

Public participation in such decisions is a relatively recent development that can be traced back only to the 1970s. Until then public participation involved social issues like civil rights; scientific and technical issues were viewed as the domain of experts. As new technologies such as biotechnology and nuclear power began to impact individuals, communities, and the environment, a notable shift took place, and numerous citizens groups were formed that clamored for the right to participate in the decision process (Peterson 1984).

Particularly controversial are siting issues involving large projects such as nuclear power plants, toxic waste disposal dumps and airports (Seley 1983). Projects to expand airports or build new ones have raised opposition everywhere, because airports have major impacts on the environment. They require large amounts of land and generate significant amounts of pollution, especially noise, which affects neighborhoods in a major way. Thus they arouse the opposition of environmental groups that tend to view with suspicion any project that involves large-scale changes in the environment. They also arouse the opposition of homeowners, because land usually has to be expropriated and because people oppose siting facilities in their neighborhoods. This is the well-known NIMBY (not in my back yard) syndrome that has become so prevalent in developed nations. In addition to such considerations, the public is legally entitled to participate in the environmental impact statement process. Thus, it should not be surprising that the issue of public participation was recognized by the planners of DIA as an important issue or that they utilized a wide variety of mechanisms to involve the public in the project.

Although the concept of public participation has attained a high degree of salience and legitimacy, there is no agreement on what participation means, how it should be carried out, or what functions it is expected to serve. As a result, participation can take place at many levels. Four are especially relevant for the DIA case.[4]

At the most elementary, power resides with the official decision-makers. The public is provided information, and there are few, if any, feedback channels or opportunities to influence the process. Much publicity is generated through various techniques such as press conferences, newsletters, articles, slide shows, speakers, public meetings, brochures, exhibits, the use of media, and so forth. The basic purpose of these activities is to persuade the public to support a project. Cynics argue that this type of participation (which can be called "tokenism") involves the manipulation of the citizenry by powerful elites, both governmental and corporate.

A higher level of citizen participation may be labeled "consultation." Here too, the public relations machinery is used to generate publicity, but citizens are provided with opportunities to present their views. Public hearings, public opinion surveys, advisory committees, and similar techniques are commonplace. The public's power, however, remains limited. At most, it can provide input into the process. But officials still make the decisions. Planning is not by the people, but for the people.

Higher levels of participation involve a greater role for the public. Two "interactive" models can be identified. The first is a "reconciliatory" model, characterized by a strong two-way flow of information between the decision-makers, the public, and various community-based groups. Common mechanisms include blue-ribbon panels, citizen advisory groups, and the like. The governmental agency, however, retains the decision-making power.

In the second model, the "participatory" model, power is shared by the public and the government. In addition to the two-way flow of information, the public participates in the actual decision-making process at the policy level, through such mechanisms as joint policy boards and planning committees. Such an approach places heavy demands on the participants and requires a significant commitment of resources by all concerned. But the public is a true partner in the process, and planning is by as well as for the people.

4. Different authors identify different levels and patterns. See, for example, Arnstein (1969) and Potapchuk (1991).

Which of these best fit the DIA case? This question is not easy to answer because the pattern of public participation changed dramatically over time. Until 1985 the public had been relatively uninvolved, apart from the Park Hill Community Group, which had been agitating for some time for an end to the noise pollution of Stapleton. In 1981, it filed a lawsuit against Denver, which Mayor Peña settled in 1985 by agreeing to close Stapleton. According to Tom Gougeon, Peña's aide, this was done deliberately to create another obstacle to any future attempt to reopen Stapleton by giving Park Hill legal standing. It was a "way to use their litigation as a vehicle to ensure that the facility be closed and stay closed" (Gougeon 1995a), an important consideration if the new airport project were to be viable. This was a prescient move because Continental Airlines sought permission from Mayor Webb to keep open its maintenance base at Stapleton.

The signing of the Memorandum of Understanding between Adams County and Denver in January 1985, however, changed the situation dramatically. Since the MOU had to be ratified by the voters of Adams County, it set into motion a series of activities with heavy public involvement.

Two months later, the New Airport Advisory Committee (NAAC) was established to serve in an advisory capacity to the planners. Its purpose was to identify the public's interests and concerns and to incorporate them into the planning process. The NAAC's membership consisted of about 100 persons representing local airports (Centennial, Front Range, and Jefferson County), homeowners' associations (Arrowhead and Eastwood Estates), real estate management firms, and various private and public organizations such as the Regional Transportation District, Rocky Mountain Tourism, Colorado Ski Country, and the Denver Audubon Society. Over time, the membership expanded to about 175 members. But attendance at each meeting declined, as did the frequency of the meetings. The participants were divided into subcommittees that addressed the concerns and issues raised, such as environmental issues (including noise), ground access, social impacts, land use, and economic development.

The New Airport Development Office (NADO) was responsible for public information. In April 1985 it launched a major effort to inform the public of the decisions that had been made, using a variety of methods. These included:

1. Media breakfasts, where reporters could meet officials and staff members involved in the project and receive briefings on such topics as the Master Planning Process, the Site Selection Study, the airfield configuration and potential noise impacts, the terms of the annexation, and the status of the terminal planning
2. Press conferences, where various officials reported on specific issues, such as the one held on March 29, 1988, in which Mayor Peña announced that the new airport would be dramatically quieter than Stapleton, or the one held on April 22, 1988, where the draft New Denver Airport Environmental Assessment was discussed
3. Speeches given to community organizations, church groups, schools, and business associations
4. Newsletters and handout materials that were not only mailed to neighborhood associations and other groups but made available through the library system
5. Television advertising campaigns that were run in conjunction with the elections in Adams County and Denver.

As noted, the environmental assessment process requires public participation, and this was handled by NADO as well. It sponsored a series of meetings in every community adjacent to the proposed site. More than 300 citizens participated and provided more than 200 pages of comments on various issues of concern, such as safety issues, air pollution, noise impacts, financing, economic impacts, and the site-selection process. These were compiled in the "Scoping Report from the Community Meetings on the New Airport Planning Process" (August 1985) and distributed to the public for further comment. In addition to these hearings, NADO also held scoping meetings with each of the NAAC's subgroups. Their concerns were forwarded to the consultants who were preparing the environmental assessment. Subsequently, additional meetings were held on the draft assessment.

The NAAC conducted numerous public hearings on other topics, such as the selection of the final runway configuration and the site-selection report. The FAA's hearings dealt primarily with the environmental issues. But the City Council held numerous public meetings on topics ranging from whether to have a referendum on the new airport to discussions of specific contracts. Altogether about

60 public hearings were held between the announcement of the agreement between Adams County and Denver in January 1985 and the Denver election of May 16, 1989.

Opponents of the project, both individuals and groups such as Vote No, used these hearings to voice their opinions and concerns. But they often encountered difficulties since the hearings were conducted and structured by those who wished to see a new facility built. As noted earlier, for example, during the important hearings held by the city council on February 2, 3, and 15, 1989, to consider whether to hold a public election, the opponents were relegated to the late hours.

It is evident that the opportunities for public participation were extensive and that the planners recognized, from the outset, the need to involve the public in the process. They accepted the right of the public to be informed and made strenuous efforts to communicate with the public and to receive, in return, public responses in the form of criticisms, observations, and recommendations. Hundreds of meetings were held with citizen groups, social organizations, church groups, and professional groups. All these activities are typical of the "consultative" approach to public participation.

Only the Adams County election provided the public with the opportunity to influence the decision—and the elites had no choice since the law mandated an election. When they did have a choice, they tried to minimize the public's role. The Denver Referendum demonstrates this point clearly, because it came about only after citizens who had doubts about the project waged a successful campaign to force the coalition to accede to an election. Even then, one can question the degree to which the people were able to exercise power since the referendum was legally nonbinding upon the city (though an adverse result could obviously be ignored only at great political peril).

Furthermore, it was not until 1985 that the public was formally enjoined to become involved in the process, even though the issue of airport capacity dated back to the early 1970s. Until then only experts, officials—and privileged socioeconomic elites—were participating. But, in fact, the final decision had already been made by the experts, and, when the project had been defined, the mechanisms for public participation were structured to obtain support. In effect, the public was not given the opportunity to participate in a very important aspect of the planning process—the formulation of the assumptions and of the alternatives that could be considered. By being excluded in the initial and most crucial stages, the public's quality and degree of involvement were severely limited. This point deserves to be emphasized, because once the problem has been defined, assumptions made, and questions set, it is difficult for the public to be truly integrated into the decision-making process.

In terms of the models discussed, the process was (with the exception of the Adams County election and perhaps the Denver Referendum) limited to the first two levels, whereby the public is expected only to be kept informed and to respond to proposals and reports that had been prepared by the "experts." Decisional power, by and large, remained in the hands of the elites, who mobilized such extensive resources that even when the public was meaningfully involved, as in the elections, it is not at all clear that the process functioned in a way that permitted the issues to be considered, evaluated, and debated to enhance the political process and the project outcome. Of course, this criticism applies to most decisions of this type across the United States, and in many other democracies as well (Feldman and Milch 1982). This does not mean, however, that the ways in which decisions are presently made on such issues should not be improved.

Failure to do so imposes heavy costs. In this case, it precluded the possibility of building flexibility into the process since it was politically impossible for the proponents to confess that their assumptions were seriously flawed and that serious mistakes resulted. At best, only marginal changes in the project could be countenanced. We return to this important point in Chapter 9. Furthermore, the effects of the structured participation and the desire to implement the project at practically any price also had important consequences. The heavy-handed way in which the project's proponents handled opponents and dissenters was destructive of the political process. Minimally, it did little to encourage a thorough discussion of the issues or the creation of a political culture sensitive to dissent and marked by trust. Ironically, the Denver experience is probably better than the norm that exists in most other cities because the Peña administration was guided by a philosophy of empowering residents and neighborhood organizations. It was often criticized by the media and the business elite for its reluctance to make decisions without extensive consultations with the affected publics (Gougeon 1995b).

Elites, planners, and technical experts tend to argue that decisions on projects should be left to them. They argue that the public possesses neither the information nor the knowledge to handle complex technological issues; its participation will not only slow down a process that is on a critical path but will produce worse outcomes. Some of those involved in the DIA case seem to share this view, which legitimizes limiting public participation to the informational and consultative levels.

But it is fallacious to assume that elites and "rational" experts will make better decisions if the public's role is limited. The issue is ultimately the quality of the interaction that takes place among the public, the elites, and the experts. Process is important, and, as noted earlier, many benefits flow from one in which the public is engaged early and instrumentally in any large project such as DIA, not least of which is the strengthening of democracy and community.

Lessons learned

- *Citizen education and participation is essential from the outset if public support is to be long-lived and democratic values respected.* In Denver, the public was brought into the process only after the "experts" had decided not to expand the existing airport, and to build a new one, and then only grudgingly. A vote of the electorate in Adams County was mandated by law, and the vote in Denver was offered only under pressure. In the end, the people pay the price of major infrastructure investment. They should have a stronger voice in policy development. Moreover, although Adams County and Denver voters were given a say, most of the metropolitan community was wholly deprived of any input. Only 25 percent of metropolitan Denver's population resides in the City and County of Denver. The higher-income suburban residents, a constituency more likely to use the airport, was wholly deprived of a role in the decision process.
- *The business and wealth interests will lobby strongly for and support growth policies that will benefit them.* In Denver, the business elite poured economic resources on the campaign to build a new airport and on politicians willing to jump on the bandwagon. Although inevitable, the pragmatic political reality must be recognized if better public policy is to be attained.

- *The media must remain an independent source of reliable, unbiased information about major infrastructure issues facing their community.* Regrettably, both Denver daily newspapers and the local television stations jumped aboard the new airport bandwagon early, and their reporting reflected it. Not until the airport was hundreds of millions of dollars overbudget did they begin to critically assess and analyze the project. Had they been more vigilant as the public's watchdog, the project would likely have been implemented far more successfully than it was.
- *If they expect to win the inevitable political battles that will be fought, local proponents or opponents must be unified, well-financed, and possess politically astute leadership.* As noted, the pro-growth wealth interests have the economic resources to fight the good fight. But in Denver, the new airport proponents were also adept in building alliances, remaining focused and unified, proficiently lobbying key officials in the federal government for support, and waging effective political campaigns.
- *Opponents of airport expansion can play a trump card with the threat of environmental opposition.* With the National Environmental Policy Act of 1969, the U.S. Congress established a cumbrous process of environmental review that can be used to tie growth interests in knots, a card Adams County successfully played in dissuading Denver from expanding onto the Rocky Mountain Arsenal.
- *Airport expansion can be a means of expanding a city's geographic boundaries.* Hemmed in on all sides, like many U.S. cities, Denver was able to persuade a suburban county to cede 53 square miles to it on grounds that a new airport would mean less noise and more jobs. The long-term economic development and tax base benefits for Denver could be enormous.
- *Campaign financing must be reformed.* It has often been noted that money is the mother's milk of politics. If political leaders are to be public servants concerned primarily with the community's well-being, the link with the business community must be transformed from one of financial dependence to one of independence.

References

Airport Update. 1987. November:1.

————. 1988. February:1.

Albin, Robert L. 1983. Clarification of chamber memo on proposed airport expansion. May 6.

————. 1995. Interview. December 8.

Amole, Gene. 1994. TV news a partner in airport follies. *Rocky Mountain News.* March 3.

————. 1995. This DIA dissenter willing to help make airport work. *Rocky Mountain News.* February 12.

Amole, Tusin. 1995. White flight made suburbs boom. *Rocky Mountain News.* September 13.

Arnstein, Sherry R. 1969. A ladder of citizen participation. *Journal of the American Institute of Planners.* July.

Barnes, James A. 1993. Cleared for takeoff. *National Journal.* September 18:2238–2244.

Brimberg, Judith and Fred Brown. 1990. Airport backers disclose war chest. *Denver Post.* March 3.

Brown, Fred. 1990. Chamber group mum on pledges, critic says. *Denver Post.* March 25.

Carnahan, Ann and Burt Hubbard. 1995a. The untold DIA story: What went wrong. *Rocky Mountain News.* February 5.

————. 1995b. Hard sell shoved DIA down Denver's throat. *Rocky Mountain News.* February 7.

City and County of Denver (Denver). 1990. Project overview: New Denver International Airport. Report commissioned by the City and County of Denver for the FAA. January.

Coates, J. 1988. Too much snow could start mayor on downhill slide. *Chicago Tribune.* January 14.

Colorado Attorney General. 1991. Report. November.

Colorado Economic Review. 1989. Business community involvement a critical ingredient to success. Third Quarter.

Cramer, Steve. 1987. Lessons learned from Atlanta. *Airport Update.* December:1.

Denver Chamber of Commerce. 1985. *Membership Directory 1984-1985.* Denver.

Denver Post. 1995. As DIA's takeoff nears, remember how we got here. February 26.

Ditmer, Joanne. 1995. Big spending doesn't make a great city. *Denver Post.* February 26.

Dwyer, Catherine and Christopher Sutton. 1994. Brown plus forty: The Denver Experience. *Urban Geography.*

Elkin, Stephen L. 1987. *City and Regime in the American Republic.* Chicago: University of Chicago Press.

Federal Aviation Administration (FAA). 1989. Record of decision for the new Denver airport. U.S. Department of Transportation, Federal Aviation Administration, Northwest Mountain Region, September 27.

Feldman, Eliott J. and Jerome Milch. 1982. *Technocracy Versus Democracy: The Comparative Politics of International Airports.* Boston: Auburn House Publishing Company.

Fink, Conrad. 1988. *Media Ethics.* New York: McGraw-Hill Book Company.

Fumento, Michael. 1993a. Peña's plane stupidity. *The American Spectator.* December:43–44.

————. 1993b. Federico's folly. *The American Spectator.* December:46–48.

Gavin, Jennifer. 1988. Airport election not necessary despite voters' wishes, Peña insists. *Denver Post.* February 27.

————. 1990. Airport donors must be revealed. *Denver Post.* January 9.

Glass, David. 1982. The social and environmental impacts of airports: The effects of the Rocky Mountain Arsenal alternative on Commerce City. Master's thesis, Department of Urban and Regional Planning, University of Colorado at Denver, October.

Gougeon, Tom. 1995a. Interview. December 8.

————. 1995b. Personal conversation. December 20.

Grelen, Jay and Jennifer Gavin. 1989. Peña aides choose order of speakers, spark furor. *Denver Post.* April 2.

Harkavy, Ward. 1994. Dark horse. *Westword.* January 18–24.

Hearings. 1987. "Authorization of the Airport and Airway Trust Fund." Before the Subcommittee on Aviation of the Committee on Commerce, Science and Transportation. United States Senate, 100th Congress, 1st Session, April 20.

Hong, Seung-Mo. 1993. Interviews with Robert Albin and Dave Ferrill.

Hornby, Bill. 1990. Four busy years for Greater Denver Corporation. *Denver Post.* November 13.

Hubbard, Burt. 1995. Denver muscled the FAA to launch DIA. *Rocky Mountain News.* February 6.

Hulteng, John. 1985. *The Messenger's Motives: Ethical Problems in the News Media.* Englewood Cliffs, NJ: Prentice Hall, Inc.

Keyes v. School District No. 1, Denver, Colo., 413 U.S. 189 (1973).

Kilzer, Lou, Robert Kowalski, and Steven Wilmsen. 1995. DIA: Dream to disappointment. *Denver Post.* February 12.

Knight, Al. 1994. Life on the airport committee. *Denver Post.* May 1.

Kowalski, Robert. 1989a. City had OK in '86 for Arsenal runway to extend Stapleton use. *Denver Post.* February 6.

———. 1989b. Project's foes, friends, battle as crowd jams city council hearing. *Denver Post*. March 2.

———. 1989c. FAA suppressed critical letter. *Denver Post*. December 2.

Locke, Tom. 1992. Disconnected? City may lose millions by awarding new airport's telephone contract to US West. *Denver Business Journal*. June 12.

McConahey, Stephen G. 1987. Why we need a new airport, now. *Airport Update*. November:1.

Molotch, Harvey. 1976. The city as a growth machine: Toward a political economy of place. *American Journal of Sociology*. September:309–332.

Moore, Scott. 1990. Growth politics in the Denver region: The 1988 Adams County airport annexation election. Unpublished paper.

Moran, Joseph. 1986. *Introduction to Environmental Science*. New York: W.H. Freeman and Co.

Petersen, James C., ed. 1984. *Citizen Participation in Science Policy*. Amherst: The University of Massachusetts Press.

Potapchuk, William R. 1991. New approaches to citizen participation. *National Civic Review*. Spring.

Rocky Mountain News. 1995a. The untold DIA story. February 5.

———. 1995b. Denver Housing: End of an era. September 13.

———. 1995c. Judge Matsch's conclusions. September 13.

Seley, John E. 1983. *The Politics of Public-Facility Planning*. Lexington, Mass: D.C. Heath and Company.

Wilmsen, Steven K. 1991. *Silverado: Neil Bush and the Savings & Loan Scandal*. Washington, DC: National Press Books.

Young, Richard. 1991. Lecture to Honors Seminar, Colorado School of Mines. November 25.

4

The economic impact of airports

"Whomsoever commands the sea commands the trade. Whomsoever commands the trade of the world commands the riches of the world, and consequently the world itself."—SIR WALTER RALEIGH

The role of economics in airport planning and decision-making has become increasingly more important to the welfare of cities, regions, and nations. Airports generate significant amounts of employment, tax, and payroll revenues, which directly benefit local and regional economies, and have become lifelines for cities in today's global marketplace. Given this degree of economic importance, and considering the amount of land airports consume, the degree of sophisticated technology involved in their construction, and their public service function, it is no wonder that airports are among the most highly energized types of projects in the world.

Accordingly, this chapter explores the economic effects associated with airports and air transportation generally, and specifically the impact of both Stapleton and the new Denver International Airport. Moving from the global to the national, regional, and finally to the local scale, this chapter examines the economic impact of DIA and how it was widely publicized and touted as a major reason why Denver and Colorado needed a new airport. As we have seen in Chapter 3, the new airport had to be sold to the public, and a powerful campaign was initiated to do so. In that campaign, supposed economic benefits were emphasized. As we shall demonstrate, the general observations regarding the role of airports in economic growth that were utilized by local planners and politicians in selling the public on the need for a new airport are highly questionable. The nature of the relation between air transportation and urban/economic growth is very complex and cannot be reduced to the simple one-way measures of causality that were commonplace in both the

Adams County and Denver campaigns. Although some evidence demonstrates that airports yield positive economic benefits, data also suggest that these are not as high as commonly believed. Accordingly, specific decisions to build new airports must be made in full awareness of the complexity of the relation and must be based primarily on the functional and financial viability of airports within national and international air transport systems. The economic impact argument alone cannot be used as the principal justification for building new airports.

Transportation, airports, and global economic development

Transportation has always been a key factor in shaping patterns of human social, economic, and cultural existence. From the invention of the wheel, to the steam locomotive, to the internal combustion engine and pneumatic tires, to the supersonic jet and space shuttle, transportation achievements have been an integral element in defining the character of great civilizations. The extensive road system of ancient Rome; the seafaring abilities of the ancient Phoenicians and Greeks, Prince Henry the Navigator and the Portuguese, Columbus, Magellan, and the Spanish; the steam engine, railroads and the industrial revolution of Great Britain; and the Wright brothers' inaugural flight at Kitty Hawk all represent tremendous achievements that profoundly influenced the course of human history.

Like communications and energy, transportation is an essential infrastructure industry that serves as a foundation for commerce, communications, tourism, and national security. It has tremendous economic, social, and cultural significance because it integrates cities, regions, and nation-states. By overcoming the friction of distance, it allows spatial interaction to occur. Today, one cannot imagine international trade, personal and business travel, or the effective exchange of ideas without an efficient transportation system.

Transportation accounts for 16 percent of U.S. gross domestic product (GDP). Internationally, transport is one of the fastest-growing sectors as nations around the world strive to make investments that are required if they are to create the kind of transportation system needed to achieve their developmental goals. The commercial airline industry transports 1.25 billion passengers and 22 million tons of cargo, about a quarter of the world's manufacturing exports based on value. Commercial aviation produces 22 million jobs (3 million directly, 7 million indirectly, and 12 million induced). Airlines are an essential component of the tour and travel industry, which generates more than $3.5 trillion in GDP, employs 127 million people, and accounts for 12.9 percent of consumer spending and 7.2 percent of worldwide capital investment (Dempsey 1995).

As air transportation has come to dominate long-distance and overseas travel, cities around the world have become increasingly reliant on air service connections to be better integrated into the international flow of commerce. Clearly, one can trace the history of world urbanization and note the role that transportation has played in fostering trade and economic growth. From the early seaports to the major rail centers, cities that possessed advantageous locations historically have benefited from the superior accessibility that their transportation connections have afforded them. Great European cities such as Athens, Rome, Venice, London, Amsterdam, and Paris all prospered as a result of their early functions as major mercantile transportation centers (Dempsey and Thoms 1986, pp. 1-2; Dempsey 1988, p. 5). Urban systems in countries around the world become organized based on local hinterland productivity and levels of connectedness within a hierarchy of lower- and higher-order centers. Today, major world cities such as Tokyo, New York, London, Hong Kong, Los Angeles, and Singapore are as reliant on superior communication networks and rapid air transport accessibility as cities from earlier eras were upon sea and rail connectivity. These days airways have replaced the oceans, and airports have replaced seaports in importance. Airlines are too numerous to be profitable in mature markets. But airports are the portals through which passengers and high-valued cargo must flow.

Within this context, airports have taken on a more important role. Major world cities and nations have devoted increasing resources to airport infrastructure to facilitate their connection to the flow of global economic activity—to put a larger cup in the economic stream, so to speak. In many ways, airports have become the hubs of today's growth centers, anchoring urban regions much the same way as shipyards or rail terminals fixed the location of economic activities in previous eras. To be sure, today's locational decisions are much more footloose but access to frequent and reliable air passenger and cargo transportation via a major international airport remains a prerequisite for most companies playing in the international field. In fact, airports have been identified as perhaps the single most important piece of infrastructure that cities and nations can build to improve their chances of success in international economic competition (O'Connor and Scott 1992). One has only to look at the current frenzy of airport development within the Asia-Pacific region to realize the profound role that air transportation is playing—and is expected to play—in the explosive growth of Pacific Rim economies. In that regard, the following section discusses the potential economic impact expected from some of the world's recent airport projects.

Recent international airports and economic impacts

Osaka's Kansai Airport

Kansai is Japan's second most significant economic region (behind only Tokyo), accounting for 2 percent of world output. The Itami Airport handled only 10 percent of Japan's cargo and 17 percent of its international passengers. But Kansai International Airport (KIA) is expected to increase the total volume to 40 percent of Japan's total. By the year 2000, the volume of international cargo is anticipated to be five times greater than at Itami, and the number of international passengers to be four times greater (Ogawa 1993).

A principal reason for construction of the new airport was a desire to boost Osaka's economic position. The Osaka/Kobe region accounts for about 20 percent of Japan's gross domestic product. Dr. Yoshio Takeuchi, president of the Kansai International Airport Co., said, "We want to make the airport the center of economic activity in Kansai. We plan to build a new 21st-century city around the airport that will embody a highly flexible electronic world." (Brown 1987).

Airport officials hope KIA will tap dormant air transport demand in western Japan, which does not fly because of the inconvenience of Tokyo's Narita Airport (Black 1994). Kansai is expected to stimulate a "revival" of the entire Kansai region. As Tetsuro Kawakami, head of the Kansai Economic Federation, said, "I feel that the effect of Kansai International Airport, together with the development of the Osaka Bay area, will let flourish the region's economic and cultural potential." (*Kyodo News* 1994).

Macau International Airport

Macau International Airport was originally planned as a facility for second-tier air carriers or for primary carriers using Macau as a complement to Hong Kong. But China gave the territory the freedom to negotiate independently bilateral air transport agreements with other countries even after it reverts to China in 1999 (*Phillips Business Information* 1994). Moreover, delays in constructing Hong Kong's new airport, the fact that it will open with only a single runway, as well as its formidable cost, have stimulated interest in Macau's airport. The airport may be able to capitalize on its potential as a regional hub for the Pearl River Delta region, a fast-growing area that includes the industrial city of Guangzhou and eight other cities (*Aviation Week & Space Technology* 1994).

Several airports have been built, or are planned, for the Pearl River basin. Guangzhou's Baiyun Airport already handles 10 million passengers a year. Guangzhou's new airport will be built in stages on 10 square miles, with the first phase completed in 2005, with capacity to handle 27 million passengers (Brower 1995). Zhuhai, on the west bank of the Pearl, opened a new airport in 1995, with a longer runway (at 4000 meters, the longest in China outside Tibet) and double the terminal space of Macau (Donoghue 1995), as well as some of the world's most modern technology (Smith 1995). Shenzhen, on the eastern bank, opened a new airport (Huangtian) in 1991, and already it is China's fifth busiest airport in terms of passenger volume and fourth busiest in aircraft movements (Donoghue 1995). Guangzhou is the provincial capital of Guangdong, which encompasses the Pearl River and its estuary, with a population of 62 million people (Donoghue 1995). China is seeking foreign investment to build or expand more than 30 airports over the next five years (*Asian Wall Street Journal* 1995b).

The airport's proponents anticipate that the new airport will assist in the transformation of Macau from a gambling haven into a business and tourist gateway to China, similar to neighboring Hong Kong (Paisley 1995).

Hong Kong's Chek Lap Kok Airport

About 20 percent of Hong Kong's enormous trade, as measured by value, moves by air. Seven million tourists fly to Hong Kong every year, spending about $7 billion. Hong Kong's existing airport, Kai

Tak, is already among the world's busiest international airports, and has reached capacity, at 24 million passengers (Mok 1993).

Skeptics claimed the new Chek Lap Kok Airport was devised as a scheme for "British colonials to make as much money as possible before 1997," when the British Hong Kong colony reverts to mainland China. Actually, Japanese firms won the largest construction contracts, although British firms were second (Becker 1995). Others note that the airport, its tunnels, expressways, bridges, and railways solve a number of transport problems, that they have been long planned, and that by packaging them together, they boost confidence in the economic future of Hong Kong even as the communist government comes to power, thereby dissuading a bail-out of capital and skilled population (Becker 1995).

As yet, it is unclear what impact the enormous cost of Chek Lap Kok will have on air traffic at Hong Kong. Nevertheless, its proponents insisted the new airport was essential if Hong Kong was to remain a commercial center (Mufson 1994).

Two other new airports will vie for Pearl River basin regional development. As described above, Macau is building its first airport 23 miles west of Chek Lap Kok, while the new Shenzhen Airport was recently opened some 23 miles to the north. Airport proponents predict that, with such an enormous regional population base, the airports will complement, rather than compete, with each other (Darmody 1993). As John Mok of the Provisional Airport Authority (PAA) observed, "We believe that the airports planned in Hong Kong and in the Southern China region will, in all likelihood, serve their own specialized markets, ultimately working in harmony rather than in competition with each other" (Mok 1993).

Kuala Lumpur Sepang International Airport

Malaysian officials perceive that the New Kuala Lumpur International Airport (NKLIA) will be a strong magnet to bolster Kuala Lumpur as an international business and travel hub. But it will compete with Bangkok to the north, which is building a new airport, and Singapore to the south, which will have a third terminal by 2000 (Hill 1993).

New Seoul International Airport

The Korean Transport Ministry predicts the new Seoul airport is "destined to become the hub of Northeast Asian air traffic" (*Airports* 1995). In 1994, Seoul's existing airport was the fastest-growing airport in the world, with a 19.6 percent growth rate (Mecham 1995).

Munich's Franz Josef Strauss Airport

It is estimated that almost 25,500 people are employed by 300 different companies linked to the new Franz Josef Strauss Airport, making it one of the largest sources of employment in Bavaria. The airport employs 2800 people directly, while Lufthansa employs 1600 (*Aviation Europe* 1995).

These new airports have promoted—and are clearly expected to continue promoting—economic growth. To determine the degree to which such expectations are justified, we turn to a consideration of the U.S. scene.

Airports and economic growth in the United States

The use of air transportation has been historically much more prevalent in the United States than anywhere else, not discounting the growing importance of aviation around the world. In fact, even today, the United States accounts for seven of the top 10, 13 of the top 20, and 23 of the top 40 airports in the world, based on terminal passenger traffic (Table 4-1). This is all the more remarkable when considering that relatively small-population cities in the United States, such as Orlando, Las Vegas, and Charlotte have airports that are among the 40 busiest in the world. Denver, with a 1990 metropolitan area population of 1.85 million people, had the eighth busiest airport in the world, serving 32.6 million passengers in 1993, an amount greater than major world city airports in Paris, Frankfurt, Osaka, Hong Kong, Rome, and Amsterdam. Among non-U.S. airports, only London Heathrow and Tokyo Haneda airports had more air passenger traffic than Denver's Stapleton.

Table 4-1. Busiest airports in the world, 1993

Airport	Terminal passengers (millions)
1 Chicago O'Hare	65.1
2 Dallas/Fort Worth	49.7
3 Los Angeles International	47.9
4 Atlanta	47.8
5 London Heathrow	47.6
6 Tokyo Haneda	41.7
7 San Francisco	32.8
8 Denver	*32.6*
9 Frankfurt	32.6
10 Miami	28.7
11 New York JFK	26.8
12 Paris CDG	26.2
13 Paris Orly	25.4
14 New York EWR	25.8
15 Hong Kong	24.4
16 Detroit	24.2
17 Boston	24.0
18 Phoenix	23.6
19 Minneapolis St. Paul	23.4
20 Osaka	23.3
21 Seoul	22.6
22 Las Vegas	22.5
23 Honolulu	22.1
24 Tokyo Narita	22.0
25 Orlando	21.5
26 Amsterdam	21.3
27 Toronto	20.5
28 Houston Intercontinental	20.2
29 London Gatwick	20.1
30 Singapore	20.0
31 St. Louis	19.9
32 New York LaGuardia	19.8

Table 4-1 *(continued)*

Airport	Terminal passengers (millions)
33 Rome Fiumicino	19.3
34 Bangkok	19.1
35 Seattle	18.8
36 Pittsburgh	18.5
37 Madrid	17.6
38 Charlotte	17.3
39 Sydney	16.8
40 Philadelphia	16.5

Source: *Airline Business*. January 1995.

Because of the greater role that airports and air transportation services have played in the growth and development of the U.S. urban system, the long-term relation between air transportation and urban growth has been the subject of increasing interest. Important questions about the nature and degree of this relation have been addressed in recent research. One study found that during the postwar era, those cities with higher air passenger/population ratios generally grew faster in subsequent time periods than those with lower air passenger/population ratios (Goetz 1992, pp. 217–238). There is a distinct pattern of high ratio/high growth cities in the U.S. South and West (e.g., Orlando, Atlanta, Dallas, Denver, Las Vegas) and a concentration of low ratio/low growth cities clustered in the U.S. Northeast and Midwest (e.g., Baltimore, Philadelphia, Buffalo, Cleveland, Detroit, Milwaukee). Another study found that a metropolitan area's hierarchical position within the networks of passenger and cargo airlines has pervasive effects on metropolitan employment growth (Irwin and Kasarda 1991), especially when they possess strong international linkages. Examples include such older gateways as New York, Chicago, and Miami, as well as newer gateways like Dallas, Seattle, and Charlotte. These findings, however, must be viewed within the wider context of the economic development benefits that result from public infrastructure investment.

Recently, vigorous debate has centered upon the existence of an "infrastructure crisis" in the United States. This debate examines the extent to which public systems, like highways, bridges, water and sewer systems, and airports, need substantial reinvestment to promote continued economic growth or whether much of this capital would be better utilized if left to the private sector (Aschauer 1989; Munnel 1990; Sanders 1993). Ongoing research is specifically examining airport investments within this framework (Gillen 1996). Other studies have noted the importance of frequent and reliable air service in the location decisions of high-technology businesses and Fortune 500 companies (Joyce 1985; Markusen et al. 1986) and the linkage between air service connectivity and professional administrative and auxiliary employment (Ivy et al. 1995).

Such findings illustrate the important role that air transportation continues to play in influencing patterns of growth within national urban systems. What much of this research indirectly implies is the continuing need for cities to ensure that their competitive positions in the national and international air transport hierarchy are central, important, and preferably dominant. Cities can employ a number of strategies to ensure that this occurs; most obvious is the continued enhancement of airport infrastructure.

Much of the research to date indicates that the nature of causality within the general relation between air transportation and urban economic growth remains unclear. In other words, cities with more and better air transportation connections tend to grow faster, but it has not been absolutely demonstrated whether the air transport activity causes more growth or growth occurring for a host of other reasons causes more air transport activity to occur. Undoubtedly a bit of both is involved—a classic case of mutual causation or bidirectionality in the relation. Cities benefit and grow from more air transport connections, and cities receive more air transport connections because they are growing. Air transport is part of the mix, but it certainly is not the only determinant of growth in an urban system. Macroscale economic cycles, structural changes in production systems, government policies, and changes in social, cultural, demographic, and environmental conditions all affect the urbanization growth process. There is more to urban and economic growth than just air transport connections.

Yet, as we have seen, most cities realize the importance of basic transportation infrastructure to encourage economic growth and thus invest heavily in airports. Denver's commitment to investment in transportation infrastructure, however, is one that reaches back to the 19th century, and understanding this background is essential because it has an important impact on the decision to build DIA.

The Denver case: Historical antecedents

University of Colorado history professor Tom Noel offered a tongue-in-cheek explanation of why Denver built the world's largest airport:

> *Why is Denver, one of this planet's smaller major cities, building one of the world's largest airports? . . . Denver suffers from "by-pass phobia." As one of the most isolated major cities, the Mile High City has always been afraid the world would pass by without noticing the little city in the middle of nowhere (Noel 1994a).*

Obviously, no city wishes to be isolated, but this concern has consistently influenced Denver's policymakers. Located hundreds of miles from any other metropolis, Denverites have always accorded transportation a high priority, some would argue to the point of an obsession.

Denver's history is inextricably linked with transportation. Since its founding by explorers canoeing along the banks of the South Platte River and Cherry Creek in 1858, Denver has possessed a strategic location, serving as a crossroads for the Rocky Mountain region and the nation. At first, Denver was no more than a small tent town housing a few settlers drawn by the lure of gold. But from its inception, its founders recognized the importance of transportation. In 1858, Denver founder General William Larimer, Jr., wrote, "Denver City is the center of all the great leading thoroughfares and is bound to be the great city" (Noel 1994b).

In 1859, nascent Denver, concerned that the city of Auraria, just across Cherry Creek, might get the region's only stage terminal, bribed the Leavenworth & Pikes Peak Express Co. with 53 lots and nine shares in the town development company to locate the terminal in Denver (Quillen 1994). Thus did Denver displace Auraria as the principal city on the Colorado Front Range.

In the 19th century, Denver served as an early trade center for the large influx of miners searching for gold in the nearby foothills. The Union Pacific Railroad laid track west from Omaha beginning in 1865, crossing into Colorado two years later. But to the chagrin of its residents, Denver was excluded from the transcontinental rail line, which was routed through Wyoming to avoid the more challenging terrain of the Colorado Rocky Mountains (Fig. 4-1).

Fig. 4-1. *The Union Pacific transcontinental railroad route bypassed Denver in favor of Cheyenne, Wyoming, leading one Union Pacific executive to describe Denver as "too dead to bury."*

The early history of Denver also illustrates that the city has always been prone to exaggerated expectations. The first territorial governor of Colorado, William Gilpin, was one of the staunchest promoters of Colorado and the West. Eschewing observations by Stephen H. Long that this semi-arid region was part of a "Great American Desert," Gilpin argued that as settlement increased, the land could be turned into a new Garden of Eden. Reflecting blind optimism, Gilpin strongly believed in the notion that "rain followed the plow," and together with the Western rail barons and land promoters, did much to encourage the rapid settlement of Colorado. Denver has characteristically been plagued by a chronic tendency to be thoroughly duped through misleading promotionalism and hucksterism in its most unabashed forms. Nevertheless, despite the cycles of extreme boom and bust, an economy historically based on resource exploitation, and the colorful characters of its past, Denver was able to grow and develop as the core center of the Rocky Mountain empire.

Abraham Lincoln appointed John Evans, a prominent Illinois citizen and founder of Northwestern University, the second territorial governor of Colorado. Evans was a builder, founding the University of Denver in 1864. He and other early town leaders understood the implications of being isolated from the main transcontinental railroad line. And Golden, Colorado, gateway to the mines of Clear Creek, was building a rail line to link Golden with the Union Pacific transcontinental line at Cheyenne, Wyoming (Quillen 1994)

Denver merchants were packing up and moving to Cheyenne, the city the Union Pacific had decided to serve. Union Pacific's vice-president Thomas Durant described Denver as "too dead to bury." Evans and Denver's other city leaders quickly banded together to create the Denver and Pacific Railway & Telegraph Company in 1867 for the purpose of building a spur line up to the Union Pacific tracks at Cheyenne. One source described it this way:

> *The year was 1868. A far-sighted group of civic leaders calling themselves the Denver Board of Trade saw their city at a crossroads and prepared to take a risk. Bypassed by the transcontinental railroad which ran through Cheyenne, Denver's leading citizens feared the city of 4000 would go the way of the ghost towns. Ignoring doubters, in a week's time they had raised the monumental sum of $300,000. The Denver-Pacific Railroad was born. Denver was on the map (Greenwald 1995).*

On May 10, 1869, the transcontinental railroad was completed at Promontory Point, Utah, and a year later, on June 22, 1870, the first trains rolled into Denver via the Denver Pacific linkage. Evans and the other leaders had assured themselves that the commercial pulse of the nation would not bypass Denver, and that Denver, rather than Golden, would be the premiere city of Colorado, and that Denver, rather than Cheyenne would be the premiere city of the West. In 1870, Denver's population was 4759, ten more people than lived there in 1860; Cheyenne had only 1450 residents, and would never catch up. Salt Lake City, however, had 12,854 residents and looked like it might lay claim to being the largest city in the Rocky Mountain West.

Just two months after the Denver Pacific completed the Cheyenne-Denver link, the Kansas Pacific Railroad reached Denver from Kansas City, and by 1872 the Denver, South Park & Pacific Railway (also started by John Evans) began serving mining towns in the nearby mountains (Dempsey et al. 1992). Later, the success of the Denver & Rio Grande Railway, started by a Civil War Union general, William Jackson Palmer, contributed to Denver's growing stature as it became the leading city of the region (Leonard and Noel 1992, p. 38).

By 1880, Denver had grown to 35,629 residents, compared with Salt Lake City's 20,768, and Cheyenne's 3456. Denver picked up the pace and never looked back (Fig. 4-2).

Yet Denver was still not on the most direct rail route from the Midwest to the West Coast. The Great Northern, Northern Pacific, Union Pacific, Santa Fe, and Southern Pacific all used different alignments that avoided the 14,000-foot mountains of the Colorado Rockies. Even Pueblo, Colorado, had a more natural east-west route up the Arkansas River Valley and over Tennessee Pass, and for a while in the early 20th century, its growth rate was a matter of concern in Denver. By the early 1900s, the leaders of Denver figured that a railroad tunnel through the mountains would provide the city with a shorter route to the West Coast. The 6.2-mile Moffat Tunnel, a massive engineering project, finally opened in 1927 with the Dotsero cutoff giving Colorado its lowest main-line crossing of the Continental Divide and proved to be an important passage for rail traffic, water diversion, and later for the emerging recreational skiing traffic in the mountains. Thus would Denver rebuff Pueblo's bid to be the premiere city of Colorado (Quillen 1994).

But by the 1920s, the seeds of rail's demise were already being sown. The internal combustion engine and its application in the suc-

Fig. 4-2. *Denver's Union Station was the transportation hub of the West during the late 19th and early 20th centuries.*

cessful Model-T Ford and other early automobiles was beginning to transform the face of transportation in this country and around the world. The airplane was the other major transportation innovation that began to compete with the railroads, and with the success of Charles Lindbergh's historic flight from New York to Paris in 1927, the aviation industry was poised for takeoff.

Denver, too, was readying itself for the age of air travel. With only a few aerodromes scattered throughout the city in the late 1920s, Denver Mayor Benjamin Stapleton proposed a new, modern airport be built to consolidate all aviation activities in a remote open area to the northeast of the city. His ambitious plan was to cost the city more than $430,000 and was met with substantial opposition, especially from the local media. Decried as "Stapleton's Folly," the plan to build a new airport in Denver was criticized as a scheme designed only to enrich a few local land owners and as a costly, wasteful boondoggle that ignored the possibility of expanding current airfields. As one critic wrote in a letter to the editor of the *Denver Post*, published March 28, 1928, "I am . . . informed that the airplane men say they will not use (Denver's new municipal airport) even if the city should make the mistake of buying it" (Hardaway 1991). But the power brokers and business elite in the city felt that a new airport was needed to better prepare Denver for the coming air boom and secured the project's implementation (Leonard and Noel 1992, pp. 432–433).

On October 17, 1929, dedication ceremonies were held and Denver Municipal Airport was open for business. At the dedication, it was boasted that the new airport hangar, with dimensions of 121 feet by 122 feet, would accommodate "not only the largest ships of today, but . . . the huge ship of the future" (Hardaway 1991). The wingspan of today's largest aircraft exceed 195 feet. Denver's airport quickly became a major stop on numerous airmail and air passenger routes operated by airlines such as Boeing Air Transport, Western Air Express, Varney Speed Lines, Continental, and Monarch (which later merged with Challenger Airlines and Arizona Airways to become Frontier Airlines). During the 1930s, with the help of the New Deal Works Progress Administration, the airport's runways were widened and extended, and a new $1 million terminal building was added. In 1944, the name of the airport was officially changed to Stapleton Airfield, to honor the mayor who originally championed the project (Leonard and Noel 1992, pp. 433–434).

In the aftermath of World War II, commercial air transport expanded rapidly. Many war veterans who had become familiar with piloting, navigating, repairing, or simply traveling in airplanes returned home to help fuel the growth of the civil aviation industry. Aircraft technology and design had greatly improved as a result of the war, and a substantial amount of surplus aircraft were now readily available for new airline ventures wishing to start up commercial operations. As a result of this favorable environment and the booming American economy, the U.S. airline industry enjoyed substantial growth throughout the early postwar period. By 1957, the number of passengers traveling by air had exceeded the total for rail for the first time ever. Air transport has remained the dominant intercity commercial mode ever since (Sampson et al. 1990).

Stapleton Airfield kept pace with the go-go years of aviation. From 12,089 passengers in 1940, the number of people traveling through Stapleton grew twentyfold by 1950 to 243,437, and grew another tenfold by 1960 to just over 2 million (Leonard and Noel 1992, p. 435). The airfield was expanded in 1959 to meet this growth in demand as well as to accommodate the new jet aircraft that were just coming into commercial production. In 1964, Stapleton Airfield was officially renamed Stapleton International Airport to reflect the city's growing interest in overseas air connections, although Denver has never really become a major direct-service international gateway. In 1969, Stapleton expanded again by extending its north-south runways onto the southern part of the nearby Rocky Mountain Arsenal. By 1970, passenger totals had grown to nearly 7.5 million, making Stapleton one of the 10 busiest airports in the world (Leonard and Noel 1992, p. 435).

The 1970s and 1980s were to witness dramatic changes in the U.S. airline industry, in the economic health of Denver, and for the future of Stapleton. Passenger traffic continued its rapid growth through the mid-1980s, and with deregulation starting in 1978, Denver soon found itself serving as a major hub for three large carriers. The 1970s energy boom contributed to massive population and economic growth in Colorado, but this was followed by a rapid decline as the price of oil, which had soared in the 1970s, collapsed in the early 1980s. Yet the transportation sector was a major industry for Colorado, and one of a very few that had grown during the period. A city seeking economic growth saw airport construction as a short-term public works project and a long-term magnet to lure business relocation to the transport mecca of the West. These events set into motion a scenario whereby Denver sought to reinvigorate its local economy by investing in another major transportation infrastructure project, a familiar strategy on which it had staked its future at pivotal moments in its past.

Yet another reason compelled either airport expansion or new airport development. Denver sits on the edge of the Great Plains and the Rocky Mountains. The weather can suddenly change, dumping snow, thunderstorms and other unstable wind and weather on the city and its airport. And its airport, Stapleton, had been poorly designed, with two parallel east-west and two parallel north-south runways laid so closely together that only one could be used in each direction during periods of inclement weather. Thus, Stapleton could be reduced to one or two runways quite rapidly. When Denver was socked with snow, thousands of travelers might find themselves stranded at Stapleton International Airport, unable to make a connection to their preferred destination. Hubbing accentuated this problem.

Salt Lake City capitalized on Denver's woes by running full-page advertisements in major national newspapers urging travelers to avoid the Denver bottleneck by connecting at the alternative western hub of Salt Lake City. Denver, which in earlier generations had fought Cheyenne and Pueblo for the right to be deemed commercial capital of the region, would have none of that. As it had built a rail line to link itself with the transcontinental route, as it had built the Moffat Tunnel to provide itself with an east-west transcontinental rail route, as it had fought to have the great east-west and north-south interstate highways built through it, Denver would build a great airport, falsely touted as equidistant between Munich and Osaka (the two major international airports being built at the time), to realize its manifest destiny as center of the travel universe. With DIA, the self-described "Queen City of the Plains" would become the "King City of the Planes." Denver would seek to continue its role as the premiere city of the West. Figure 4-3 reveals the historical growth rates of each of the towns and cities discussed.

Thus has Denver become the self-proclaimed inland seaport for the "Rocky Mountain Empire" (Quillen 1994). If the South Platte River could be made navigable, no doubt the state's political leaders would insist that Denver be made an inland port. But realistically, from stagecoach to rail to air service, Denver's political and business

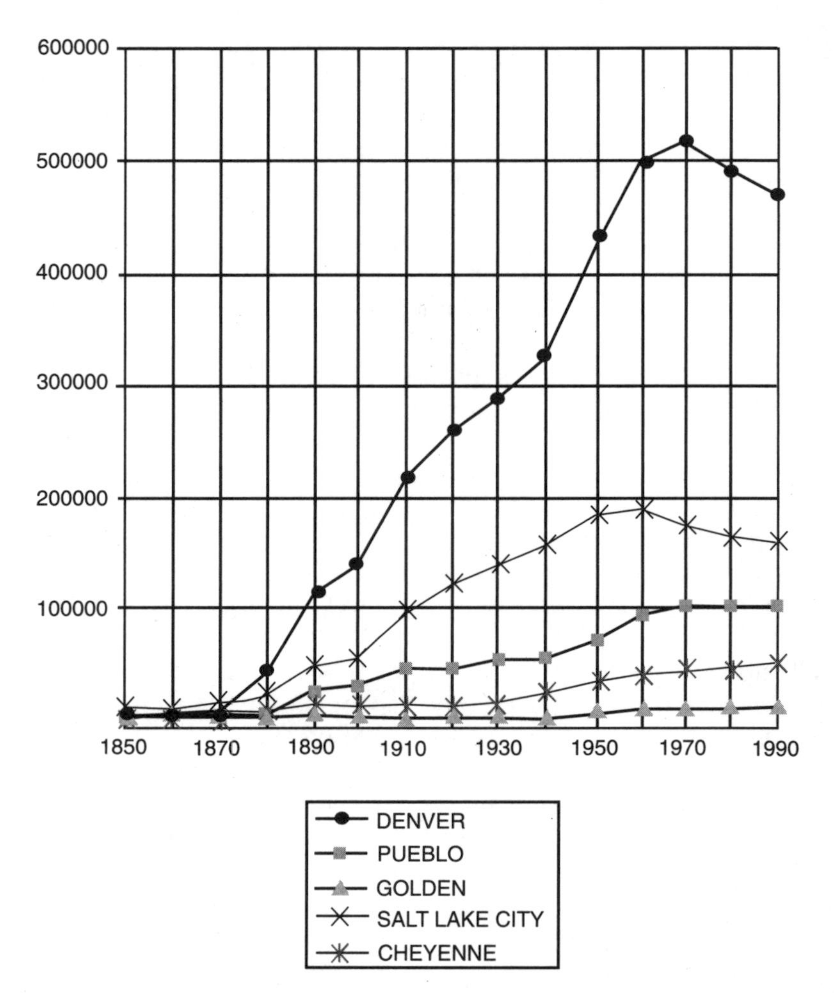

Fig. 4-3. *Population of selected western U.S. cities from 1850 to 1990.*

leaders have taken it upon themselves (with especially vibrant zeal in some cases) to ensure Denver's place among the nation's leading trade centers. As early as the 1940s, they were touting Denver as the nation's aviation center, a city destined to be truly important in the coming "air age." More often than not, these strong affirmations of purpose were heard when major decisions were being made about expanding airport capacity. It is no wonder then that many of these sorts of arguments were revisited during the planning and selling of Denver International Airport.

Economic impacts and the selling of DIA

One of the driving forces behind the push to build a new airport in Denver during the mid- and late-1980s was the expected positive effect that increased air passenger and cargo service, especially to international destinations, would have on the local and regional economy. The processes of economic globalization, including increased international trade, direct foreign investment, joint ownership and marketing agreements between foreign and domestic companies, and opportunities for increased international tourism, were clearly becoming more important to the U.S. economy overall and particularly so for Denver and Colorado.

Recognizing the need for economic analyses, the Colorado Forum, a group of Colorado business leaders, was asked by the City and County of Denver in 1985 to lead a Regional Task Force on the Economic Impact of Stapleton Airport (Regional Task Force 1986). The task force assisted the city in commissioning three reports that focused on different questions of economic impact. Booz-Allen & Hamilton conducted a study on economic impact in Colorado from Stapleton and future airport development (Booz-Allen 1986); Browne, Bortz & Coddington examined land-use impact around Stapleton Airport (Browne, Bortz 1986); and Sussman and Gray considered the implications of new airport development in Denver by surveying the economic and land-use impacts in other cities occurring as a result of their recent airport projects (Sussman and Gray 1986).

A principal finding of the Booz-Allen & Hamilton study was the realization that many of the key growth industries in Denver and Colorado were actually quite strongly linked to the airport and the air service it provided. One of the most important basic industries in the state is tourism and recreation: skiing during the winter months, and camping, hiking, and sightseeing during the summer (Fig. 4-4). More than 1.2 million skiers visited Colorado during the 1985–86 ski season, with 60 percent arriving via Stapleton airport. These visiting skiers pumped more than $400 million into the state's economy and supported more than 25,000 jobs in the ski industry. More than 19,000 jobs were dependent upon summer tourism in 1985, and there were an additional 1300 jobs for hotel and restaurant workers near Stapleton (Regional Task Force 1986). The continued expansion of airport infrastructure in Denver was seen as vital for allowing the important Colorado tourism industry to continue to grow in the future.

A considerable amount of other economic activity in the Denver area was found to be attributable to the airport. In addition to the 49,000 tourism-related jobs, there were nearly 21,000 direct jobs at Stapleton in 1985, mostly airline and airport service employees. Another 13,000 were induced workers, or jobs created by firms that provide goods and services to the airlines and airport. Finally, another 57,000 related jobs in firms that relied on frequent and reliable air service—in sectors such as services, finance, trade, and high technology—were estimated to be attributable to Stapleton Airport (Regional Task Force 1986).

All told, approximately 140,000 jobs were identified by Booz-Allen & Hamilton to be directly or indirectly related to the role of the airport. This airport-related employment was similar in magnitude to the 168,000 Denver-area workers employed in the booming energy industry at its peak in 1980 (Regional Task Force 1986). Transportation had proven to be among the most stable of Colorado's industries vis-à-vis oil, gas, minerals, agriculture, heavy industry, or high-tech. In the late 1980s, United and Continental Airlines together employed as many Coloradans as the state's largest employer, USWest, a regional

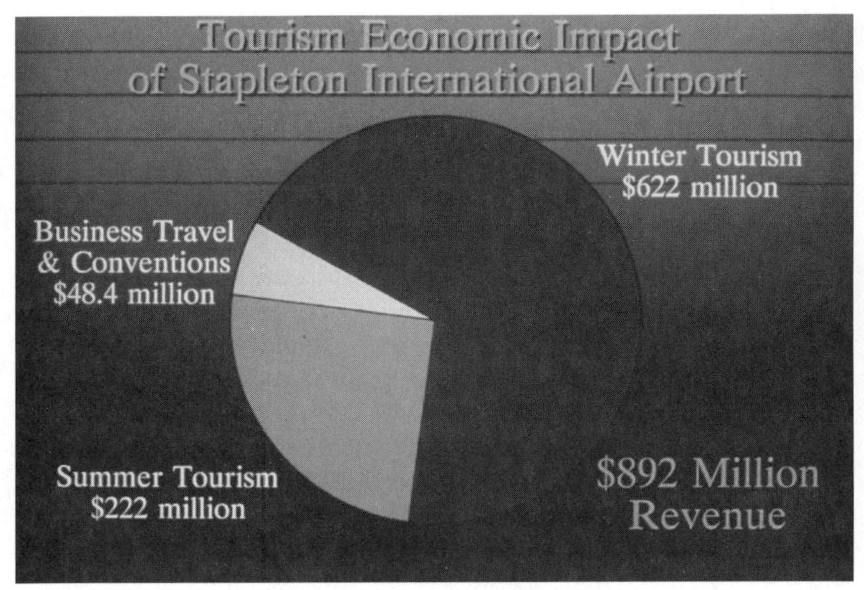

Fig. 4-4. *With its magnificent mountains (52 of which exceed 14,000 feet in elevation), tourism is one of the most important sectors of Colorado's economy. This chart reveals the relationship between Stapleton Airport and Colorado's tourism industry.*

telephone conglomerate. Aviation's importance to Colorado was reflected in the assertion that one of every 10 jobs in the state was airport-related (Regional Task Force 1986). The implication of these findings was that if Denver and Colorado wanted to ensure the continuation of a significant number of jobs, and in fact, increase these numbers, a new airport with expanded capacity would be needed because Stapleton would not be able to handle additional increases, and these air service-related activities would find other locations to be more accessible and amenable.

The Sussman and Gray report, along with another study commissioned by the Colorado National Banks, examined the business impact from several other recent U.S. airport projects (Colorado National Banks 1989). These studies focused particularly upon the economic impacts realized in the cases of Dallas/Fort Worth, after building DFW Airport in 1974, and Atlanta, after Hartsfield International was reconfigured in 1980. Business and community leaders from these two cities felt strongly about the role of the airport in stimulating economic growth in their regions. In fact, the airport was referred to as the single most important economic asset to the region in each case (Regional Task Force 1986). The experiences in Dallas and Atlanta soon became the rallying cry for Denver's "growth machine."

Especially compelling in the Dallas and Atlanta cases was the role that their airports, and their increased air service, played in stimulating international business activity. DFW International Airport was estimated to have produced $5.5 billion in spin-off growth and the relocation of major Fortune 500 corporate headquarters, such as J.C. Penney and American Airlines, to Dallas. In the Las Colinas development near DFW airport, some 750 companies, including Panasonic, Hitachi, NEC, Abbott Laboratories, and Honda, had regional or national facilities in operation. Atlanta Hartsfield International Airport's 1980 reconfiguration was deemed to have created $7 billion in spin-off growth, corporate relocation (e.g., Georgia-Pacific and United Parcel Service), and an explosion in foreign direct investment. In 1978, about 150 foreign corporations were doing business in Georgia. A decade later, 1000 foreign firms were employing 75,000 Georgians. More conservative estimates by Atlanta representatives showed that more than 800 international businesses were attracted to the region during the 1975–1985 period, representing an investment of over $3.3 billion and the creation of 54 000 new jobs (Regional Task Force 1986). In an increasingly international economic environment, in which balance of trade and generation of basic employment were fundamental concerns, these findings represented extremely compelling evidence of the role that airports played in stimulating international business activity.

The City of Denver quickly latched onto the significance of these findings and began to publicize them widely. In fact, when passenger enplanements at Stapleton began to drop in the late 1980s, the city switched its principal strategy in promoting the need for the new airport. It de-emphasized the new airport's importance in alleviating bottlenecks in the nation's air traffic system and increasingly began to sell the new airport on the basis of its potential local and regional economic stimulus (Fig. 4-5). In an economy that had experienced a massive capital withdrawal in the aftermath of the energy boom, a major public works project promising many new jobs and increased investment sounded too good to pass up for many people in the region who were struggling to make ends meet (Fig. 4-6). The airport fit with the strategy of Federico Peña and the Greater Denver Chamber of Commerce to emphasize public works as a stimulant for a moribund economy. Furthermore, there was much to be gained symbolically if Denver and Colorado were to show the world that they were indeed "open for business" and were welcoming new growth by means of a new airport, unlike previous years when this region was perceived as hostile to growth.

Denver: Is it really halfway between Osaka and Munich?

In emphasizing the economic potential of the new airport, its supporters highlighted Denver's central location in the emerging global marketplace. The goal was to persuade local residents that Denver would become a major international air service hub if a new airport were built. Promotional materials produced by the city showed Denver as having an ideal geographical location to be an international center, supposedly halfway between two other world cities with major new airports: Osaka, Japan, and Munich, Germany. Denver, Osaka (or Tokyo), and Munich were portrayed as the global air hubs of the 21st century, with Denver ideally located midway between the other two.

This slick piece of geographical advertising was incorrect in two fundamental ways. First, the city's promotional map distorts the actual

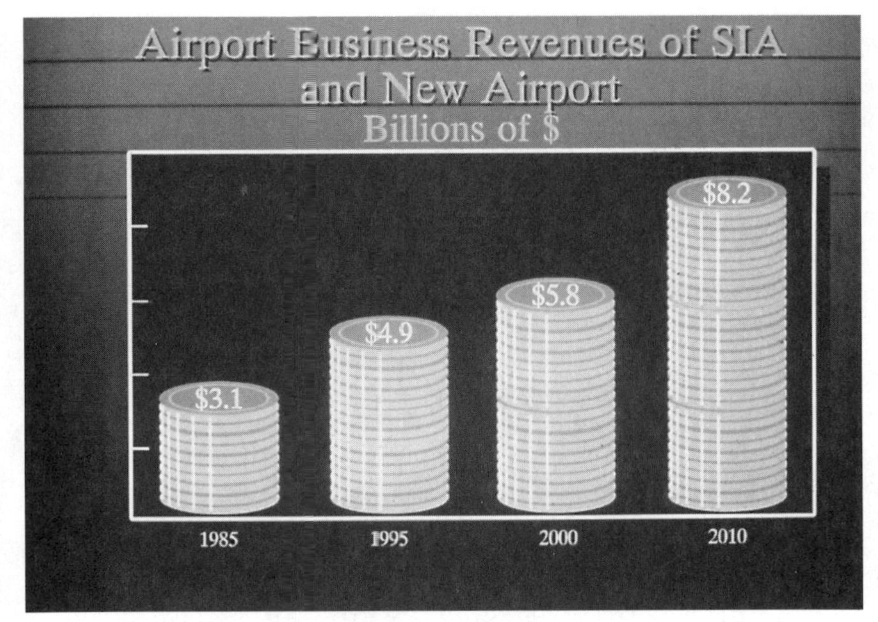

Fig. 4-5. *The new airport itself was deemed to be a venue for increased economic activity, as this chart reveals.*

distances, especially between Denver and the Far East. From the map, it looks like Tokyo is closer to Denver than either London or Munich. In fact, even though Denver is in the western United States, it is more than 500 miles farther to Tokyo than to Munich, and nearly 1000 miles farther to Tokyo than to London. Second, even though most of us tend to perceive distances from a flat-world map perspective, air travel distances are measured based on Great Circle distances. The halfway point between Tokyo and Munich is actually at the Russian town of Matylka located in the West Siberian Plain. Denver does form one corner of a nearly equilateral triangle joining the three cities, but nearly any city in North America could also make the same claim. In fact, Phoenix is the closest major U.S. city to being equidistant from Tokyo and Munich (Flynn 1992a).

Aside from the city's embarrassing display of geographic illiteracy, these promotional errors had a practical relevance. Although one could reasonably discuss direct flights to most European destinations from Denver, there were both technological and economic reasons why direct flights to the Far East were unlikely even if a new airport were to be built. A Boeing 747-400 departing during the warm summer months would have difficulty reaching Tokyo directly because

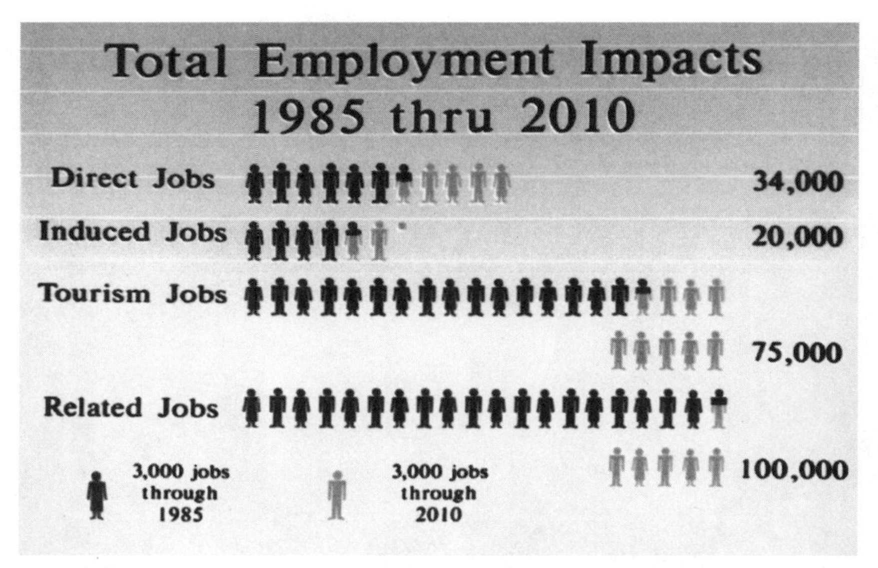

Fig. 4-6. *The construction of a new airport and the new economic activity expected to be stimulated by the existence of a new airport were projected to create a significant number of new jobs for the Colorado economy.*

of the extra fuel that would be required to lift off in Denver's thinner and less humid atmosphere (Flynn 1992b). Recognizing this environmental constraint, the city included in its airport master plan a 16,000-foot runway to facilitate long-distance takeoffs. With this extra-long runway, many, but not all, locations in the Far East would be within range weight restrictions. It would, however, limit passengers, baggage, or cargo on longer flights taking off during hot summer afternoons. The city wanted the extra-long runway by DIA's opening day, but those plans had to be postponed. Subsequently, the city provided a totally different rationale for the sixth runway, arguing that it was essential for the smooth operation of the airport in poor-weather conditions and for reducing adverse noise impacts.

The other obstacles that Denver faced regarding its envisioned status as an international air hub were regulatory and economic. Despite increased globalization in international air service, these routes are still governed by bilateral air service agreements negotiated between nation-states (Dempsey 1987). The agreements specify exactly which cities are allowed to receive direct international service from points abroad. Although there has been a considerable loosening of restrictions in awarding international direct service to more cities, Denver has not been considered a high-priority international destination by either the U.S. Department of Transportation (which, with the State Department, handles bilateral negotiations) or by the airlines. Denver has never been a major international gateway. Historically, it had regular international service to only a few destinations in Mexico and Canada, and, Continental operated a direct seasonal service to London. But Continental's 1994 decision to dismantle its hub operations in Denver also meant the loss of that London flight. Denver has been trying hard to reinstitute direct international service to London or another European destination but with little success. Denver became a leading member of U.S. Airports for Better International Airport Service (USABIAS) which sought to persuade the U.S. DOT to make international route awards on the basis of city impact, not airline equity. In a recent round of international route awards, however, a British Airways flight from London was approved to Phoenix rather than Denver. So, Denver continues to struggle to attain an international air-service presence, but despite its best promotional efforts, the outlook for substantial direct international service in the near future looks bleak (as of 1996, Denver enjoyed nonstop international service only to several cities in Canada and Mexico and weekly summer service to Amsterdam).

DIA as a catalyst for regional economic growth

Equally problematic was the claim that Denver International Airport would stimulate the Colorado and Rocky Mountain regional economy. Improved air passenger and cargo service would allow other communities in Colorado and the region to be better connected into the national and international air transportation system. Businesses would then be more inclined to locate in this region because of the superior accessibility made possible by DIA.

What seems to be occurring so far is that DIA has actually inhibited economic activity in the state and region. This is due largely to the events described in Chapter 2. Since United Airlines has become the dominant carrier at Denver, it has systematically raised its fares to Denver from all points in which no other airline competition exists. When Continental operated a hub out of Denver, many smaller and medium-sized communities in Colorado and the Rocky Mountain region had another alternative that worked to keep fares lower. With

Continental's departure, many cities and towns found themselves either with only one carrier providing service at substantially higher fares, or in some cases, no service at all.

This was especially troublesome for some of the state's major ski areas. One of the original justifications for building DIA was to ensure that the state's ski areas continued to receive a high volume of out-of-state skiers through the airport. Instead, the bad publicity surrounding DIA and its infamous baggage system has caused potential skiers to skip Colorado and to ski elsewhere, most notably in Utah. In fact, the Utah ski and tourism association has used DIA's problems as a key part of its own promotional advertising, asking tourists whether they want to risk having their baggage mangled in Denver. An additional concern has been the case of the Aspen, Colorado, airport, which had relied on Continental Express service to bring skiers from around the country through Denver. After Continental's pullout, Aspen experienced a dramatic decrease in its skiing volume and has been scrambling ever since to find another carrier willing to provide this air service. Other ski towns have been trying to upgrade their service directly from other cities in the United States, bypassing Denver. Hayden Airport in northwest Colorado, for example, has the capacity to handle jet aircraft and has been experiencing increasing numbers of direct flights from cities such as Los Angeles, Dallas, and Chicago during the winter months. There are limits to this type of service, however, as most mountain airports do not have the capacity to handle large jets, especially during troublesome weather episodes.

In addition to the ski areas, a number of communities located on Colorado's western slope (west of the Continental Divide) have been outraged by the high fares and reduced service now being provided by United Airlines and its code-sharing affiliate, United Express, through Denver. In August 1995, Governor Roy Romer responded to the criticisms and organized a meeting in Grand Junction, Colorado, to discuss the issue of air service on the western slope. In this meeting, Romer felt the sting of criticism for his role in helping to promote DIA, which is now perceived as the reason why air service in the rest of the state has suffered. He also was stung at criticism that he did nothing to help Continental keep its maintenance base at Stapleton Airport. The most obvious strategy to remedy the air service situation, as well as the high-fare problem that Denver area travelers face, is to entice additional carriers to serve

Denver and these regional communities. Unfortunately for Denver and the towns dependent on air service from Denver, there has not been much interest among other airlines to initiate such service. Much of that stems from discriminatory code-sharing by the dominant hub airline, which refuses to share connecting traffic with independent jet carriers thereby relegating small communities on the Colorado western slope (and throughout the Rocky Mountain and Great Plains region) to high-cost monopoly turboprop service from Denver.

One place in Colorado that has benefited from DIA is Colorado Springs and its regional airport. Shortly after DIA opened, a new low-cost carrier, Western Pacific Airlines, began operations using Colorado Springs as its hub. It offered very low fares to several large cities throughout the West, including Los Angeles, San Francisco, Phoenix, Seattle, and Kansas City. In order to match this competition, United radically cut its fares out of Colorado Springs but kept its Denver fares high. In a large number of cases, the same fare class would cost as much as $700 or $800 more out of Denver. This fare differential led many Denverites to drive south to Colorado Springs (75 miles away from downtown Denver) to take advantage of the lower fares. Ironically, many of these United flights out of Colorado Springs stopped in Denver before going on to their final destination. This sequence of events caused passenger traffic to increase sharply at the Colorado Springs airport while depressing DIA traffic. The economy of Colorado Springs has benefited greatly as a result.

In short, instead of acting like an economic fountain spraying benefits over the communities in the state and the region, DIA has acted more like a vacuum cleaner, sucking money away from these communities and from Denver itself, to United's headquarters in Chicago. United is quite defensive about the allegation that it was reaping monopoly rents on origin and destination (O&D) traffic at Denver. Its CEO, Gerald Greenwald, claimed its rents at Denver had soared 600 percent, leading to an annual bill of $180 million (compared with $35 million at Stapleton), even factoring in the increased efficiencies enjoyed at the new facility (Greenwald 1995). It should be noted that some of these rent increases were due to United's demands for larger and more expensive facilities at the new airport. United vice-president Roger Gibson contended that United had raised prices only 15 percent in 1995, although conceding that business travel fares had risen 46 percent (Gibson 1995). But whether caused by high DIA costs, or the United monopoly, or both, Denver travelers felt the pinch of sharply higher ticket prices, a malady not conducive to the economic growth DIA was built to stimulate.

Local land-use impact from airport development

Another purported benefit of building DIA was the large amount of local land development that would occur for both the City of Denver and Adams County in the areas around the new airport. Such development would, of course, benefit land developers who happen to own large tracts of land either in or near new airport sites (see Chapter 3).

During the time that Stapleton International Airport was in operation, a variety of land-use activities emerged within its local area. A number of hotels, motels, and restaurants came to line Quebec Street just west of Stapleton. Cargo distribution operations, warehouses, and truck and rail terminals dominated much of the area to the north, east and south of Stapleton. United operated its training facilities at Denver, while United and Continental both operated maintenance bases at Stapleton. This sort of land-use activity is quite common for airport environs. Most older airports have similar land uses in their immediate vicinities, although there are variations depending upon the community.

Studies initiated by the Colorado Forum and the Colorado National Bank in the late 1980s analyzed land-use impact around recent airport projects in the United States, such as Dallas/Fort Worth, Atlanta, Kansas City, and Washington Dulles (Sussman and Gray 1986). These studies discovered that newer forms of development had occurred, especially large mixed-use planned developments, incorporating commercial, office, and residential land uses. The prototype of this development was the Las Colinas area near the Dallas/Fort Worth Airport, which opened in 1973. It is described as a 12,000-acre master-planned development of commercial, residential, educational, recreational, and retail land uses designed to harmonize with the natural environment. In 1989, Las Colinas contained 13 million square feet of office space, 7 million square feet of warehouse space, 900 companies, and 8552 residential units. One-half of the 7000 acres proposed for development was designated for open spaces, parks, lakes, greenbelts, biking and pedestrian trails, country clubs, tennis clubs, schools and museums (Colorado National Banks 1989). This type of development was very different from airport-area land uses of the past. It was a more upscale, less industrial-based type of development that was more akin to the suburban mixed-use centers mushrooming on the outskirts of metropolitan areas around the country. These developments were typical of those found today in "edge cities," or suburban downtowns, which now rival the traditional central business districts (CBDs) in terms of economic activity.

Such developments around airports obviously depend upon the viability of the airport itself. Yet because of the fickle nature of airline management and their dismal economic performance since deregulation, many airlines have abandoned hubs at cities that invested large sums of money in airport infrastructure, thus leaving the airport with excessive capacity and a high debt burden. Kansas City built a new airport for TWA's hub. But TWA relocated to St. Louis. For a time, Braniff and then Eastern Airlines tried to make a hub of Kansas City, but both found their way to bankruptcy instead. The same scenario was played out in Dayton, Ohio, which built infrastructure for Piedmont (absorbed by USAir), as well as San Jose, California, and Raleigh/Durham, North Carolina, both of which built hubs for American Airlines, which turned them over to Reno and Midway Airlines, respectively. In other words, under deregulation airports cannot rely on the commitment or ability of airlines to maintain hubs even where major investments are made in airport infrastructure to accommodate them. Witness Denver, which built an entire concourse for Continental Airlines, which subsequently folded its Denver hub and today needs only three gates. In other words, there appears to be no "build it and they will come" magic to investment in airport infrastructure.

This point was implicitly recognized in the Task Force Summary Report, which identified several factors as key to the pace and quality of development in new airport environs (Colorado Forum 1986). One such factor is how well the airport fits into the overall pattern of urbanization within a metropolitan area. Since most airports are not located directly in the path of the most-developed sector of the city, they tend to be found near the major transportation or industrial wedges emanating from the CBD. In the cases of Dallas/Forth Worth and Atlanta, the airports were found to actually redirect growth toward their site locations even though little development had been based there previously. The effects for Kansas City were not as significant, mainly because the airport was perceived to be very far away from the rest of the metro area. The same reason also was cited for the lack of development around Washington's Dulles Airport in its early years. But now the rest of the metro area has grown farther out into northern Virginia, and a substantial amount of development can be found in that airport area.

Other factors identified by the task force included

1. An adequate transportation system with provision for 360-degree access to the airport site.
2. A comprehensive and implementable land-use plan for the airport environs.
3. Special zoning that prevents residential development in noise-sensitive areas.
4. Perceived distance to the airport—the shorter the better.
5. Cost of land and extent of real estate speculation prior to opening of the airport.

In the case of Denver, several of these "preconditions for growth" factors are proving to be problematic. First and foremost, Denver International Airport is perceived to be very far away from the rest of the metropolitan area. In fact, it actually is very far away. When the new airport was still being debated in the late-1980s, the City and its propaganda machine were touting that the new airport would only be 17 miles away from downtown. This compared relatively favorably with Dallas/Fort Worth (17 miles from Dallas; 15 miles from Fort Worth), Kansas City (17 miles) and Washington Dulles (26 miles). This 17-mile distance, however, was the distance from downtown Denver to an earlier site, not the eventual location of the airport terminal. Yet that 17-mile figure continued to be used and promoted during the 1988 and 1989 airport referenda even though the decision to move the airport site further out had already been made. The 24-mile distance to the actual site obviously did not compare favorably with other recent airports and was extremely unfavorable when compared to Stapleton, which was only 7 miles away from the Denver CBD. The city's promotional materials in the 1990s (after the referenda and bond sales) show the distance as 23 miles from downtown. Thus, Denver bears the distinction of having its only airport located farther away from its CBD than any other city.

Another factor that is not yet in place around DIA is transportation access in a 360-degree direction. Ground access is discussed in

Chapter 6; suffice it to say that DIA has only one main entrance, Peña Boulevard (named for the father of DIA, Federico Peña), which connects the airport to I-70.

The narrow strip of land along Peña Boulevard that is under the jurisdiction of the City and County of Denver has been included as part of Denver's Gateway Land Use Plan (Figs. 4-7 and 4-8), an ambitious and visionary plan for future development along the only entrance into the new airport. It includes plans for high-quality mixed-use developments, livable neighborhoods, town squares and activity centers, balanced transportation, environmental protection, ample parks and trails, and quality infrastructure (Denver 1991). It has clearly been influenced by DFW's Las Colinas and similar developments around other recent airports, and attempts to prevent the sprawling strip development that typifies the outskirts of many U.S. cities.

Like any plan, however, it is just a plan unless land owners and developers subscribe to it. Before the airport opened, no substantial land-use development occurred in the Gateway area or anywhere along Peña Boulevard. The delays and problems that the airport encountered as it sought to open undoubtedly had a chilling effect on potential developers' interest in the airport area.

Shortly after DIA opened on February 28, 1995, however, a number of specific land-development projects were announced (Raabe 1995a). Bill Pauls, one of the major developers of the successful Denver Technological Center in southeast Denver, announced plans for a multimillion-dollar business park on a 1200-acre site located near the intersection of Peña Boulevard and I-70. Dallas developer Baruch Properties was interested in a 442-acre parcel for a mixed-use development just west of Peña Boulevard between 48th Avenue and 56th Avenue. A 161-room Marriott Fairfield Inn at 69th Avenue and Tower Road was built and opened in 1996. Groundbreaking on a 202-room Marriott Courtyard hotel in the same vicinity started in October 1996. Lead developer L. C. Fulenwider announced plans for a major high-quality, mixed-use development on its parcel of land within the Gateway area. By the summer of 1995, the City of Denver had come to an agreement with Westin to build a new hotel on the airport site itself, right next to the terminal building (by 1996, this project had been delayed). A number of office, retail, recreational, lodging, and other real estate developments were either under construction or proposed as of October 1996 (Conley 1996). By the

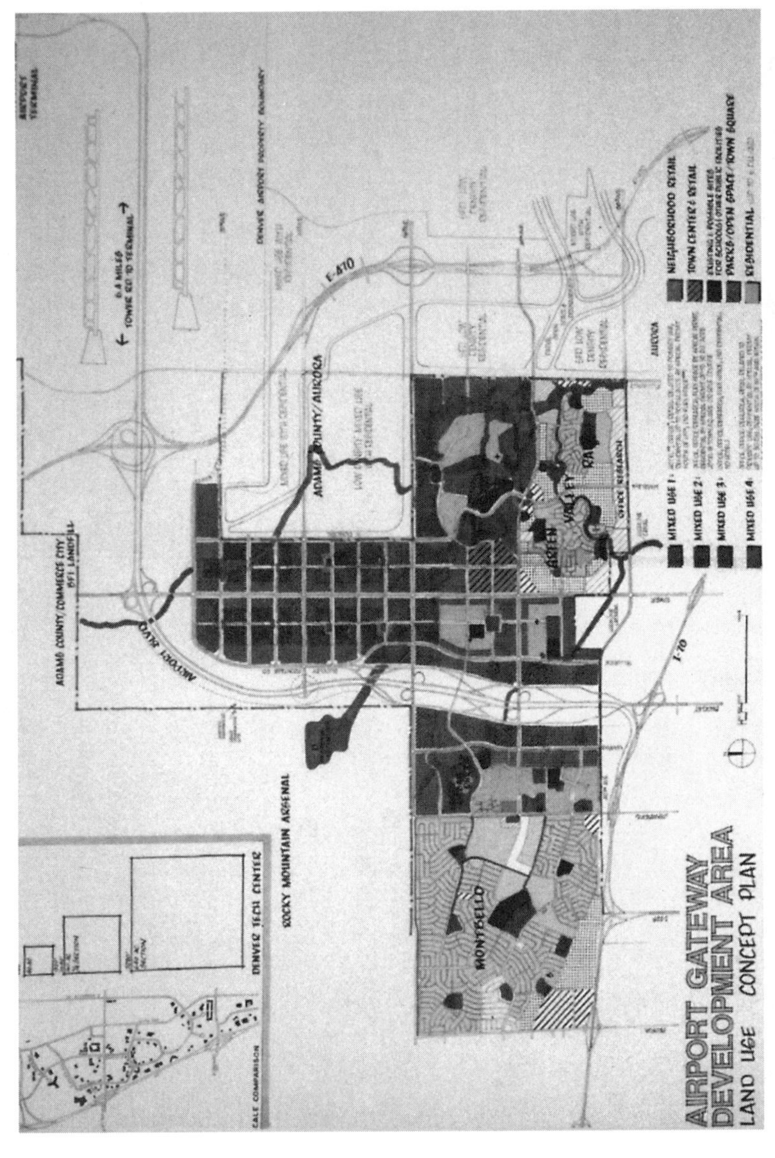

Fig. 4-7. *Several new airport-related developments have emerged, including a "Gateway Development Area" on both sides of Peña Boulevard.*

summer of 1996, only the Marriott Fairfield and a Conoco gasoline station had been built. But if all of the proposed projects actually occur, the area along Peña Boulevard to the airport will undergo a dramatic transformation. What is now open farmland or rangeland is expected to become the site of billions of dollars in residential and commercial development over the next 10 to 20 years (Conley 1996; Raabe 1995b).

Most of these aforementioned developments are within the jurisdiction of the City of Denver, either on the airport grounds itself or in the Gateway area. The nearby city of Aurora has also been working to realize spin-offs from the new airport. The Aurora Economic Development Council has been operating a DIA marketing program since early 1994. After DIA opened, it mailed 10,000 brochures to targeted companies asking them to consider relocating or expanding at Aurora sites near the airport. Aurora has a number of existing business parks near the new airport but so far has not announced any significant new ventures. Aurora and Denver are negotiating to cooperate on joint marketing for the area (Raabe 1995c).

Adams County also has jurisdiction over substantial amounts of land near the airport, and it too wishes to reap the benefits of airport-related economic development. Perhaps the only major selling point of the airport during the Adams County election was the economic spillover effects that would occur, providing jobs for the residents and revenues for the county. In fact, an economic development "corridor" was purposely created in Adams County by moving DIA further north and east (see Chapter 6). As part of its intergovernmental agreement with the city of Denver, Adams County insisted that limits be placed on the amount of development that Denver could have within its airport environs, so that more of it could be located within Adams County (Brown 1991). To date, neither Denver nor Adams County has witnessed much airport-related development, but, as we have seen, Denver has projects currently in the planning stage, and a few have been built. A principal factor in Denver's edge is Peña Boulevard, the only access road into the new airport, which goes right through the Gateway plan area and is entirely within Denver's jurisdiction. Adams County will need to press for additional access points into the airport and the E-470 beltway to spur development interest on its land. But aside from the original agreement, there does not seem to be much cooperation between Denver and Adams County regarding land-use planning in the immediate airport area.

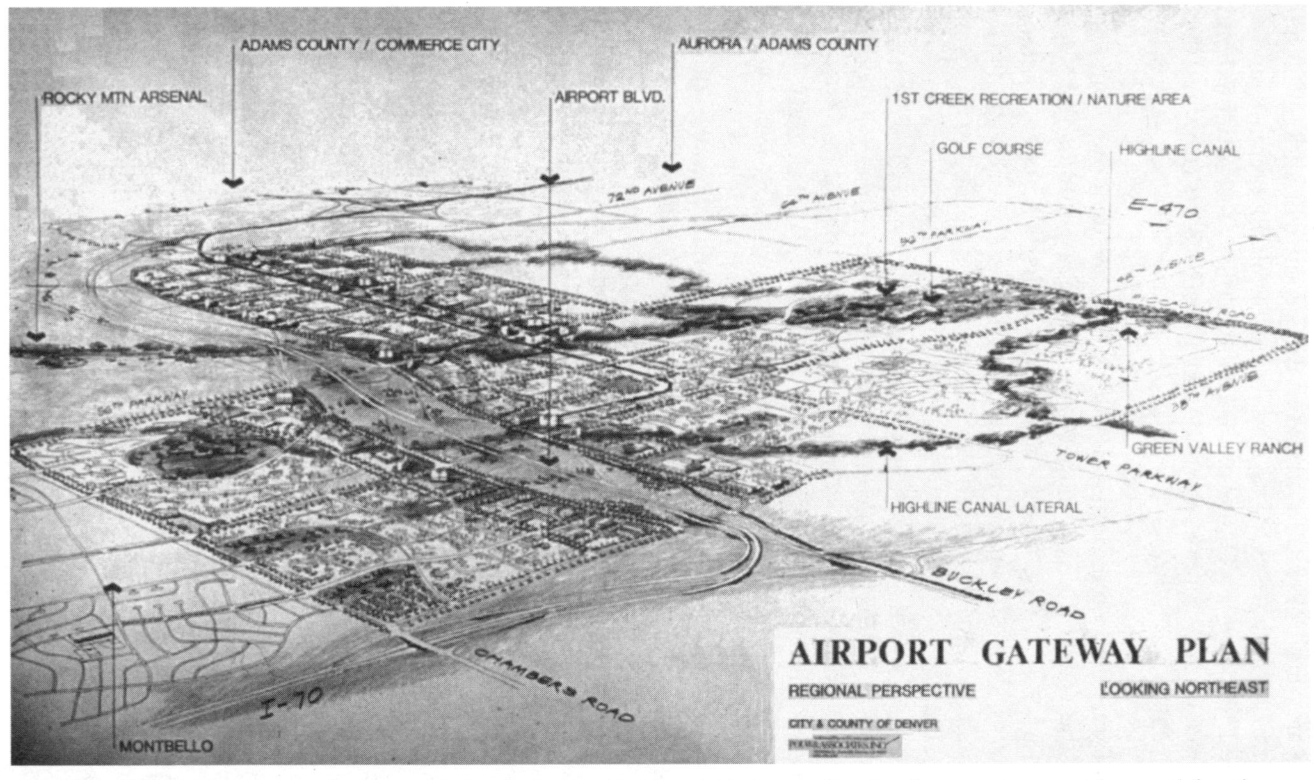

Fig. 4-8. *This artist's rendering of the schematic in Fig. 4-7 shows the development zone planned for the area surrounding Peña Boulevard.*

These two governmental entities have resumed their traditional adversarial relationship after an ever-so-brief period of détente.

The lack of cooperative, or at least coordinated, land-use planning will have several ramifications for the future development profile of the area. With each jurisdiction pursuing its own plans, and individual developers within jurisdictions doing likewise, there is a distinct possibility that the area will become a hodgepodge of development and will resemble the urban sprawl that so many urban planners and landscape architects detest. Developers may be able to play one economic development agency off another and may receive tax breaks, subsidies, and other incentives for locating their developments in certain jurisdictions.

Another negative ramification from the lack of coordinated land-use planning stems from the multitude of development plans that call for increasing numbers of residential units within the airport environs area. One of the original justifications for having to build an entirely new airport was that Stapleton had become crowded in by residential development, and these residents were complaining about aircraft noise. The solution to the problem was to move airport operations to a greenfield site where noise impacts were expected to be very low (more about actual noise impact at DIA in Chapter 6).

To prevent the same residential encroachment from happening, Denver and Adams County agreed to set noise buffer zones to protect existing and future residents from noise. Denver was allowed to annex 45 square miles for the airport itself and an additional 8 square miles for the noise buffers (Brown 1991). Denver rezoned the buffer areas to prohibit residential development in areas with higher than 65 LDN (level day/night—roughly equivalent to an average of 65 decibels over one day/night period) readings. (See Chapter 6 for further explanation of noise level measurement.) Adams County has an even more stringent zoning requirement, prohibiting residential development within the 60 LDN contour. There will be no residences within 5 miles of the end of any built or planned runway, and no residential development within a 2-mile buffer zone around the airport land (Brown 1991).

With all of these residential development safeguards in place, it might seem that the issue has been addressed. Unfortunately, the science of noise monitoring is not as precise as the agreements suggest. Actual noise readings may not necessarily conform to the proposed LDN noise contours. Furthermore, auditory perception of noise as measured by reported complaints may not exhibit much correspondence with projected LDN contours. What results is an imperfect awareness of the noise-impact surface around an airport and the real threat of future noise litigation that may hound DIA just as it did Stapleton. Allowing any residential units anywhere near the new airport is a risky proposition. But so many development plans have included residential complexes in areas where residential use is not prohibited that there is bound to be increasing residential encroachment around the new airport. The lack of coordination among competing jurisdictions will exacerbate the residential growth problem because these entities will likely be very accommodating to developers and will allow them to do essentially what they want without much restriction. After all, the economic development agencies and developers will note, wasn't that the reason why the new airport was built in the first place?

In the midst of this local economic development competition, another potential advantage that Denver has over its rivals is the redevelopment of the defunct Stapleton airport site. One of the key motivating factors behind Denver's drive to relocate its airport was to enlarge its territorial boundaries and open up the 4700-acre Stapleton site for additional development. Former mayor Federico Peña emphasized the development potential of both the DIA and Stapleton sites as a deciding factor in going forth with the new airport (*Rocky Mountain News* 1995):

> *I strongly believe that five, 10, 15, 20 years from now, people will look at the airport, will look at the land that was annexed, will look at the redevelopment opportunities of Stapleton and say this is one of the best investments and economic decisions for the city that was made, many many years ago.*

In the period leading to the two airport referenda, the redevelopment of Stapleton was held out as another bonus of the plan. But as the years after groundbreaking at DIA began to slip by, there didn't seem to be many real land-use alternatives emerging for Stapleton. The only intriguing possibility for Stapleton during this time was its potential use as an annex to the Smithsonian Air and Space Museum, housing extra aircraft that had been accumulated by the museum over previous decades. Although it had some merit, the idea eventually was abandoned when it became evident that the museum aircraft could be housed at Washington Dulles Airport, a location that appealed to the Smithsonian and several influential congressmen from northern Virginia. Under increasing pressure, the City of Denver maintained that eventually something would be found for Stapleton, preferably an activity that could make some use of the old airport terminal building, concourses, hangars, and runways. The city began to shift some of its personnel and resources from Gateway toward the Stapleton redevelopment project. A private group, the Stapleton Redevelopment Foundation, was formed and began working with the city to identify potential developers for Stapleton.

Once Stapleton was closed and DIA opened in February 1995, development interest picked up somewhat at Stapleton, although not as much as in the Gateway area at DIA. The Union Pacific Corporation had earlier acquired a 270-acre parcel of Stapleton land from the city in exchange for a key parcel along I-70 that the city needed in order to build Peña Boulevard. This Union Pacific parcel on the eastern margin of Stapleton is now slated to become a $125 million business park called the Stapleton Business Center, containing 4 million square feet of warehouse, hotel, and retail space (Raabe 1995b). Developers Bill Pauls (who is also developing a business park at DIA) and Paul Powers are in charge of the marketing and development of this project. A 155-acre parcel located in the northeast corner of Stapleton is being purchased by the King Soopers grocery chain to serve as a distribution center (Raabe 1995b). Recent interest has also been focused on using some of the aircraft hangars and related property at Stapleton for a motion picture studio. In October 1996, United Airlines announced it would be expanding its pilot training facilities at Stapleton by spending about $14 million on a new building that would house additional flight simulators (Leib 1990).

The Stapleton Redevelopment Foundation, in conjunction with the City and County of Denver and the Citizens Advisory Board, produced the Stapleton Development Plan in March 1995. Incorporating many innovative urban-planning designs, the plan charts out future development for the Stapleton area. Envisioning "a network of urban villages, employment centers, and significant open spaces, all linked by a commitment to the protection of natural resources and the development of human resources" (Stapleton Development Plan 1995), it represents a blueprint into the future not only of the Stapleton area, but of the region as well. It is uncertain, however, when—or to what degree—the plan will become a reality.

In a matter of just a few years, Denver was able to realize a windfall of extra land coming from Stapleton's closure, the opening of new DIA land, and several other sites including closed military bases and renovated railyards and warehouse districts. With all of this extra land on the market, as well as the parcels in suburban jurisdictions, the Denver area appears to have a glut of available land just waiting for offers from developers. Although there has been much more airport land interest since the new airport opened, it remains to be seen how all of these land parcels will eventually be developed and whether these uses will be compatible with the public's vision of the nature of future growth in the region. Will airport-related development be economically and environmentally sustainable in the long run?

Conclusion

From the early- to mid-1990s, the economy of Denver, the metro area, and Colorado has experienced relatively substantial growth. It appears that the city, region, and state have overcome the depths of the energy industry withdrawal, and that this new growth seems to be less dependent on any one sector. Telecommunications, computers, and a variety of information-related service activities are driving the current growth cycle. Of course, much of the growth has also been fueled by rapid population migration from California and elsewhere, so it is difficult to assess how long this current boom will last.

A series of intriguing questions could be raised that considers the role of Denver International Airport in this current growth cycle. Was the construction of DIA a factor behind this growth in Denver and the region, or would the growth have occurred regardless? How much extra economic benefit (or loss) has Denver and the region received from DIA as opposed to Stapleton? Will businesses locate in Denver because of DIA that otherwise would not if Stapleton was still the main airport? Was the cost of DIA worth the investment from an economic development perspective? These questions will need to be addressed in future years, and their answers will ultimately become important parts of the final judgment of this project. The Denver case also offers some insights into the theoretical relation between airports and urban/economic developments, as expressed in the following lessons.

Lessons learned

- *Growing economic integration across the world will mean that expeditious and effective communication and transportation will be important in the global economy.* An increasing amount of interaction will occur via improved telecommunications technologies, and, to some degree, this will substitute for a portion of air-passenger traffic. Nevertheless, important decisions and meetings will continue to be conducted in face-to-face settings, and that requires quick and efficient air service. The way for cities and nations to ensure that type of service is to be situated as a major air transport center within national and international airline networks. The provision of quality airport capacity at low fees is a contributing factor in the realization of this level of service.
- *Despite the documented relation between airports and urban growth, airports are not a panacea for lagging economic development.* The relation between air transportation, airports, and urban/economic growth is complex and cannot be treated as a simple one-way causation process. There are cases where cities have "built it," but "they did not come." Recent events occurring in the ever-turbulent airline industry have conspired to depress traffic, raise fares, and call into question the global, national, and regional economic impacts that it was supposed to be providing. DIA has spurred some local development interest in some parcels near the new airport, as well as a few at redeveloping Stapleton. But by and large, the hoped-for impacts have not been occurring. This suggests that airport projects must fundamentally be supported on the basis of their role in the national and international air transportation system. To some degree this was true in the Denver case. In the long run, DIA's virtually unlimited capacity will allow Denver to handle more traffic, and perhaps more economic activity, as other city's airports become increasingly congested. But many DIA supporters "hyped" the economic impacts argument when traffic at Denver declined in the late 1980s, and the Denver community will find the inflated costs of DIA difficult to absorb. Justification for building additional airport capacity

must be based on solid air traffic needs, not as a panacea for lagging economic development.

- *Land-use planning around airports should be coordinated and consistent among affected jurisdictions.* In the rush to realize economic development impacts from new airports, there is a tremendous temptation to encourage developers by providing incentives and allowing them to build what they want, even if it doesn't conform to a master plan. If planning is to be effective, jurisdictions must agree on coordinated and consistent land-use plans around airports.
- *Place insurmountable barriers to residential development near airports.* Allowing residential development to occur anywhere near airports is asking for trouble and will sow the seeds for the airport's future demise. No matter how lucrative the development seems, land-use planners and politicians must resist the temptation to allow residential development to encroach upon the airport's environs. We know too little about the impact of airport noise and its effective measurement to base development plans on projected noise contours (see Chapter 6). It is better to be safe than sorry.

References

Airports. 1995. Korea seeks further U.S. participation in new airport. June 20:245.

Aschauer, David A. 1989. Is public expenditure productive? *Journal of Monetary Economics.* 23:177–200.

Asian Wall Street Journal. 1995b. China seeks investors for airports. July 19.

Aviation Europe. 1995. Munich airport large source of jobs. June 15:6.

Aviation Week & Space Technology. 1994. U.S. funds feasibility study for second stage of Macau airport. June 6:36.

Becker, Stuart. 1995. Airports: Chinese opera. *Far Eastern Economic Review.* April 6:54.

Black, Alexandra. 1994. Trade and tourist boom from new airport. *Inter Press Service Network.* September 9.

Booz-Allen & Hamilton, Inc. 1986. The Regional Economic Impact of Stapleton International Airport and Future Airport Development.

Brower, Helen. 1995. Mainland adds, upgrades airports. *Travel Weekly.* August 17:24.

Brown, David. 1987. Japanese building international offshore airport to serve Osaka. *Aviation Week & Space Technology.* July 13:38.

———. 1991. Denver airport: International hub of the future; Denver aims for global hub status with new airport under construction. *Aviation Week & Space Technology*. March 11.

Browne, Bortz, & Coddington, Inc. 1986. A Land Use Analysis of the Environs of Stapleton International Airport.

City and County of Denver (Denver). 1991. The Gateway Plan.

Colorado Forum. 1986. The Regional Economic Impact of Stapleton International Airport and Future Airport Development: Summary Report.

Colorado National Banks. 1989. Ready for Takeoff: The Business Impact of Three Recent Airport Developments in the U.S. Prepared by Coley/Forrest, Inc.

Conley, Steve. 1996. Growth around DIA taking off. *Denver Post*. October 27.

Darmody, Thomas. 1993. The design and development of world class airports. Paper presented to the IBC International Conference on Airport Development & Expansion, Hong Kong, October 28.

Dempsey, Paul. 1987. *Law & Foreign Policy in International Aviation*. Irvington-on-Hudson, NY: Transnational Publications.

———. 1988. *The Social & Economic Consequences of Deregulation: The Transportation Industry in Transition*. Westport, CT: Quorum Books.

———. 1995. Airlines in turbulence: Strategies for survival. *Transportation Law Journal*. 23:16–17.

——— and William Thoms. 1986. *Law & Economic Regulation in Transportation*. Westport, CT: Quorum Books.

———, Andrew Goetz, and Joseph Szyliowicz. 1992. The University of Denver Center for Transportation Studies: Education at the crossroads. *Transportation Law Journal*. 21:4.

Donoghue, J. A. 1995. The Pearl-Y gateways. *Air Transport World*. February 1:75.

Flynn, Kevin. 1992a. New-airport backers look to Europe before Asia. *Rocky Mountain News*. April 20.

———. 1992b. Denver to be left high and dry? *Rocky Mountain News*. April 20.

Gibson, Roger. 1995. United Airlines decries guest editorial. *Denver Business Journal*. October 6.

Gillen, David W. 1996. Issues in airport economic development. Paper presented at Transportation Research Board annual meeting. Washington, D.C. January.

Goetz, Andrew R. 1992. Air passenger transportation and growth in the U.S. urban system, 1950–1987. *Growth and Change*. 23.

Greenwald, Gerald. 1995. Future bright for DIA and Denver. *Denver Post*. February 28.

Hardaway, Robert. 1991. What will airport haters carp about now? *Denver Post.* July 1.

Hill, Leonard. 1993. Asia's newest "dragon." *Air Transport World.* September 1:66.

Irwin, Michael D. and John D. Kasarda. 1991. Air passenger linkages and employment growth in metropolitan areas. *American Sociological Review.* 56:524–537.

Ivy, R. L., F. J. Fik, and E. J. Malecki. 1995. Changes in air service connectivity and employment. *Environment and Planning.* A 27:165–179.

Joyce, E. 1985. Testimony from The Economic Impact of Federal Airline Transportation Policies on East Tennessee: Hearing Before the Senate Committee on the Budget, 99th Congress, 1st Session.

Kyodo News International. 1994. Kansai International Airport inaugurated. September 5.

Leib, Jeffrey. 1996. United to expand Stapleton training site. *Denver Post.* October 31.

Leonard, Stephen J. and Thomas J. Noel. 1992. *Denver: Mining Camp to Metropolis.* Niwot, CO: University Press of Colorado.

Markusen, Ann, Peter Hall, and Amy Glasmeier. 1986. *High-Tech America: The What, How, Where, and Why of the Sunrise Industries.* Boston: Allen and Unwin.

Mecham, Michael. 1995. Seoul ranks no. 1 in airport growth. *Aviation Week & Space Technology.* April 10:34.

Mok, John. 1993. The development of Hong Kong's new international airport. Paper delivered at International Conference on Aviation & Airport Infrastructure, Denver, Colorado, December 8.

Mufson, Steven. 1994. Accord boosts Hong Kong airport. *Washington Post.* November 5.

Munnel, Alicia H., ed. 1990. *Is There a Shortfall in Public Capital Investment?* Conference proceedings. Boston: Federal Reserve Bank.

Noel, Tom. 1994a. From a jumped claim to DIA, Denver has never lacked visionaries. *Denver Post.* April 2.

————. 1994b. By-pass phobia leads Denver to spin giant webs of steel and concrete. *Denver Post.* May 14.

O'Connor, Kevin and Ann Scott. 1992. Airline services and metropolitan areas in the Asia-Pacific region 1970–1990. *Review of Urban and Regional Development Studies.* 4:241.

Ogawa, Zenjiro. 1993. Kansai International Airport projects. Paper delivered at International Conference on Aviation & Airport Infrastructure, Denver, Colorado, December 8.

Paisley, Ed. 1995. On a wing and a prayer. *Far Eastern Economic Review.* May 5:76.

Phillips Business Information. 1994. New Macau airport set to open July 1995. January 28.

Quillen, Ed. 1994. DIA follows Colorado tradition. *Denver Post*. April 5.

Raabe, Steve. 1995a. Deals flying at DIA parcels. *Denver Post*. March 20.

————. 1995b. Developers to market airport site. *Denver Post*. September 2.

————. 1995c. Denver, Aurora to tout DIA. *Denver Post*. September 2.

Regional Task Force on the Economic Impact of Stapleton Airport. 1986. The Regional Economic Impact of Stapleton International Airport and Future Airport Development, Summary Report.

Rocky Mountain News. 1995. Peña defends his role in new airport, slams media coverage. January 29.

Sampson, Roy J., Martin T. Farris, and David L. Schrock. 1990. *Domestic Transportation: Theory and Practice*, 6th ed. Boston: Houghton Mifflin Company.

Sanders, Heywood T. 1993. What infrastructure crisis? *The Public Interest*. 110 (Winter):3–18.

Smith, Craig. 1995. Zhuhzi airport opens to uncertain future. *Asian Wall Street Journal*. May 31.

Stapleton Development Plan. 1995. Prepared by Stapleton Redevelopment Foundation, City and County of Denver, and Citizens Advisory Board.

Sussman, Gennifer and Frank Gray. 1986. Implications of the Construction of Major New Airport Facilities for Economic Development in the Metro Denver Region.

5

Financing the field of dreams

"We were lied to about the bonds, we were lied to about the risks, we were lied to about the benefits of the project; but the bonds are not defaulting."—GORDON YALE, SECURITIES EXPERT

Modern airports are expensive. They are large and complex institutions, cities unto themselves, through which millions of people and tons of cargo pass annually. Such activities require very large investments. In the 1970s a new airport (Dallas/Fort Worth) cost $700 million; by the 1980s new airports were costing billions of dollars. Such tremendous outlays have not discouraged communities and governmental agencies from investing ever larger sums in new facilities. Driven by the need for additional airport capacity, a major boom in airport construction is underway throughout the world. The International Air Transport Association (IATA) and the International Civil Aviation Organization (ICAO) estimate that, worldwide, $250 billion will be spent for airports between now and 2010, of which $100 billion will be required for the Asia-Pacific region alone (Dunham 1993). Europe will have to invest between $30 and $40 billion by 2000 (Ashford and Moore 1992, p. 57).

The cost for the new airports ranges from a low of $0.9 billion at Macau to $14.4 billion at Kansai and more than $20 billion at Hong Kong. Several are in the $3.5 to $7 billion range.

Whatever the cost, the sponsors of any airport project must arrange for the financing. This issue of financing is a crucial element in any airport planning effort, because each step requires financing, and the entire program is based on the ability to obtain the necessary funds. Of course, the construction of additional airport capacity is of direct concern to the primary tenants, the airlines. For them, airport expansion has positive and negative components.

On the positive side of the ledger, the growth in flights and passengers can create congestion on the land side (in terms of surface access), air side (in terms of runway, tarmac, and air space), and in the terminal. Some of that can be resolved with better utilization of scarce resources, such as technological advances in aircraft navigation or peak-period pricing. But, ultimately, congestion can cause airports to expand terminals and add runways, and new airports to be built. New infrastructure can enhance carrier efficiency and productivity in serving a growing customer base. It reduces congestion and delay, leading to enhanced utilization of aircraft and labor and reduced consumption of fuel. The U.S. Federal Aviation Administration (FAA) predicts that, absent infrastructure expansion, serious delays at more than 30 of the nation's largest airports will cause $1.1 billion in additional airline costs by the year 2001 (FAA 1993).

On the negative side, most of the cost of new and expanded infrastructure must be borne by the airlines (in the form of landing fees, terminal fees, aircraft parking fees, gate and hangar rental, ground handling services, air traffic control charges, and fuel taxes) and their passengers (in the form of passenger facility charges, parking, and tolls). Contributors to the revenue stream include passengers, taxpayers, and concessionaires, and the sale and lease of real estate.

From the perspective of the airports, user costs are a relatively modest portion of airline operating expenses—a mere 4.1 percent of total average annual operating costs since 1978 (Dunham 1993). But from the airlines' perspective, whose net profit margins in the United States ranged between 2 and 3 percent before deregulation and collapsed to less than 1 percent since, even a modest economic burden is an onerous one.

During the 1980s, airline user charges constituted between 70 and 90 percent of airport revenue (although other sources insist that passenger carriers pay only about a quarter of airport costs, about the same as concessions). ICAO predicts an average 9 percent annual increase in airport landing and associated charges and an average 12 percent annual increase in route facility charges through the end of this decade. What is clear is that in recent years, airport and other charges imposed upon airlines have grown faster than most other operating expenses and the ability of airline operating revenue to digest them (Tompkins 1993).

And while airport capital-equipment needs will total between $250 billion and $350 billion by 2010 (with much of that paid, directly or indirectly, by the airlines), airline capital needs worldwide (mostly for new aircraft) will, by some estimates, total $815 billion by the year 2000 (Meredith 1995). Given the inadequate profitability of the U.S. airline industry since deregulation, these capital requirements will be difficult to achieve.

Economic recession dampens passenger demand, thereby relieving some pressure on the infrastructure and squeezing airline profits, making it more difficult for carriers to bear the cost of airport development. It is said of airlines that they order aircraft in good times and take delivery in bad. Of airports, it can be said that construction is begun in good times and completed in bad. Another compounding difficulty is the accuracy of the forecasts. Airports are built to meet future expected demand. If that demand is not forthcoming due to poor projections (a common problem, as we saw in Chapter 2) or other factors, then the financial viability of the project can be endangered. If it is a public project financed by the state, then the airport's losses are covered by the treasury. If it is financed by private bonds, however, the project may default. We shall return to this point in more detail later.

The financing process

The U.S. pattern differs from that which prevails in many other countries because of the ways in which private money is channeled into public projects. In the United States, the costs of an airport are everywhere borne largely by the airlines and concessionaires and is financed largely through long-term debt. Essentially airlines pay for what they use; they contract to pay an amount that will cover most of the operating costs of the facility. The U.S. pattern also differs in the nature of the ownership arrangements. In the United States more than 4500 airports are public facilities, owned and operated by local or municipal governments, sometimes through an airport authority whose members are either elected or appointed. It is the owner of the airport who makes the decisions concerning the construction, financing, and operation, although airlines often obtain a majority-in-interest clause in their lease agreements, which effectively gives them a veto power over future expansion. DIA's planners fought successfully to have the city retain control. As shown in Chapter 3, the federal government also plays an important role because it can be an important source of funds.

In the United Kingdom, all major airports since the Airports Act (1986) are private companies, many of which have shares owned by regional authorities. The British Airports Authority, a public company, controls such major airports as Heathrow and Gatwick and accounts for about 75 percent of all passenger traffic in the United Kingdom. In France, Italy, Scandinavia, Ireland, and much of the rest of the world, the major airports are owned by the state, although a trend toward privatization is evident in many countries (Ashford and Moore 1992, pp. 1–3, 56).

In the United States, airport projects can be funded in various ways, including federal grants and local surpluses, but almost all have to borrow money through the issuance of bonds of some kind. These take two general forms, although there are many variations. General obligation bonds are supported through the full faith and credit of the issuer and are secured through property taxes, sales taxes, use taxes, and other sources of revenue. If necessary, property taxes will be raised to pay these bonds. The other general category, revenue bonds, are paid off by income derived from the operation of the enterprise. DIA was financed with revenue bonds, which means it was *not* backed by the full faith and credit of the City and County of Denver.

Revenue bonds do not impose claims on general tax revenues or on the general funds of local governmental units. In essence, a bond is an agreement to repay a borrowed sum at the end of a specific period, say 25 to 30 years. Bond proceeds are used for other purposes besides construction since many other costs are involved, such as interest payments that are due before the airport generates revenues, paying off outstanding bonds, and setting funds aside to meet unexpected contingencies (Peat Marwick Mitchell 1980).

In the early days of aviation, airport construction projects were financed by general obligation bonds. The industry was in its infancy, and airports were not in a position to generate significant revenues. By the 1930s, however, some airports were producing enough income to suggest that their expansion could be funded through airport system revenue bonds (ASRB). Indeed, an early example of this type of financing occurred in Denver when the city issued such a bond to pay for some facilities for Continental Airlines. By the 1980s, revenue bonds had become commonplace—between 1984 and

1988, for example, $8.3 billion of bonds for various airport projects were issued (Bell 1991, p. 93). Table 5-1 shows the degree to which the major U.S. airports are financed through revenue bonds (Graham 1992, p. 194).

Of course such financing is not limited to airports. In the 1980s, the financing of electric power plants, and later of health-care facilities, represented a large share of the market. In recent years, however, transportation facilities have emerged as the leading area. This sector has been marked by numerous megaprojects, of which the Denver International Airport is the biggest. Judging the degree to which it (or any other project) will operate as planned and generate adequate revenues to be self-sustaining and meet the interest and principal payments is a complex and difficult task. Despite detailed analyses, many large projects have generated losses rather than profits for investors, as the Washington Public Power Supply System (WPPSS) debacle demonstrated (Petersen 1994). (In this case, Washington State's planned nuclear power plants never became operational due to technological and environmental problems, and the bonds went into default.) We are not aware, however, of any airport bond default.

Underwriters are supposed to be a check on issuers because they may be legally liable if they do not exercise "due diligence." Not only must the underwriter take reasonable care that all the material facts are disclosed, there is also an obligation for the underwriter to come to a reasonable belief that the bonds will meet their assigned interest rate. Theoretically, therefore, highly speculative projects should not find financing because of the risk that the underwriters would assume. As a practical matter, however, that system does not function in this manner because the securities industry is driven by commissions and transactions. Therefore, as a general rule, deals get done not because they should get done but because they can be done, and compromises of varying degrees occur every day. The bond industry has not been noted for its ethical behavior, and it has been loosely regulated, although, as we shall see, this situation may be changing, partly because of the DIA case.

The existing pattern also has important political ramifications since it creates an environment replete with powerful actors. One of these, the federal government and its agencies, was discussed in detail in Chapter 3. Two other actors, however, play a prominent role—the

Table 5-1. Airport bonds of large U.S. hub airports 1990[1]

	Type of bond, %				
	Revenue	General obligation	Special facility	Other	Total
Detroit Metropolitan	100				100
Honolulu	100				100
St. Louis/Lambert	100				100
Los Angeles	100				100
Miami	100				100
San Francisco	100				100
Seattle	100				100
Tampa	100				100
Washington Dulles	100				100
Washington National	100				100
Orlando	98		2		100
Philadelphia	95	5			100
Pittsburgh	91	9			100
Las Vegas					
McCarran	90	10			100

Houston Intercontinental	90		10		100
Denver Stapleton	87		13		100
Salt Lake City	75	10	15		100
Kansas City	71		29		100
Dallas/Fort Worth	70		30		100
Atlanta Hartsfield	61		39		100
Chicago O'Hare	60		40		100
Charlotte	59	13	28		100
Phoenix	38	26		36	100
Memphis	21	11	51	17	100
Minneapolis	17	83			100
New York JFK				100	100
New York LaGuardia				100	100
New York Newark				100	100
Baltimore				100	100
Boston Logan				100	100
San Diego		100			100
Airports using these bonds	81	29	32	23	

1. In most instances, these are consolidated bonds issued by a port authority or other transportation authority.

Source: Graham 1992.

bond firms that sell the bonds that raise the capital and the rating agencies that assess the credit quality of the bonds.

Municipal bonds are a multibillion dollar business, generating huge profits for the bond houses granted the right to sell them. Such profits come in two ways—commissions and the trades that a company carries out on its own behalf. The underwriter can buy a portion of the issue at a discount and later sell it at the market price. Thus, bond underwriters are powerful external actors, competing with each other for the right to participate in a bond offering. In the case of a huge project like Denver's airport, bond house managers practically salivate, because the potential profits are enormous—by 1995 more than $35 million had been earned in fees. Given such a context, it should not be surprising that the bond houses are politically active or that the competition to become an underwriter is intense. In short, the municipal bond business was "immensely profitable, intensely political—and largely unregulated" (Hedges et al. 1993).

Also possessing great influence are firms, such as Standard & Poor's and Moody's, that rate the bonds. The conclusions they reach move markets; the higher the rating accorded to an issue, the higher its perceived quality and the lower the interest that has to be paid to potential buyers. These companies monitor the project and may change the ratings of the bond on the basis of developments. The lowest-ranked bonds, "junk bonds," carry the highest yield. In December 1994, for example, AAA-rated bonds yielded 6.62 percent, while BAA bonds, a lower investment-grade rating, yielded 7.17 percent (*Moody's Bond Record* 1994). In the case of a project the magnitude of the new Denver airport, where billions of dollars are involved, the difference between a yield of 6.0 percent and one of 6.5 percent has enormous implications for the debt burden and the financial well-being of the project. The rating of any project is based on a careful analysis of the degree to which a project is financially sound—the degree to which its potential cash flow can meet the financial obligations that have been incurred. Hence two issues are critical: the cost and the potential revenues.

Outside the United States, airports have traditionally tended to be public sector enterprises, although their ownership ranges from the central government to local governments. All, however, have been financed through governmental sources, primarily through treasury grants or cheap loans, though occasionally through private sector loans guaranteed by the government. This pattern, however, has changed dramatically in recent years as governments everywhere have moved to privatize state enterprises of all kinds, including airports (Ashford and Moore 1992, p. 56). Still, the new airports are largely financed through some governmental entity rather than long-term debt.

Privatization can take various forms, including management contracts, turnkey or joint venture arrangements, public incorporation with share ownership either retained by a government entity or placed in the hands of the public, management buyouts, or outright sale (Ashford and Moore 1992, p. 58). Although private developers usually bear a higher cost of capital vis-à-vis the government and lack the government's eminent domain powers, private firms, driven by a profit motive, often produce a product (here, airport services) with fewer employees and greater economy and efficiency. The privatized British Airports Authority has proven that real estate and concessions can be developed into a significantly enhanced revenue stream. Nonetheless, airports are a monopoly bottleneck, and unless regulated, have the ability to extort monopoly rents from their customers (primarily the airlines).

The specific financing arrangements for the new airports that are being built around the world are as follows:

1. *Munich*. Munich's airport was built by the Munich Airport Authority. The Free State of Bavaria provided 51 percent of the funding, the government of Germany 26 percent, and the city of Munich 23 percent (Mordoff 1988). All three governmental institutions must approve improvements, and all three audit the airport's expenditures (*Aviation Week & Space Technology* 1991). By 1991, the airport was anticipated to cost 1 billion Deutsche marks ($590 million) for land and 7.5 billion Deutsche marks ($4.41 billion) for construction. Had it been built 20 years earlier, in 1971, the full cost for land and construction would have been 4 billion Deutsche marks. The additional cost represents the delay caused by environmental opposition and litigation (*Aviation Week & Space Technology* 1991). Revenue of 600 million Deutsche marks ($380 million) was anticipated for the first full year of operation. However, high mortgage and interest payments will cause gross losses for several years (Hill 1993). The new airport reported a net

deficit after depreciation of 100 million Deutsche marks ($64.5 million) in 1993, and 95 million Deutsche marks ($61.3 million) in 1994 (*Aviation Europe* 1995).

2. *Osaka.* The Kansai International Airport Company was formed in 1994 to supervise construction and manage the airport (Moorman 1994). Kansai International Airport (KIA) was Japan's first airport to have been built and operated by a private firm and financed commercially through long-term debt (Brown 1987). Kansai was originally projected to cost 1 trillion yen, or about $7.7 billion (Ott 1993). The national government was to invest $546 million, and local governments and private sources invested $137 million, with $6 billion anticipated to be borrowed (Brown 1987).

   But Kansai cost 40 percent more than was originally projected (*Aviation Week & Space Technology* 1994). Ultimately, the government of Japan provided two-thirds of the funding, local governments provided one-sixth, and about 1000 local businesses contributed one-sixth. KIA is saddled with $4.4 billion in debt (and rising) and needs about $2 million a day to service debt (Moorman 1994). It lost more than $200 million in its first six months of operation (*Aviation Week & Space Technology* 1995).

   Cost overruns at Kansai, stimulated principally by the fact that parts of the island on which the airport was built began to sink during construction, led the airport to impose charges of about $24 a ton as a landing fee, or nearly $9673 for a 747 (Moorman 1994), which is the same level as Tokyo Narita (the world's most expensive landing-fee airport) and about five times as much as landing fees at New York's John F. Kennedy International Airport (Gross 1994). Jetway and baggage-handling fees will also be among the highest in the world, as are rental rates on concessionaires (it was anticipated a cup of coffee might cost $10; Lassiter 1994). Airside charges (e.g., landing fees) are 10 percent higher than Tokyo Narita, while landside charges (e.g., concessionaire rentals) are up to 300 percent higher (*Aviation Week & Space Technology* 1994). Office space and baggage-facility rental fees are 331 percent and 750 percent higher at KIA, respectively, than at Narita; cargo office, first-class lounge, and check-in counter space at KIA are 332, 347, and 330 percent higher, respectively, than at Narita

(*Moorman* 1994). The passenger airport departure tax is $26 (the world's highest); the automobile toll across the two-layer truss airport bridge is $17, while a cargo vehicle pays $120; a taxi ride from Osaka station costs $190 (*San Diego Union-Tribune* 1994); and parking a car for three days at the airport costs $145 (Black 1994). KIA is an extremely expensive airport at which to do business.

This situation stems from the amount of debt KIA carries. Debt service will cost 200 million yen ($2.04 million) a day (*Aviation Week & Space Technology* 1994). The airport was projected to sustain a $400 million loss during its first year of operation (Moorman 1994). It is operating on the basis of a "5-9-23" concept, whereby the airport would earn a profit after five years, eliminate its deficit after nine years, and redeem its debt in 23 years. That is quite different from projections made in the 1980s that the airport would reach the breakeven point by 1998 and pay dividends by the year 2003 (Brown 1987).

3. *Seoul*. Korean Airport Construction Authority (KOACA) will raise will raise $3.84 billion (55 percent) from loans, bonds, and sale of land at the prepared site. The Korean government will provide $2.58 billion in subsidies (37 percent). Private sector investment will provide $489 million (7 percent) [(Shin 1993)]. KOACA funding will be used for airport facilities. The government's funding will be used principally for airport access and site preparation. Private sector investment will be used to construct the cargo terminals and other secondary support facilities (Shin 1993).

4. *Macau*. The $912 million Macau International Airport was financed through an agreement with the Banco Nacional Ultramarine, the Banco Comercial de Macau, and financial institutions in New Zealand and Australia. Landing fees will be an average of the five busiest airports in the region (*Phillips Business Information* 1994). It is anticipated that the landing fee for an Airbus A320 will be $730; the passenger departure tax will be $16 (*Aviation Daily* 1995).

   The Macau Airport Company (CAM), which holds the 25-year concession on the airport, is owned by the Macau government (51 percent), Hong Kong's Stanley Ho's STDM (33.3 percent). Ho also owns all nine Macau casinos, the hydrofoil shuttle from Hong Kong, the sea terminal, and accounts for 25 percent of Macau's gross domestic product), and other investors (22 percent) [Donoghue 1995].

One concern in both Macau and Hong Kong is reversion of the colonies to China in 1999 and 1997, respectively. The 25-year concession granted CAM was an effort to ensure some stability. Mainland Chinese approval of airport financing of MIA was without incident, unlike the reservations Beijing raised with respect to Hong Kong's new airport (*Asian Wall Street Journal* 1994). Moreover, with Hong Kong reverting two years earlier, Macau will be able to assess how China's ascendancy to political dominance affects the economic environment. And MIA will be operational before reversion, while Chek Lap Kok will be open only after Hong Kong reverts to China (*Asian Wall Street Journal* 1994).

5. *Hong Kong*. The new airport at Chek Lap Kok is expected to cost $8.7 billion (Mok 1993). However, with the additional suspension bridges, tunnels, highways, high-speed rail, and ocean terminal facilities being built, the total cost will exceed $20 billion.

The new airport was originally planned with the Provisional Airport Authority (PAA) working in conjunction with the departments of Civil Aviation, Highways and Territory Development, as well as the Mass Transit Railway Corporation, the Hong Kong Legislative Council, the Airport Committee of the Joint Liaison Group (with China), the Airport Consultative Committee, and airlines (Jansen 1994). Although 100 percent owned by the Hong Kong government, the PAA operated as an independent corporation operating under commercial principles (Mok 1993).

To pave the way for financing the airport terminal, the Hong Kong legislature converted the provisional body that oversees airport construction into an independent corporate body named the Airport Authority. The Mass Transit Railway Corporation, a semi-governmental body that runs the subway system and is building a railway to the airport, is already incorporated. The bill followed an agreement between the United Kingdom and China, concluded in the summer of 1995, which resolved the dispute on financial support for the airport and its railway (*Asian Wall Street Journal* 1995). China had objected to what it perceived to be excessive reliance on borrowed capital, which would cause a heavy burden of indebtedness after it assumes sovereignty over the British colony in 1997 (*Orlando Sentinel* 1995). China also resisted on grounds that it had neither been consulted on the desirability of a new airport nor been included in the decision-making process (Becker 1995). It objected to what it perceived to be an excessively costly gold-plated facility devised by the British, for whom China has had long-standing antipathy. Moreover, Beijing was unhappy about Governor Chris Patten's proposals for enhanced democratic representation in the Hong Kong local government (Mufson 1994).

Chek Lap Kok was originally scheduled to open in July 1997, but the lack of a financial agreement with China slowed the schedule. Although the airport's construction is scheduled for completion in 1997, the airport is not scheduled to open until April 1998 to allow completion of commissioning and trial operations. A rail line to the new airport is scheduled to open in June 1998 (*Airports* 1995).

As noted, only 40 percent of the $20 billion price tag is for the airport itself. The remaining funds are being spent for the rail and roadway suspension bridge, a 21-mile airport express rail line, and a third harbor tunnel for Hong Kong Island. Approximately 30 percent of costs will be raised from tenants, franchises, and such accounts as air traffic control. Aeronautical charges, expected to be triple those of Kai Tak, will provide 40 percent of the airport's revenue (*Aviation Week & Space Technology* 1994).

The Hong Kong government put up $4.5 billion in cash for the airport and $2.9 billion for the airport railway. The airport and railway each will borrow $1.4 billion from banks, to be repaid in 2001, four years after the airport is scheduled to open (*The Economist* 1994). The Chinese government insisted that three-fourths of the airport debt be retired by its opening date (Mecham 1995).

6. *Kuala Lumpur.* The Sepang airport is managed by the Public Works Department of the Malaysian federal government (Jansen 1994). The airport is being financed primarily by government-guaranteed unsecured bonds. It also can access a 61.5 billion yen Japanese government loan (*World Airport Week* 1995).

The DIA case

The DIA's approach to financing was typical of all U.S. airports in terms of both its reliance on long-term debt and its detailed financial planning. This planning was closely linked to the various stages in the overall planning process since each stage has financial implications. These were spelled out in detail, specifying the particular expenses that were involved and the source of the funding. This rigorous financial framework permitted upgrade analyses to be carried out as conditions changed.

The first preliminary plan for a new airport was prepared by a consulting firm for DRCOG in 1980. It identified the key factors, determining the overall cost as:

1. Land acquisition and site development
2. Airfield facilities based on forecasted demand
3. The terminal concept, whose "importance . . . to the overall cost of the airport cannot be overemphasized"
4. Cost escalation due to inflation.

It then estimated that a new airport would cost between $1 and $2 billion, not including financing costs (Peat Marwick Mitchell 1980). The study also considered the advantages and disadvantages, from a financial viewpoint, of various administrative arrangements for a new airport. These ranged from Denver being the sponsoring agency to a regional authority to a state aviation agency.

As discussed earlier, large projects are subject to unanticipated changes. As is evident from the above summaries, airports are no exception. As DIA took shape, the precise financial plans had to be revised in light of new conditions. By 1990 the airport was projected to cost $2.1 billion, essentially at the high 1980 estimates of between $1 billion and $2 billion. As events developed, however, these estimates were first scaled down to $1.9 billion and then escalated to $3.3 billion. The city's original cost estimates included only the cost of construction, while subsequent upward revisions more realistically included the cost of land and other items. We return to the cost issue in Chapter 9.

DIA was to be financed at no cost to the Denver taxpayer. This was an important consideration, as we have seen, in gaining the popular support essential to launch the project. Such a commitment could be made because the financing would be based on the issuance of airport revenue bonds, for which the taxpayers could not be held liable.

The original financing plan was drawn up by the Wall Street firm of Smith Barney, Harris Upham and Company, which acted as financial advisor to the city. It identified the various sources of funding that would be used to pay for the construction of DIA. The most important, by far, was the sale of revenue bonds, which is discussed in detail later. The others consisted of the following sources (ASRB 1990).

1. Federal funds. As noted earlier, Denver successfully campaigned for federal funds from the Airport and Airway Trust Fund and was granted approximately $500 million, to be disbursed over a nine-year period. In May 1990, the FAA formally notified the city, through a Letter of Intent, that it would do so, provided that Congress appropriate adequate funds. Through 1995, the city had received $319 million and expected that figure to rise to $437 million by 1999. It had also received an additional $20 million for runway lighting and site preparation for its planned international runway (ASRB 1995).
2. The disposition of Stapleton. The city hoped to realize $100 million from the sale of land occupied by the former airport in the years from 1994 to 2000, but it did not incorporate this amount in its plan because of the uncertainties involved (ASRB 1990). However, the GAO stated that the city expected to generate $89 million from this source (GAO 1991, p. 37).
3. Capital fund. The city had available its capital fund, from which it drew $43.9 million in 1989 for equity financing. It also drew out additional funds for various other purposes so that by 1995 it had spent $175 million from this source. Proceeds from bond sales and federal grants were used to repay $155 million (ASRB 1994).
4. 1985 project account. This account was generated by the sale of bonds in 1985 to make improvements at Stapleton. The unspent monies were allocatable to the new airport.

These were supplemented, over the years, by the following two sources.

1. The passenger facility charge (PFC). In October 1990, Congress passed the Aviation Safety and Capacity Expansion

Act, which permitted the Secretary of Transportation to grant requests by airports to levy up to $3 per head tax on passengers and use the facility charge for approved projects. The funds are to be used for projects that improve safety, security, protect the environment, and to pay the debt service on bonds issued for such purposes. Denver promptly submitted a formal request that was approved by the FAA in May 1992. Collection began on July 1, 1992. Between 1992 and 1995, the city collected $97 million and expected to raise another $255 million by the year 2000 (ASRB 1994). This unanticipated event may well have kept the project competitive with other airports.

2. Civil aviation fuel tax. In 1988 the city imposed a 2-cent tax on aviation fuel and used some of these resources to help defray the costs incurred by the numerous delays.

The DIA bonds

The success of the new project depended on the city's ability to raise more than $2 billion through bond sales. Originally it was estimated that a total of $1.746 billion worth of parity bonds would be required, plus another $550 million of variable rate subordinate bonds. These were to be issued through two offerings each in 1990 ($700 million) and 1991 ($840 million), and three in 1992 ($735 million) (Denver 1990). However, as the project costs escalated, more funds were needed, and Denver issued additional bonds until the total reached almost $3.5 billion. The specific issues to date are as follows:

Series 1990A	$700,003,000
Series 1991A	$500,003,000
Series 1991D	$600,001,000
Series 1992A	$252,180,000
Series 1992B	$315,000,000
Series 1992C	$392,160,000
Series 1992D-G	$135,000,000
Series 1994A	$257,000,000
Series 1995A-B	$329,290,000
Total	$3,480,647,000

Source: ASRB 1995.

Additional bond sales were scheduled for November 1995 and January 1997 to fund the capital improvement program, which totals $205 million (Denver 1995a).

The bonds are to be redeemed at various times ending in 2025. Payment of interest and principal is to be made each year and the DIA's financial viability is dependent on its ability to generate enough revenues to pay that sum each year. It was anticipated that DIA would have to achieve net revenues of between $145 million and $177 million annually between 1994 and 2000 for the debt service (GAO 1991, p. 38). Because of the many changes and the delays in opening the facility, which added an additional $361 million in costs, the debt service cost climbed to $344.5 million per year (GAO 1994, p. 37). The probability that DIA will be able to handle its debt service is discussed later.

Airport revenues

If revenue bonds are to sell, there must be evidence that the cash flow of the project will enable it to repay the borrowed principal and interest. Accordingly, the financial plan included a detailed analysis of the debt service requirements and the revenues that would be generated at the new airport.

Airports can be highly profitable enterprises because, as regional monopolies, they have the potential to generate large revenues. Indeed, the profitability of large airports increased steadily in the 1980s. In 1988 the ratio of revenues to expenditures for London's airports stood at 1.69 for Heathrow and 1.49 for Gatwick, 1.38 for Nice, and between 1.10 and 1.15 for Marseilles, Amsterdam, Frankfurt, Copenhagen, and Vienna. Although profitable, U.S. airports cannot earn such profits because they have fewer international passengers who tend to be big spenders in the duty-free shops and because of the agreements that they have with the airlines (Graham 1992, p. 3). Further, U.S. federal law prohibits the expenditure of airport revenue for nonairport purposes, thereby diminishing the incentive for local governments to treat airports as profit centers. Still, Stapleton turned a significant profit, with net revenues of about $50 million a year in the 1985–1988 period, $63 million in 1989, $74 million in 1990, and $80 million in 1991. It was projected to earn $83 million in 1992 and $95 million in 1993 when it was to be closed (ASRB 1990; ASRB 1992).

How profitable DIA will be remains to be seen. But its potential is limited by the concessions the city made to the airlines. DIA is obligated to allocate to them, as credits against their financial commitments, 80 percent of the revenues that remain after legal bond obligations and the operating budget items are met. The percentage drops to 75 percent in Fiscal Year (FY) 2000 and to 50 percent in FY 2005 up to a maximum of $40 million (ASRB 1995). The larger the airport's net profit, the more flexibility it has to pursue infrastructure expansion without incurring additional debt.

Figure 5-1 shows the flow of revenues.

Airports have several sources of revenue. In the United States most of these are categorized as "aviation" and "nonaviation" income. The former consists of income generated by the airlines—passenger facility charges (PFC), lease agreements with the airline for gates and other facilities, landing fees, and apron and aircraft parking charges. The latter is derived from concession fees inside the terminal, parking revenues, and charges for ground transportation. These are sometimes subdivided into "inside concession revenues" and "outside concession revenues." Other sources of income are generated through interest charges, fees for consulting and services, and leases to commercial buildings on airport land.

Stapleton earned about 60 percent of its total income from airline landing fees and terminal lease charges. It was a profitable operation, even though the number of passengers had decreased.

The new airport was expected to enjoy the same sources of revenue, albeit in different proportions, as Fig. 5-2 shows.

Now, almost 70 percent of the income was to be generated through airline operations and another 18 percent through concessions of all sorts. Thus the projections of the airline revenues became, more than ever, a matter of vital concern.

Airline revenues are usually calculated on a per enplaned passenger cost basis. To the uninitiated it would appear that the higher the figure, the greater the income. In reality, however, the higher the figure, the higher the costs to the airline and the less likely it will be a happy tenant. Higher airport costs must ultimately be borne by airline passengers in the form of higher ticket taxes, which reduces demand for air transportation. The ideal for airport operators is to identify a figure that will enable them to generate revenues adequate to cover their capital and operating expenses (Fig. 5-3) as efficiently as possible so

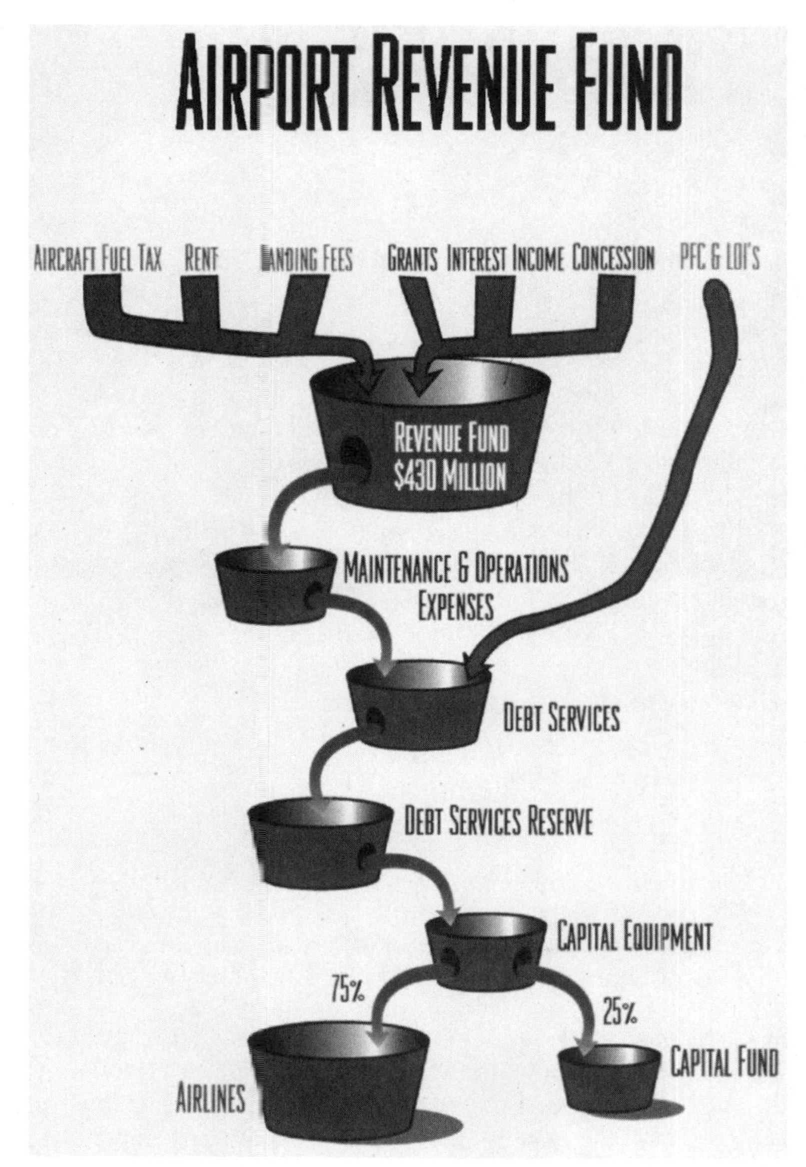

Fig. 5-1. *For its first year of operation, DIA's expenditures were projected to be more than $400 million, more than half of which is attributable to debt service.*

as not to impose too heavy an economic burden on the airlines and their passengers. The two critical variables, of course, are the number of passengers using the facility and the costs that have to be covered.

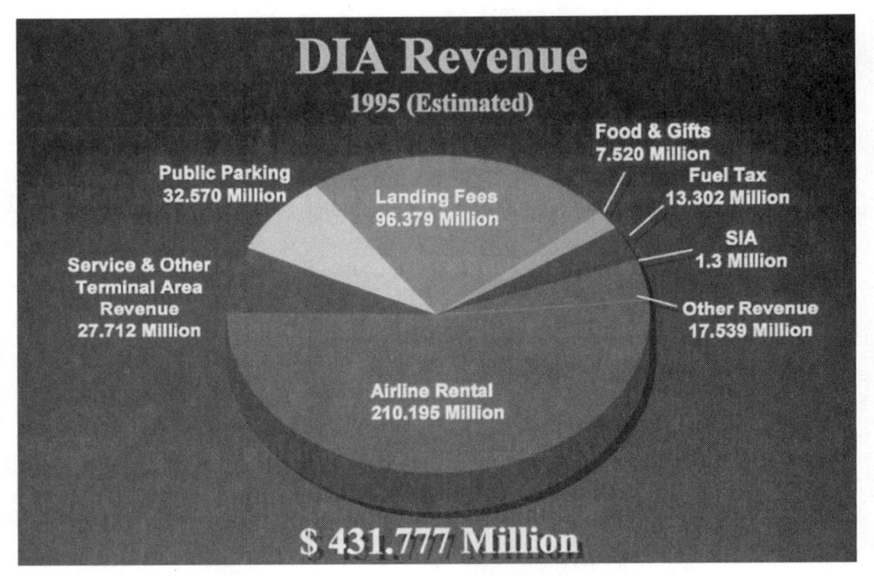

Fig. 5-2. *Revenues were projected to exceed $430 million in 1995. Of that, airlines would pay more than $300 million in gate leases and landing fees.*

As we have seen, at DIA the costs went up. Unfortunately, the passenger projections did not go up as well; on the contrary, they had to be adjusted downward as the assumptions concerning growth and hubs that had driven the early wildly optimistic forecasts became increasingly untenable. Accordingly, the city continued to refine its traffic estimates to more plausible levels. Table 5-2 shows how the estimates for traffic and cost per enplaned passenger changed over time.

Even at the low 1990 estimate, the per-passenger cost figures were very high both in comparison to Stapleton—$5.74 in 1989—and to the rates at other airports. Recognizing the difficulty, the city argued that the comparable cost at DIA was about $12 to $13 and that this figure was competitive with what the costs would be (an estimated $10) at other major airports such as JFK, O'Hare, and Dulles when their development projects were completed (ASRB 1990). In addition, the airlines' operating costs would be cut because of the operating efficiencies that the new airport would provide. Two years later, the report added the following "In the final analysis, the 'reasonableness' of these costs is a judgment that must be made by individual airlines in deciding to serve the Denver market" (ASRB

Table 5-2. Traffic and cost per enplaned passenger

	Traffic (millions)		Enplaned passenger total cost (future dollars)	
	1995	**2000**	**1995**	**2000**
1990	17.0	20.26	13.63	12.31
1991				
Low	11.0	12.1		
Medium	14.0	16.6		
High	17.0	20.7	13.12	11.88
1992				
Two hubbing carriers	16.5	19.6	14.50	14.35
One hubbing carrier	13.0	14.4	20.00	21.65
1994	16.0	18.2	18.15	17.20

Sources: Denver 1990, Table 17; Denver 1991 Table 15; Denver 1992, Table 15; Denver 1994, Table 15.

1992). Unfortunately, Continental made the decision to pull out, turning DIA into a fortress hub for United and creating new difficulties for the bond sales, as we shall see.

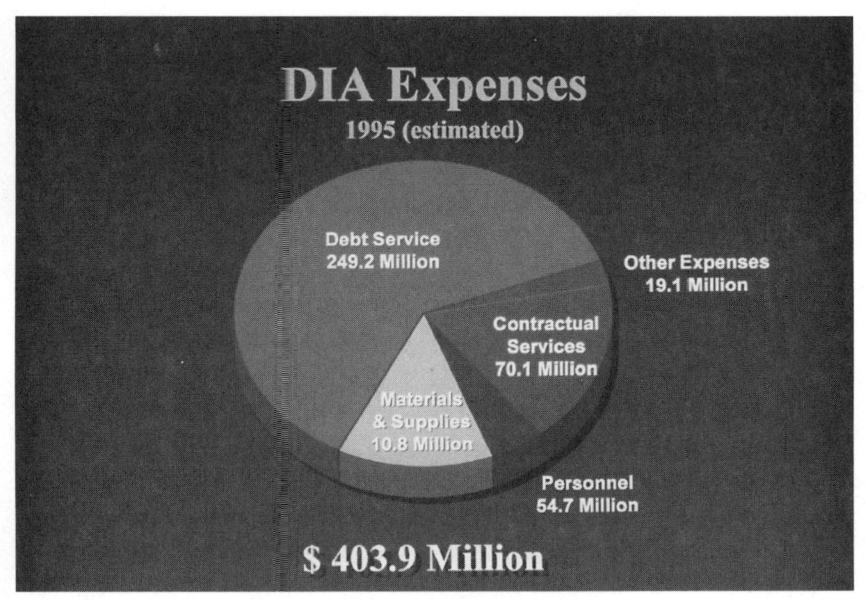

Fig. 5-3. *The revenue stream at DIA is divided among several expenditure items. Of the net "profit," 75 percent is returned to the airlines.*

The bond sales

The results of the first offering were awaited with trepidation, but the bonds sold extremely well. To understand the reasons, it is necessary to analyze the nature of the bond market and the role of the underwriting firms.

Until 1986, municipal bonds were owned largely by rich investors. The 1986 Tax Reform Act, however, equalized the tax advantages that had accrued to many investments, thus making bonds attractive to many middle-class investors. As a result, the demand for municipal bonds increased rapidly. The boom produced two consequences. First, the importance of ratings increased. Second, many investors turned to professionals who managed diversified tax-free funds. The 1990s witnessed an explosion in such funds and in their assets. Thus, although there are more individual investors in municipal bonds than ever before, the concentration of power is also greater than ever before (Petersen 1994).

It is the institutional buyers who drive the market, because they have huge amounts of cash to invest. They are ranked by performance, that is, the yield that they achieve for the funds that they manage. Accordingly, the prospect of high yields can offset a concern about potential risks. Furthermore, the risks are indeterminate because they lie in the future and no one knows for certain what will happen. Hence, assessments of the risks involved sometimes conflict; it becomes a matter of judging which is valid and how much risk one is willing to tolerate.

Many factors enter into such judgments. The short time horizon that drives many investment decisions tends to increase the willingness to accept high yields, since any particular money manager might well have moved to another firm when problems arise with the issue. In the meantime, the manager will have achieved a high yield for the fund and been handsomely rewarded for an above-average performance. A more specific consideration that entered the calculation in regards to the DIA bonds was the widespread belief that DIA could not be allowed to fail.

Underwriting firms help achieve a positive outcome in many ways, even if the bonds are actually poor investments. Many well-known firms have marketed millions of dollars of such securities, sometimes knowingly. The bond houses prosper precisely because they know how to market any kind of issue at the best price, even in the face of a falling market, reluctant investors, and rating changes. As the Public Securities Association's own manual states, "Successful underwriting depends on trading and sales personnel who know the market well and can reach prospective buyers quickly" (Monsarrat 1991).

The success of the DIA bonds was due in no small part to the skill of its underwriters. The first issue came to market amidst investor uncertainty and doubts about the risks involved. Yet, to the surprise of many, the issue was oversubscribed at a yield of 9.185 percent. Goldman Sachs, the lead underwriter, actively promoted the bonds and risked—and made—millions of dollars trading them. One trader put it this way: "You have to take your hat off to Goldman; they did a tremendous job pleading their case to the (investor) community and offering price concessions to get the job done. They were involved intensely on the local political scene" (Monsarrat 1991).

The politics of underwriting

The DIA bonds attracted national attention because of the size of the offering. Firms always struggled to secure a share of promising issues and used whatever means they could to influence the outcome; given the potential profits of this issue, the competition among underwriters was extremely fierce. Sometimes the issuing entity would arrange a competition with precise criteria, but unless the bidding process ended with the best price being selected (a process naturally hated by underwriters because it minimizes profits), the issuer could find ways to exercise a subjective judgment. Politics thus becomes a factor not only in terms of political contributions but within the underwriting firm and in its relationship with the issuing entity.

In some cases the underwriter is in a strong position and can, at the 11th hour, impose more favorable terms upon a deal. This time it was the other way around. The city had great leverage because of the magnitude of the fees and also because it was a challenging deal. Whoever got this deal done would be an airport hero because everyone considered the issue as a difficult one since it was unlikely to receive a high-grade investment bond rating. Furthermore, the city made it very clear from the beginning that it would run an orderly process and that the underwriters would be rotated. That had a profound impact on everyone's behavior because the firms believed that if they did not toe the line, they would be kicked out.

Regional firms naturally develop close relations with public officials in their community and nurture these carefully. This was true of one of the most prestigious local investment firms, Boettcher and Company, which had played a prominent role in financing Denver and Colorado's growth for decades. Unfortunately, by the early 1980s, the firm had suffered a series of major reverses from bad real estate and underwriting deals that had involved junk municipal bonds. Furthermore, in the late 1980s, the bond business was in the doldrums. Now DIA emerged—the largest debt financing in Colorado's history. Every major firm was interested in participating in the deal, but Boettcher was desperate to win the leading position. It drafted a proposal to the city which, if accepted, would earn it renewed prestige as well as millions of dollars in underwriter's fees and at least as much again through trades.[1]

Boettcher carried out a preliminary analysis of the project's financial viability in late 1989. The bond analyst looked at the history of traffic forecasts and concluded that although the estimates were on the high side, Denver was prudently dropping its estimates and providing reasonable ones. Despite the relatively positive conclusions, Boettcher, concerned with the possible political ramifications, decided not to publish the study. The following table shows how the prospectus incorporated traffic estimates that were reduced markedly from the optimistic projections (discussed in Chapter 2) that had been used in the past:

	1995	**2000**
EIS (1986)	33.5	
PMM (1986)	25.8	30.7
FAA (1987)	27.8	34.3
Bond prospectus (1990)	17.0	20.26

Sources: See Table 2-2; 1990 Bond Prospectus.

Realism and objectivity also drove the choice of underwriters. Because of the complexities of the issue, the city's selection process was quite objective. It retained Smith Barney as its financial advisor. Smith Barney also wanted to be the lead underwriter, but the city properly viewed this as a conflict of interest; Smith Barney, which

1. In most instances, these are consolidated bonds issued by a port authority or other transportation authority.

chose to be the financial advisor, did not want Merrill Lynch to get the business and recommended Goldman Sachs, a well-known, though slightly lesser player in the bonds industry. The city accepted the recommendation because it recognized that a successful bond issue would have to be handled by a prestigious, experienced Wall Street firm. Its decision was clearly validated by the result. Boettcher and Company, however, was named co-senior manager, an achievement that was promptly greeted with a celebration at Boettcher headquarters.

Although the bonds had sold well, events in the Middle East (Iraq's invasion of Kuwait in August 1990) profoundly affected the airline industry. On the one hand, jet fuel prices had doubled; on the other, the rise in energy prices had hurt the U.S. economy and air traffic dropped precipitously. Hence, the airlines were under enormous financial pressure, especially Texas Air's Continental, which, under Frank Lorenzo's stewardship, had assumed enormous debt to acquire People Express and Eastern Air Lines. It now looked as though the traffic that would pass through the airport could well be 14 million enplanements instead of the projected 17 million. Such a drop would have a significant impact on the per passenger cost to the airlines.

As noted earlier, this indicator—airline costs per enplaned passenger (ACEP)—is commonly used to indicate the revenues that would be derived from the airlines. Given a cost of $2.1 billion for the airport (the figure used in the 1990 prospectus), the city estimated that the ACEP would be $13.63 in 1995. If the number of passengers drops by almost 20 percent, however, then the cost per passenger increases correspondingly to $17.50. The ACEP issue was a fundamental one, for the higher the cost, the less attractive the facility to the airlines. The projected ACEP of $13.63 was significantly higher than that which existed at Stapleton—$5.74 (in 1989). The city argued that the new ACEP would not be a deterrent to the airlines because the ACEP would rise to $10 at other major airports where large development projects were underway and because airport operating costs are a small percentage of an airline's costs.

Because of such considerations, Boettcher's analyst, Gordon Yale, set to work on a new research report in August 1990. The report discussed the many uncertainties surrounding the project—the opposition to the project, the lack of an agreement with UAL, the events in the Middle East and the financial difficulties of Continental (which made it likely that it would cut back)—and concluded that the DIA bonds would probably be downgraded. The report was prepared with trepidation because it was clear that the Peña administration dealt harshly with its critics.

Now the same issue arose. Should the report be published? Numerous meetings were held. The author had powerful support, including the director of research in Chicago and Mike Bell, Boettcher's senior airport investment banker, because such a report was needed; Boettcher was already being sued over local bond defaults. Accordingly, the company's counsel argued that the report was essential.

If it were to be published, how should it be styled? This question proved even more troublesome than the decision to publish. Some argued that Boettcher should make the strongest case possible for its position, that nothing should be held back. After all, the city would not react kindly, regardless of how the analysis was presented. The chairman and chief executive officer, on the other hand, did not like the idea of a highly detailed analysis at all since he was attempting to enhance Boettcher's relations with the city so as to secure more underwritings. At most he wanted a one-page summary. The struggle raged for six weeks within the company.

In late November, the *Rocky Mountain News* ran a story stating that Boettcher believed that the bonds would be downgraded. Someone had leaked an early draft of the report to the paper. The resulting storm was predictable as the city rushed to defend the project. Mayor Peña criticized Boettcher for its weak internal controls, defended the project, and attacked the report, which was labeled in the *Rocky Mountain News* story as the work of an unqualified analyst.

The leak sparked a furor within Boettcher. The identity of the person who had leaked the report became an object of great speculation, and the author came under suspicion. He volunteered to take a lie detector test, but his superiors declined the offer. Still he remained under a cloud, and his standing within the firm was further eroded when the CEO declared to the *Wall Street Journal* that the research was flawed and that the official report, which would include additional information (the leaked version had not included a section analyzing the impact of the new federal legislation authorizing passenger facility charges), would present different conclusions.

The analyst, who naturally felt that he had been disavowed, went to the general counsel to offer his resignation, but it was not accepted. Discussions now ensued at the highest levels of Kemper, which had taken over Boettcher. Finally the decision was made to stand behind the report because it would be unseemly for the firm to appear to bow to the will of politicians. Thus, as one national story put it, "Boettcher & Co. on Friday released a final report warning that bonds for the $2.3 billion Denver International Airport could be downgraded—three days after a draft to that effect angered project officials" (Racine 1990).

Events continued to affect the project, notably Continental's filing for Chapter 11 bankruptcy, and new analyses were carried out at Boettcher. One focused on the possibility of mothballing the project. The study found that this was a feasible option; the debt could be serviced by revenues from Stapleton coupled with the income from the passenger facility charges, although there is some dispute as to whether airport and tax revenue could be used for this purpose.

Denver had spent or committed about $200 million and had contracts for another $500 to $800 million. These contracts, however, had an exit clause that permitted the city to cancel them, if it paid a 5 percent penalty. The calculations showed that the city could take all unspent proceeds and repurchase the bonds at par—though the city might have to threaten the bondholders with default. Essentially, the city had created this difficulty for itself. Bonds are contracts, and many carry clauses that permit the issuer to call the bonds at a specified price, usually par, with a small premium. The underwriters had suggested that such a clause be included in the original Denver prospectus when its details were under discussion in the spring of 1990. The city had refused because it did not want to give the airlines, who were opposed to abandoning Stapleton, any reason to believe that the project could ever be halted.

This study was taken to the city with the hope that its officials would give serious consideration to these conclusions. Such an assumption was, of course, quite naive. As was noted in Chapter 3, a powerful political coalition had been built that was fully committed to implementing the project, whatever the obstacles. Hence the official reaction was swift and predictable. A meeting was promptly arranged between Mayor Peña and Kemper officials. As a result, the bond house was urged to fire the author and not to publish or discuss the analysis. Boettcher decided to place the analysis in a safe and not distribute it.

But the project was in a state of crisis. Continental had entered Chapter 11 and United Airlines was refusing to sign airport leases. Standard & Poor's had placed the bonds on credit watch and Goldman Sachs held an inventory thought to be $100 million of the bonds it had bought in order to prop up the price, which had fallen by 12 percent (Racine 1991).

Mayor Peña and his aides were not deterred. They moved swiftly and effectively to overcome these difficulties. They drew up a new plan to deal with the situation, announcing that in light of "weakened economic conditions in the airline industry, uncertainties relating to Continental . . . and the lack of agreement with United, the city has reduced the size and number of certain of the facilities . . ." The original project (which was to have cost $2.1 billion) had consisted of a terminal with three modules and three concourses (with a total of 92 to 100 gates) and five runways (Denver 1990). "The Current Program" was to consist of a terminal with two modules, four runways, and three concourses with 70 to 80 gates, at a cost of $1.9 billion. Furthermore, the city was prepared to make additional changes (eliminate another concourse and another runway) that would drop the cost further, down to $1.7 billion (Denver 1991).

Denver's flexibility in dealing with a potential worst-case scenario apparently reassured Wall Street, though for a price—higher interest coupons. Moody's retained its BAA1 rating, and the $500 million issue was sold within a day. The crisis seemed to have passed; now there was reason for optimism, though Standard & Poor's downgraded the bonds to BBB-. Further good news came in June, when the city announced that an agreement had been reached with UAL. The price of the bonds rose. The agreement, however, was a costly one, ballooning the project's cost, according to the city, by $600 million, for a total of $2.7 billion. Obviously, the debt burden also increased significantly, but the financial planners had carried out detailed analyses of the project's ability to handle the additional debt. As a result, the city signed the UAL agreement only after a new revenue stream—the passenger facility charge—became available to help defray the additional interest charges.

Some members of the City Council who were anxious about the status of the airport learned of the existence of the Boettcher report and, after considerable discussion, made a decision to hold public hearings. Before the hearings, a city representative called Boettcher officials, requesting a copy of the study. Boettcher refused unless it was permitted to update it. Shortly thereafter, Boettcher was informed that if it failed to produce the document, it would no longer be able to participate in future underwritings. A copy was furnished to the city. The hearings, however, were superficial. Gennifer Sussman, the airport's chief financial officer, was called to testify. She presented an analysis that responded to the council's questions and argued against mothballing the project. Though Boettcher representatives had indicated their willingness to testify, none were called. The council voted unanimously to proceed with the airport (Denver 1992).

Over time, the DIA bonds continued to enjoy a bumpy ride as the project's implementation did not proceed smoothly, notably in getting its baggage system to work. Their ratings rose and fell with the news, dropping from investment grade A- to a borderline BBB- in early 1991, some of the darkest early days of the project. This rating gave Denver the dubious honor of earning the lowest grade ever accorded to the bonds of a major U.S. airport (Racine 1992). Subsequently they were restored to BBB when the city reached agreement with UAL, but they hit a historic low in May 1994 when baggage system breakdowns led Standard & Poor's to place them in the junk category (BB). The bonds' oscillations are listed in the following table:

Date	Standard & Poor	Moody
To May 10, 1990	A-	A
May 10, 1990	BBB	Cond. BAA-1
March 12, 1991	BBB-	Cond. BAA-1
February 1992	BBB	Cond. BAA-1
May 4, 1992	BBB	Cond. BAA
May 16, 1994	BB	Cond. BAA

These ratings reflected the news that came out concerning the project. And news was plentiful. For example, the following events affected the price of the bonds (Todd 1994):

September 8, 1993—Speculation of opening delay to spring 1994 begins

October 26, 1993—Official delay to March 9, 1994

December 17, 1993—Continental says Denver hub not profitable

January 22, 1994—Continental cutbacks

March 1, 1994—Delay opening to May 15, 1994

March 15, 1994—UAL board approves delay funding

May 3, 1994—Opening delayed indefinitely

May 12, 1994—S&P downgrade

July 12, 1994—Baggage testing problems

These developments led to such significant price fluctuations that the bond's behavior was compared by many professionals to that of stocks. DIA "is traded almost like an equity," commented one authority. "As news comes out, the price of the bond fluctuates wildly, depending on the news." Not surprisingly, Rudy Andrus, associate vice-president of Dain Bosworth, a national underwriting firm, remarked "If you don't like volatility, you don't like these bonds." Knowledgeable traders could make, or lose, millions in this situation. The bonds' initial dollar price was $98.25; they subsequently traded as high as $103. They moved down to $95 in December 1990 when Boettcher produced its report highlighting potential risks. When Continental entered Chapter 11 in January 1991, the 8.5 percent bonds maturing in 2023 fell from 97 percent of par to 91 percent of par. Over time the price continued to fluctuate as the bonds were downgraded to junk status and restored to investment grade and as the project evolved through its eventful history. For example, in October 1992, the 6.75 percent bonds maturing in 2022 traded at 95 cents; in November 1994 they fell to 81 cents (Rebchook 1995).

Such gyrations did not delight many of the bond holders, who had purchased the bonds expecting stability. When this proved to be illusory, several bondholders turned to the courts for redress.

The bondholders' lawsuits

Bondholders have legal standing, and as bond ownership has become more widespread, stricter rules concerning disclosure have been promulgated. Inadequate disclosure is the basis of several lawsuits filed against the City of Denver. The plaintiffs charge that the city withheld information concerning the status of the baggage system and thus misled them as to the risks associated with their investment. In the words of the suit, filed in Denver U.S. District Court by Bonnie Sonnenfeld of Tampa, Florida: "As a result of defendant's materially misleading misstatements and omissions, the market significantly overvalued the bonds, and plaintiff and members of the class were induced to purchase the bonds—both in new issues and on the secondary market—at artificially inflated prices. If the true state of affairs had been disclosed, the securities would have been priced much lower than they actually were" (Rebchook 1995).

Denver is, of course, defending itself vigorously against these charges. The city attorney, Dan Muse, declared that the suit had little validity. Although mistakes were made, the city insisted these were not deliberate and are typical of large projects, which always experience large delays and cost overruns. "They bought the DIA bonds, they got their invested rate, there has been no default on interest rates. They have no beef," Muse said (Rebchook 1995).

The central issue, however, is not whether DIA was a typical megaproject or whether Denver paid interest on the bonds, but whether it made full disclosure to potential bond buyers. Colorado Senator Hank Brown has voiced his doubts in this regard, pointing explicitly to an unpublished 1990 report by a consulting firm, Breier Neidle Patrone Associates, which concluded that an automated baggage system could not be completed by October 1993, the scheduled opening. "It's my belief," Brown said, "that if the purchasers of the revenue bonds had been fully apprised of the . . . contents of the report, that it would have been very difficult to sell the bonds without correcting that problem." City officials have responded that the city acted properly and that numerous other reports had reached a different conclusion (Kowalski 1994).

Another area where doubts about adequacy of disclosure have been raised involves the amount of money that the airlines would save by operating at the new airport. The city claimed that because DIA's configuration was more efficient than Stapleton's, the airlines would have fewer delays, thus saving between $50 million and $100 million per year. The airlines' calculations indicate far smaller savings. In 1988, Continental, then a major hubbing carrier, estimated a savings of about $10 million and United now estimates its savings at a mere $15 million. The city's figure was based on a consultant's report that used obsolete 1989 traffic projections and current delay values. Some believe that the use of this report reflected a consistent pattern by the city to use the most optimistic data available so as to put the most favorable spin on its bond offerings. Such a practice has been severely criticized by at least one financial analyst, Gordon Yale, who noted: "You can't just make stuff up and put it in the bond offering documents. . . ." (Carnahan 1994).

City officials deny vigorously that they ever did so. The airport's former chief financial officer has pointed out that delay costs can be estimated in various ways and that the analysis came up with the best estimate from the available data. (The airlines stopped providing reports during the period of conflict.) Furthermore, this issue was treated the same way as all other financial matters. Every piece of information was tested and reviewed, and great care was taken to ensure that the city's publications outlined all risks accurately (Sussman 1996). However, lengthy and well-documented reports by Steven Paulson, an Associated Press reporter, suggests that there were numerous instances where the city was repeatedly informed that the scheduled opening dates were unrealistic but failed to reveal this publicly (Paulson 1995a and b).

This controversy awaits resolution in the courts since all these issues are captured in the class-action suit filed by Larry Rabinowitz against the City of Denver, Lazard Freres, Merrill Lynch, Prior McClendon, Counts, Goldman Sachs, and KMPG Peat Marwick. The suit charges (Rabinowitz v. City of Denver et al. 1995):

> *During the class period, defendants . . . issued, sold and/or encouraged trading in Denver Airport Revenue System Bonds . . . . These Bonds were issued under false pretenses and pursuant to materially false and misleading offering materials or Official Statements and other statements of defendants which failed to fully and adequately disclose the risks associated with investing in the Bonds.*
>
> *Specifically, defendants misled investors by failing to disclose (1) the Airport's projected revenues and costs (2) the Airport's ability to cover its operating costs and outstanding debt (3) problems with the Airport's baggage handling system (4) potential safety hazards associated with runway construction (5) Denver's projected transportation needs and (6) disputes with major lease holders, including Continental Airlines.*

Defendants were aware of certain of these problems as early as 1990 and did not disclose them . . . . Rather . . . defendants established a so-called "truth squad" . . . to deflect media criticism and smooth over the bond market's concerns about the Airport.

As a result of defendants' conduct as alleged herein, the market significantly over-valued the Bonds, and plaintiff and members of the Class were induced to purchase the Bonds—both in new issues and on the secondary market—at artificially inflated prices . . . .

Moreover, defendants' conduct has resulted in numerous investigations . . . thereby further adversely affecting the value of the Bonds . . . .

This and the other legal actions have important implications for the city. Although Denver is not liable for any shortfalls in revenues that prevent the airport from paying off its debt, this immunity does not extend to issues of fraud and, if Denver officials are found to have misled the bond purchasers by not providing adequate information in the bond prospectuses, then Denver could be subject to a variety of sanctions. These range from a warning to criminal indictments, as a result of which taxpayers could find themselves responsible for the monetary reparations. That Denver takes these suits—and the SEC investigation into whether the city made adequate disclosures—very seriously indeed is indicated by its willingness to seek prestigious (and high-priced) legal expertise. To date, significant sums in legal fees have been spent to counter the lawsuits and related investigations, which led the city auditor to remark (Denver 1995b):

I didn't realize when everyone talked about how DIA was going to mean full employment that what that would mean was full employment for lawyers. I never dreamed that when the airport was completed that we would exchange construction workers for lawyers.

The political linkages

One of the suits involving the Securities and Exchange Commission is directly related to the issue of financing, particularly the degree to which underwriters were selected on the basis of political considerations. This practice has deep roots since public finance has always linked bond houses to elected officials. Prior to 1975, the municipal bond market was completely unregulated; but a scandal led to ineffective reforms, which created the perfectly legal "pay to play" system. Donations to politicians was an accepted way of doing business. The increased costs of campaigns coupled with the increased use of negotiated arrangements provided powerful incentives for the emergence of "pay to play." In the absence of such contributions, bond houses would seldom be seriously considered for participation in the lucrative issuing consortium (Hedges et al. 1993; Petersen 1994).

The practice of "pay to play" became blatant and widespread. Although its scope cannot be defined precisely, a *U.S. News and World Report* study revealed that the 10 leading municipal bond houses contributed more than $5 million to congressional campaigns and the presidential race in 1992. The top two underwriters, Merrill Lynch and Goldman Sachs, donated $2.2 million. They were followed by Lehman Brothers, $524,337; First Boston, $397,890; Smith Barney, $412,611; Prudential, $311,329; PaineWebber, $394,847; and J.P. Morgan, $45,610 (Hedges et al. 1993). These figures do not, of course, include all the contributions that these firms made to local races.

The temptation to "pay to play" was practically irresistible, given the ways in which contracts for bond sales are awarded. They are not usually awarded on the basis of competitive bids. In Denver, the decisions were made by the mayor, on the basis of recommendations made by a broadly based committee. Given the degree of freedom that political officials enjoy in this area, it should not be surprising to find that any firm interested in participating in the consortium—and given the commissions and potential profits from trading the DIA bonds, practically every major Wall Street firm would be—contributed to local Denver campaigns. The roster included such financial luminaries as Goldman Sachs, Lehman Brothers, Merrill Lynch, Smith Barney, and Lazard Freres, as well as many prominent law firms.

With all the ingredients in place, it is not surprising that serious abuses proliferated; scandals in the form of secret payments and influence peddling were uncovered in 1993. That action had to be taken to eliminate what came increasingly to be regarded as widespread corruption, and the larger bond houses decided in 1993 to eliminate political payments (Kowalski 1994). Then a blatant conflict-of-interest case in Massachusetts prompted the SEC, FBI, and Internal Revenue Service to investigate Lazard Freres and Merrill Lynch, and, in 1994, the SEC issued a new rule. Now bond houses that made campaign contributions to potential clients could not work for them for two years.

Since Lazard Freres had served as a financial advisor to the City of Denver and Merrill Lynch had participated in six consortia between 1990 and 1992, DIA naturally became one of the many projects to which the SEC extended its investigation. It issued three separate subpoenas to Denver officials. Two in January 1994 were limited to the work done by Merrill Lynch and Lazard Freres. The third subpoena, in March 1994, covered all the firms that had been selected as underwriters for DIA (Flynn 1994).

Thanks to the work of Robert Kowalski, an enterprising *Denver Post* reporter, fairly precise data are available on the New York-Denver connection. The available records, which are flawed, show that bond firms contributed, in 1990 and 1991, $13,900 to Federico Peña's re-election campaign. When he decided not to run, these firms strongly supported Wellington Webb, donating almost $100,000 to Mayor Webb's two mayoral campaigns, about 10 percent of the total that he raised. Furthermore, about $65,000 was donated after Webb was elected mayor and before decisions about who would issue the next series of bonds had been made. These donors also contributed about $24,000 to the campaigns of the city auditor and city council members who were involved in selecting the underwriting group.

Between 1990 and September 1994, the bond house donations to elected officials who played a role choosing underwriters by companies and individuals with an interest in underwriting the bonds were as follows:

- Mayor Wellington Webb $96,000
- Former Mayor Federico Peña $13,900
- Councilmember Ted Hackworth $2,350
- (Former) Councilmember Stephanie Foote $4,050
- Councilmember Ramona Martinez $550
- Councilmember Happy Haynes $500
- Councilmember Tim Sandos $10,025
- Councilmember Bill Scheitler $850
- City Auditor Bob Crider $5,700

Of course, many have denied that these contributions influenced the decisions that were made. Councilmember Happy Haynes, chair of the council's airport committee and member of the panel that chose the underwriters, argued that such an implication was unwarranted: "I'd say absolutely not," she said. "I'm not sure anybody has a clue who contributed anything, much less to whom." (Kowalski 1994). Peña insisted, "The process was thorough, fair, and objective and demonstrated to the public that the city's business went to the best financial teams . . . . It was a model of good government, a model and a process that I am proud of" (Flynn 1994). Still, it is noteworthy that all the bond issues were handled by syndicates, the first comprising 15 separate underwriting firms. This practice was commonplace. It has been defended on the grounds that it broadens the base of potential bond buyers and reduces an underwriter's risk. Nevertheless, by distributing profits among many firms, the number of companies that can be expected to contribute to political campaigns also increases. Practically every single firm that participated in the bond issues felt so strongly about local politics that they all made contributions, especially to mayoral campaigns.

They not only donated to political campaigns, they were expected to do so. Rita Kahn, for example, was a fundraiser for Wellington Webb when he ran for mayor in June 1991. She worked the financial district, visiting the bond firms. Her activities have been described as follows: "When they saw Kahn, the underwriters got the message . . . . 'She would come in and ask for contributions," says one underwriter, "and people would show up at fund-raisers or send in their checks'" (Hedges et al. 1993).

Until 1993 when it abandoned the practice, the most generous was the firm of Lehman Brothers, a subsidiary of Shearson Lehman Hutton. Lehman was named the lead institution for the 1991D series ($600 million), the 1992A series ($253 million), the 1992B series ($315 million), and the 1994A series ($257 million). For its services it received $3.6 million. It contributed $19,500 to Mayor Webb's campaign committee between 1991 and 1993, most of which came after Webb became Mayor in June 1991. An additional $6500 came in the summer of 1991 in the form of contributions from three subsidiaries, even though the mayor had stated that he would reject any contributions over $2000. Jim Cain, an assistant vice-president with Shearson, noted that the donations "were always made according to industry practice and in accordance with all state and local laws."

When asked if the contributions to the Webb campaign were intended to help gain the firm a role in the DIA bonds sales, Cain said: "Oh, I can't comment on that" (Kowalski 1994).

Goldman, Sachs was also a generous contributor to local campaigns. Beginning with Mayor Peña's administration, it was the lead underwriter for four sales (the 1990A series, $600 million, the 1990B-E series, $200 million, the 1991A series, $500 million, and the 1995 Series A-B, $329 million) and participated in four other offerings. It earned $5.8 million in commissions and fees. Its PAC and officials donated, in the summer of 1992, at least $11,000 to Webb's campaign. Similarly, Pryor, McClendon, Counts & Co. and Prudential Securities, which earned $3.1 million and $1.2 million, respectively, sponsored a reception for Mayor Webb and Atlanta's mayor during the Democratic National Convention in New York City in July 1992. PMC (a minority-owned firm) also made generous contributions to Webb, about $28,000 between 1990 and 1993 (Kowalski 1994). Such generosity apparently did not go unrewarded: the firm (which had been selected for the original team by the Peña administration and was eligible for rotation) became lead underwriter for the 1992 DIA bond issue even though they had not been selected by the screening committee (Flynn 1994). It should be noted that PMC (and the other firms too) were playing by the rules—its contribution was disclosed and consonant with Colorado law. PMC officials also claimed that it did not "pay to play" since they supported a candidate who was not, at the time of the donation the leader in the race (Weiser 1995).

Lawyers are essential participants in any bond issue, and they too were eager to participate in the financial bonanza of the DIA bonds. Those with linkages to Mayor Webb—and they were many—prospered. Six separate firms worked on each of the issues, earning more than $4 million. Each of them contributed to the mayor's campaign and some were close personal friends. One noteworthy example is Linwood T. Holt, who had served as Webb's personal lawyer and whose firm, soon after the election was hired to work on the bonds. Subsequently, Mr. Holt was temporarily suspended from legal practice by the Colorado Supreme Court (and by the city). He left the firm. His former partners formed a new firm, which continued to do legal work on the airport (Hodges 1994).

Another example is provided by the law firm now known as Trimble and Nulan, which served as co-bond counsel for bond sales beginning in 1990, earning $800,000. Between 1991 and 1993 the firm contributed $3600 to the mayor's campaign. Its head, King Trimble, was also a friend of the mayor's wife, Wilma Webb. A former member of the state house of representatives, he resigned his seat to run for city council. Wilma Webb was appointed to fill the vacancy. Subsequently, she was reimbursed for travel expenses incurred when participating in the opening of Munich's new airport in 1992, from her campaign fund (to which Trimble had contributed), even though she was not seeking reelection. When the secretary of state filed a complaint, Wilma Webb agreed to return the donations, which were then paid into a different fund. Between 1991 and 1993, Trimble and members of his firm also donated $3600 to Mayor Webb's campaign (Kowalski 1994). We return to this point in Chapter 9. Suffice it here to note that such practices were commonplace in the world of municipal financing.

The SEC's attempt to introduce a measure of probity into the process by limiting political contributions and other gifts was well received by the major underwriters since it universalized their own decision. However, it ran into opposition from smaller, especially regional, firms. They had always nurtured close political ties to local officials and feared that the big firms would find ways to circumvent the rule. Minority groups also objected. Not surprisingly, the recipients of the political largesse, ranging from treasurers to council members to mayors, also opposed a rule, which deprived them of significant political income (Petersen 1994). Still, given the need for public confidence and the potential for political corruption, the SEC's decision was long overdue and may well contribute to the creation of a healthier structure for public financing.

The power of debt

The system of bond financing that exists in the United States affects local decision processes in more subtle ways than through the kind of political influence that underwriters and others hope to achieve through campaign contributions. The very fact that a city has issued large amounts of bonds gives the holders of the bonds and the rating agencies power to influence its decisions. As noted above, the ratings that are accorded to bonds are of great importance to the municipality that has issued them, especially when it wishes to issue more. As the bond rating sinks, the interest rate that the municipality has to offer rises to offset the higher risk. Threats of downgrading bonds, therefore, have potency.

This is precisely what happened in the case of MarkAir, a low-cost airline that was willing to fly out of DIA. When Continental abandoned its hub, United Airlines turned DIA into a fortress hub and raised fares dramatically. Concerned about the empty gates and the impact of high fares on traffic and on the community (many persons were driving to nearby Colorado Springs to take advantage of low fares), city officials wooed numerous other airlines. One target, low-cost Southwest Airlines, refused to operate at DIA because of the high per enplaned passenger cost.

MarkAir, on the other hand, was willing to move to DIA MarkAir's financial position was, however, precarious. It had declared bankruptcy and wanted a five-year, $30 million loan from the city. The State of Alaska had turned down a similar request from the Anchorage-based airline. Although there was a risk that the airline would not survive, Denver pursued the arrangement, believing that MarkAir's operations would generate more than enough revenues to cover the loans. On learning of this development, Moody's told Denver that if it approved the loan to MarkAir, the ratings on its bonds would be dropped to junk-bond status. The threat sufficed to bring the negotiations to a halt—even though the citizenry would benefit from a more competitive environment that would lead to lower fares. Another example of the rating agency's power is provided by the reappointment of Jim DeLong as Denver's aviation director (see Chapter 9).

A more subtle example of how the debt load that a municipality accepts by issuing revenue bonds influences its decisions is provided by Denver's relationship to United Airlines. As noted, the original financial design for DIA was based on the assumption that at least two airlines would be hubbing at DIA. Further, UAL proved to be a hard bargainer, gaining important concessions before it agreed to sign a leasing agreement, an agreement that was vital for the airport's financial credibility with investors. In the agreement, Denver agreed to release United Airlines from the terms of its lease if the gate fees exceed $20 per enplaned passenger in 1990 dollars. As we have seen, the cost per enplaned passenger turned out to be far higher than projected in the financial forecasts and now runs about $18 in current dollars.

If Denver ever needs to increase revenues, it can only do so to a significant degree by charging United more. But, if forced to pay more than the aforementioned contractual ceiling, United would be released from its leases and could reduce the number of gates (and flights) it operates at DIA, a move that would obviously have disastrous consequences for the airport's future. Hence Denver must generate enough income (primarily through fees) to pay off the debt, but it cannot do so by increasing the charges to its biggest client significantly above current rents.

Furthermore, the very fact that Denver must charge high fees to generate the revenues to pay off the debt's interest and principal provided the airlines with an excuse to hike their fares by about $40, a move that has obviously imposed a burden on the community. This point is discussed in detail in Chapter 11.

Nor can one ignore the degree to which the financing was responsible for the cost overruns, which are discussed in Chapter 9. The city felt itself to be under tremendous pressure to open the airport by its scheduled date because of the payments that it had to make to the bondholders. Accordingly, it consistently tried to avoid schedule slippage and placed many elements on a fast track to ensure that the deadline be met. Ironically, these steps led to many more serious problems and an even later opening date than would have been the case if the city had adhered to a more realistic timetable.

DIA's financial future

Despite the tortuous path that DIA has traveled, there are good reasons to believe, though the information is preliminary and incomplete, that it may well have a reasonably sound financial future. The 1995 bond prospectus revealed that, even though passenger traffic for the first four months of 1995 was 9.6 percent below the comparable period at Stapleton in 1994, the financial picture was more promising. DIA's income reached $29.5 million (annualized) in 1995, well above predictions. Traffic is expected to rise by 2.4 percent each year until the year 2000, and the forecast shows that DIA will probably net over $30 million in 1996. Furthermore, the 1995 bond sale went well so that the $233 million of 1984 and 1985 bonds, which had carried interest rates of 8 to 10.5 percent, were now replaced by bonds paying only 6 percent. The savings to the city are enormous—the airport's debt service is reduced by $121 million for the 1995–1999 period (ASRB 1995; GAO 1994; Leib 1995a).

Still, troublesome financial issues continue to cloud DIA's future. Several of these have been discussed and involve the many investigations and lawsuits that are under way, as well as the city's weak position vis-à-vis UAL. Also of concern are the continuing difficulties with the airlines over fees and charges, especially the prospect that DIA might not receive the anticipated federal AIP (Airport Improvement Projects) grants ($95 million), and the future of Continental's gates. The prevailing financial projections include both these sources of revenues. Continental's original agreement with the city called for payment of $273 million for 20 gates until the year 2000, an agreement that has since been renegotiated down to half the number of gates and allows Continental to sublease several of those.

Ultimately, the future of the airport depends on the traffic that is generated. It appears that DIA will process enough passenger enplanements so that the cost per passenger will not be exorbitant. This conclusion is based on a recent study by the General Accounting Office. It carried out a careful analysis of the city's latest projections to determine the impact of passenger enplanements on DIA's financial well-being. These are more conservative than the FAA's figures, as the following table shows:

	1995	2000
FAA	17 million	21.9 million
City	16 million	18.3 million

Source: GAO, October 1994.

The GAO then developed its own lower and higher projections. The former assumed that traffic would be 10 percent lower than the city's forecast, or 14.6 million in 1995, and would grow by only 1 percent annually, rising to 15.5 million enplanements by 2000. The latter assumes that the total number of enplanements will reach 90 percent of the FAA's projections, 19 7 million. It then subjected these figures to a probability analysis and found that there was a 90 percent chance that the 15.5 million figure would be exceeded and a 10 percent chance that the 19.7 million would be surpassed. It also analyzed the per passenger enplaned cost and concluded that (in 1990 dollars) the peak would be reached in 1996 ($15.57 in 1990 dollars) and subsequently decline to about $13 (in 1990 dollars) in the year 2000. The GAO then fed these data through a financial risk analysis model to determine the probability that DIA would be unable to service its debt which, for 1996, is about $290 million. It found that "Projected traffic levels would need to be nearly 20–25 percent lower before . . . there would be significant risk that DIA's estimated costs might exceed revenues" (GAO 1994, pp. 41–45).

From a financial perspective, therefore, DIA appears to be in reasonable shape, even assuming slow growth for the immediate future. Still, one should view the future with only cautious optimism, because the possibility of unexpected difficulties remains very real (Leib 1995b). We explore potential troublespots in Chapter 11.

Obviously the project's financial well-being is not the only consideration that should be taken into account when one assesses the project. Accordingly, in the concluding chapter, we discuss the costs and benefits to the city and the region.

Lessons learned

- *Honesty is the best policy.* This ancient saying has proved its validity over the centuries. It behooves officials to fully disclose all salient facts when issuing bonds. Even the appearance of not doing so will seriously damage the community's reputation and subject it to potential litigation. If a city is found guilty in court, the consequences for its financial well-being can be extremely serious.
- *Act ethically.* Even if it is legally permissible to create large consortia of underwriters and to hire numerous lawyers so as to generate political support, doing so is often not in the best interests of the community. Only as many lawyers and underwriters as are truly needed should be hired, and they should be selected on the basis of professional criteria and cost. Even the appearance of corruption by elected officials degrades the democratic process.

  When planning the financing, maintain flexibility and monitor developments constantly. A sound financing plan should contain detailed feasibility and risk analyses and reserves to cover emergencies. Some unanticipated events can be beneficial, as in the case of the passenger facility charge, which provided Denver with additional revenues (about $100 million between 1992 and 1995), but these are far less common than events that have negative impacts on the financial arrangements. Accordingly, the basic financial framework should permit ongoing analyses and revisions.

- *Subject bond underwriters to the same competitive bidding process as other city contractors.* Negotiated agreements do possess certain advantages, especially when large projects like DIA are involved and a firm possesses unique expertise. At the same time, the issuer pays substantially more fees to the underwriter. Accordingly, the costs and benefits of each approach should be carefully weighed, and the burden of proof should rest on those who favor a negotiated agreement.
- *Keep the community fully informed of all financial developments.* The airport planners failed to make clear just what the original cost figures included and the degree to which subsequent escalations reflected major changes in scope. As a result, much negative publicity was generated, and misunderstandings of the actual situation became widespread.

References

Airports. 1995. U.K., China reach agreement on Hong Kong airport financing. July 4:261.

Ashford, Norman and Clifton Moore. 1992. *Airport Finance.* New York: Van Nostrand Reinhold.

Asian Wall Street Journal. 1995. Hong Kong passes bill to set airport body. July 20.

————. 1994. Airport to be Macao's gate to region. October 31.

ASRB. 1990–1995. See City and County of Denver.

Aviation Daily. 1995. Macau's first airport to open Nov. 9. July 28:149.

Aviation Europe. 1995. Munich airport still in the red. February 2:6.

Aviation Week & Space Technology. 1991. Germany struggles to meet airport needs. August 26:41.

————. 1994. Growth outpaces Asian airports. August 29:57.

————. 1995. Kansai losses. July 17:19.

Becker, Stuart. 1995. Airports: Chinese opera. *Far Eastern Economic Review.* April 6:54.

Bell, Michael. 1991. Airport financing. *Airport Regulation, Law and Public Policy.* Robert Hardaway, ed. New York: Quorum Books.

Black, Alexandra. 1994. Trade and tourist boom from new airport. *Inter Press Service Global Information Network.* September 9.

Brown, David. 1987. Japanese building international offshore airport to serve Osaka. *Aviation Week & Space Technology*. July 13:38.

Carnahan, Ann. 1994. Efficiency estimates at DIA in dispute. *Rocky Mountain News*. December 12.

City and County of Denver (Denver). 1990. 1990A Bond Prospectus.

————. 1990. Airport System Revenue Bonds (ASRB), Series 1990A-B. May 10.

————. (Denver). 1991. 1991A Bond Prospectus.

————. (Denver). 1992. 1992A Bond Prospectus.

————. (Denver). 1994A Bond Prospectus.

————. 1992. ASRB, Series 1992A, B, D-G. May 6.

————. 1994. ASRB, Series 1994A. September 1.

————. 1995. ASRB, Series 1995A, B. June 1.

————. (Denver). 1995a. 1995A-B Bond Prospectus.

————. (Denver). 1995b. Office of the Auditor news release. February 28.

Donoghue, J. A. 1995. The Pearl-Y gateways. *Air Transport World*. February 1:75.

Dunham, Oris. 1993. Infrastructure Constraints—Deeds Not Words. 7th IATA High-Level Aviation Symposium 109, Cairo, Egypt.

The Economist. 1994. Fasten your seat belts. November 12:42.

Federal Aviation Administration (FAA). 1993. Aviation System Capacity Annual Report 5.

Flynn, Kevin. 1994. Inquiry could hurt Webb. *Rocky Mountain News*. July 14.

General Accounting Office (GAO). 1991. New Denver airport: Safety, construction, capacity, and financing considerations. September.

————. 1994. New Denver airport: Impact of the delayed baggage system. October.

Graham, Anne. 1992. Airports in the United States and Canada. *The Airport Business*, Rigas Doganis, ed. New York: Routledge.

Gross, Neil. 1994. Japan wanted an airport, it got a real mess. *Business Week*. June 6:50.

Hedges, S., W. Cohen, and A. Martinez. 1993. The politics of money. *U.S. News and World Report*. September 20.

Hill, Leonard. 1993. Beyond expectations. *Air Transport World*. June 1:182.

Hodges, Arthur. 1994. Gentlemen prefer bonds. *Westword*. May 18–24.

Jansen, Peter. 1994. Legal issues in airport construction. Paper delivered at IBC Conference on Asia Pacific Airports '94, Singapore, July 25.

Kowalski, Robert. 1994. DIA firms gave $96,000 to Webb campaign. *Denver Post*. June 19.

Lassiter, Eric. 1994. Japan to open much-delayed Kansai airport. *Travel Weekly*. August 29:4.

Leib, Jeffrey. 1995a. DIA projections healthy, woes also cited in bond report. *Denver Post.* June 6.

————. 1995b. Doomsday scenario for DIA? *Denver Post.* December 3.

Mecham, Michael. 1995. Airport fees stable for Asian carriers. *Aviation Week & Space Technology.* August 28.

Meredith, John. 1995. Room to boom. *Airline Business.* January:38–41.

Mok, John. 1993. The development of Hong Kong's new international airport. Paper delivered at International Conference on Aviation & Airport Infrastructure, Denver, Colorado, December 8, 1993.

Monsarrat, Sean. 1991. Balancing the market's hazards is key to successful underwriting. *The Bond Buyer.* April 26.

Moody's Bond Record. 1994.

Moorman, Robert. 1994. Osaka to me. *Air Transport World.* October 1:62.

Mordoff, Keith. 1988. Air Transport Munich's new international airport expected to begin operations in 1991. *Aviation Week & Space Technology.* February 22:92.

Mufson, Steven. 1994. Accord boosts Hong Kong airport. *Washington Post.* November 5.

Orlando Sentinel. 1995. Hong Kong builds $20.3 billion airport. April 9.

Ott, James. 1993. Kansai sets highest fees. *Aviation Week & Space Technology.* October 25:21.

Paulson, Steven. 1995a. Ignored warnings chronology. *Associated Press.* February 19.

————. 1995b. Denver may have misled investors about delays at new airport. *Associated Press.* February 23.

Peat Marwick Mitchell. 1980. Assessment of general financial requirements. File memorandum prepared for Denver Regional Council of Governments. January.

Petersen, John E. 1994. A guide to what's ahead for the municipal bond market. *Governing Magazine.* December:59.

Phillips Business Information. 1994. New Macau airport set to open July 1995. January 28.

Rabinowitz v. City of Denver et al. 1995. United States District Court, March 30.

Racine, John. 1990. Boettcher warns downgrade is possible on Denver International Airport. *The Bond Buyer.* December 3.

————. 1991. Moody's affirms Denver airport's conditional Baa1 ahead of April issue. *The Bond Buyer.* March 22.

————. 1992. DIA bonds restored to BBB status by Standard & Poors. *The Bond Buyer.* January 10.

Rebchook, John. 1995. Investor sues Denver, alleges fraud; City attorney promises fight, says timing of lawsuit on eve of opening is 'no coincidence.' *Rocky Mountain News.* February 28.

San Diego Union-Tribune. 1994. Japan opens airport on man-made island. September 5.

Shin, Jong-Heui. 1993. Airport developments in Korea. Paper delivered at International Conference on Aviation & Airport Infrastructure, Denver, Colorado, December 8.

Sussman, Gennifer. 1996. Interview. January 21.

Todd, Gregory. 1994. Airport bonds ride the bad news roller coaster. *Rocky Mountain News.* September 11.

Tompkins, Robert. 1993. Infrastructure Capacity Financing Through User Charges. Unpublished address at IBC Conference in Hong Kong, October 28.

Weiser, Benjamin. 1995. In the minority, and mad: Black-owned investment firms say SEC's "pay to play" curbs are unfair. *Washington Post.* January 29.

World Airport Week. 1995. NSIA considers opening with extra runway. July 1.

6

Location, location, location: Site selection, environmental impacts, and ground access

"The solution to pollution is dilution."—ANONYMOUS[1]

Location, location, location—the mantra of developers and agents—has become the stock reply concerning the three most important factors in real estate. Though perhaps a *bit* exaggerated, this expression does convey a basic truism regarding any development project. Certainly, if one wishes to initiate a commercial, industrial, or office business enterprise, or to purchase a residential housing unit, one of the first concerns in the decision calculus must be location.

In the realm of airport planning, the locational question takes on even more importance. Since they consume a considerable amount of land, contribute to noise, air, and water pollution, and are usually located relatively near major population centers, airports represent a very visible and controversial urban land use. Airports are subject to the NIMBY—Not in My Back Yard—syndrome, as most people would prefer not to have airports located near where they live, but nevertheless expect them to be close enough to use without having to drive too far to reach them.

Accordingly, this chapter compares the locational aspects of airports around the world with Denver International, paying particular attention to site selection, environmental impacts, and ground access. The DIA site selection process and the factors that contributed to the ultimate selection of the final site are identified and analyzed. A discussion of the environmental impact assessment procedures that are

1. Quoted by James (Skip) Spensley, environmental attorney and director of the New Airport Development Office for DIA.

now required for all major federal projects, including airports, provides the appropriate background for consideration of the DIA case. Of particular importance are the environmental and ground access implications of locating a new airport 24 miles northeast from the city center on a 53-square-mile parcel of relatively open farmland in a metropolitan area notorious for air pollution problems and in a semi-arid region where provision of water resources to accommodate new growth is a chronic concern. Specific environmental issues surrounding the new airport were noise impacts at the new site compared to Stapleton, effects on metropolitan air quality from more vehicle-miles driven due to the airport's remote location, potentially harsher meteorological conditions at the new site, and construction problems due to expanding soils. Finally, ground access planning at DIA is discussed, including the issues of parking and provision of alternative transportation modes.

Location and site selection

Contributions from the study of location theory inform us that any locational decision is dependent on a complex set of historical, geographical, social, economic, engineering, and political considerations. Many of these can be assessed by means of technical analyses, but as with any human behavioral process, not all decisions can be explained by the workings of objective rationality.

Site selection analysis for airports involves numerous factors, including the availability of land for expansion and alternative runway configurations, presence of other airports and airspace congestion in the area, surrounding obstructions, atmospheric conditions, accessibility to ground transport, economy of construction, and availability of utilities (Horonjeff and McKelvey 1994, p. 193). These and other factors are addressed in site selection studies and are reanalyzed in environmental assessments or other screening procedures put in place by local or national governments. Objective weighting and scoring mechanisms are developed in these studies, which are used to evaluate each potential site. Based on results from these technical assessments, optimal sites are selected and recommended for airport project implementation (Deem and Reed 1966).

What might seem to be a fairly straightforward, rational process of site selection at some point becomes subject to political factors. Once initial sites are identified and meet basic threshold criteria, other "nontechnical" factors tend to become more important in the actual site selection process. Yet, despite efforts to choose sites that maximize positive benefits and minimize negative impacts, actual sites are usually less than optimal and many times seem to invite a fair share of criticism and negative fallout. Because the ultimate selection depends on both technical and political decision-making, it is not always certain which takes on more importance in the overall site selection process and thus which bears more responsibility for locational decision-making. The challenge to airport planners and decision-makers is to structure a process that addresses both technical and political factors in an open manner that does not result in poor decisions emanating from technical evaluation failures, from ineffectual political negotiations and bargaining, or from "under-the-table" political patronage and cronyism.

Comparative airport locations

In the early days of aviation, airports were relatively small in size and, given the more confined area of cities at that time, located much closer to city centers. But the combined effect of post-World War II automobile-induced suburban expansion and the need for larger airports due to the technological and commercial growth of aviation created land-use conflicts across most metropolitan areas. It was more difficult to expand older airports because they had become circumscribed by urban development, and residents near those airports began to object more strenuously to the increased noise coming from larger and louder jet aircraft. The obvious result was to build new airports farther away from the metropolitan area but not too far so as to make the new facilities inconvenient. Thus, any airport site selection exercise has to balance these two fundamental, and opposing, locational forces, in addition to considering the numerous other factors cited earlier, such as environmental impacts, airspace congestion, and the like.

When considering the sizes and locations of existing and new airports in the United States and around the world, it is obvious that newer airports tend to be larger and located farther away from city centers. Table 6-1 shows that most older airports are usually anywhere from 3 to 15 miles (5 to 24 kilometers) away from city centers, while newer airports tend to be anywhere from 15 to 40 miles (24 to 65 kilometers) away, with the exception of some recent airports built on landfill in adjacent waters.

Table 6-1. Locations for major airports of the world

Airport	Date opened	Distance from city center
Hong Kong Kai Tak	1929	3 mi (4.8 km)
Osaka Kansai	1994	4 mi (6 km)
Frankfurt Rhein/Main	1936	5 mi (8.1 km)
Denver Stapleton (old)	1929	7 mi (11.3 km)
Vancouver International	1930s	7 mi (11.3 km)
Munich Riem	1939	7 mi (11.3 km)
New York LaGuardia	1939	8 mi (12.9 km)
Atlanta Hartsfield	1925	9 mi (14.5 km)
Paris Orly	1961	9 mi (14.5 km)
Paris Charles de Gaulle	1974	12 mi (19.4 km)
London Heathrow	1946	15 mi (24.2 km)
New York JFK	1948	15 mi (24.2 km)
Chicago O'Hare	1955	15 mi (24.2 km)
Hong Kong Chek Lap Kok Airport	1997 (est.)	15 mi (24.2 km)
Munich Franz Josef Strauss	1992	16 mi (25.8 km)
Toronto International Malton	1937	17 mi (27.4 km)
Kansas City International	1963	17 mi (27.4 km)
Houston Intercontinental	1969	17 mi (27.4 km)
Dallas/Fort Worth International	1974	17 mi (27.4 km)
Bangkok Nong Kgu Hao International	2000 (est.)	19 mi (30 km)
Los Angeles International	1930	20 mi (32.3 km)
Denver International	1995	24 mi (38.7 km)
Washington Dulles	1962	27 mi (43.5 km)
Seoul Yongjong International	2000 (est.)	32 mi (52 km)
Montreal Mirabel	1975	40 mi (64.5 km)
Tokyo Narita	1978	41 mi (66.1 km)
Kuala Lumpur Sepang International	1998 (est.)	41 mi (66.1 km)

Sources: Allen 1979; Colorado National Banks 1989; Feldman and Milch 1982.

Recent international airports: Locational and environmental aspects

The following discussion highlights some of the locational and background aspects associated with several new airport projects around the world.

Munich's Franz Josef Strauss Airport

As early as 1954, the Munich Airport Authority was contemplating a replacement for Riem Airport (Mecham 1992). Political leaders and airport officials began to study seriously the need for a new airport after a midair collision over Munich in 1960 (Mordoff 1988). Safety, noise, and capacity considerations drove the need to build a new airport and retire the old one. Yet noise, pollution, ground traffic congestion, and other environmental concerns posed the most formidable obstacles to new airport development.

In 1963, studies began to determine the best location for a new airport. Twenty sites were evaluated. In 1966, the list was reduced to two—Hofolding Forest and the Erding-North/Freising area. In 1969, the Bavarian government selected the latter for the new airport. Munich's new airport was originally planned for 5066 acres and three runways; litigation reduced it to 3427 acres and two major runways (Upton 1992). At 5.4 square miles, Strauss is one-tenth the size of DIA. Nonetheless, the new airport is five times the size of the airport it replaces. Half the land is set aside for future development (Hill 1991). The new airport is located 16 miles (25.8 kilometers) northeast of downtown Munich, beside the Isar River. That, of course, is 8 miles closer to downtown Munich than DIA is to downtown Denver.

Osaka's Kansai Airport

One of the fundamental problems confounding public officials in Japan is the inability to condemn land under eminent domain power and purchase it for fair market value. Traditional farmland is valued by its owners well beyond its economic value, and many farmers resist all efforts to purchase their property. Moreover, as is especially true in densely settled countries, Japan is noise-sensitive. Efforts to expand the existing airport were resisted by local residents for the same reasons people everywhere resist additional airport infrastructure—noise and fear of accidents (Brown 1987). Thus, the land, safety, and noise problems could only be resolved by building an airport offshore.

Japan's Ministry of Transport began to study the possibility of building a new international airport in the Kansai region in 1968. The Cabinet approved the initial Kansai airport plan in 1984. Noise pollution was a driving force for the decision of the Ministry of Transport to authorize construction in 1986 (Moorman 1994). Construction began in 1987, and reclamation of the enclosed area began in 1988. Construction of the passenger terminal began in 1991 as the bridge trusses were completed. Kansai International Airport (KIA) opened on September 4, 1994 (Moorman 1994).

Kansai was originally planned to be only 1 kilometer offshore, on solid seabed. But fishermen protested, and the airport island was moved an additional 5 kilometers, above a seabed consisting of 20 meters of alluvial silt and mud (*The Economist* 1995). This soft seabed caused parts of the manmade island to sink, thus complicating the engineering task and adding appreciably to the construction cost. The island on which the airport sits is shaped like an aircraft carrier and linked to Osaka by a 4-kilometer double-decker six-lane highway/railway truss bridge (Black 1994).

Macau International Airport

An artificial island was built adjacent to the island of Taipa, one of two islands adjacent to the peninsula of Macau (*Travel Weekly* 1995), using hydraulic sandfilling (*Aviation Week & Space Technology* 1994a) to expand the tiny Portugese colony's usable real estate by more than 20 percent (Donoghue 1995). Four hundred hectares were reclaimed between the islands of Taipa and Coloane, which will be the venue for the airport and airport-related businesses (*Asian Wall Street Journal* 1994). By virtue of the fact that the airport is in the bay, noise impacts on residents are anticipated to be modest, creating the possibility of 24-hour departures (*Travel Weekly* 1995).

Hong Kong's Chek Lap Kok Airport

Hong Kong's Chek Lap Kok Airport (CLKA), the world's most expensive by a large margin, will replace the crowded and slot-constrained Kai Tak Airport, one of the world's busiest (Mufson 1994), with a single runway jutting out into the bay. Kai Tak is one of the world's most congested airports, with a land mass of only 310 hectares serving more than 22 million passengers a year (Darmody 1993). It was built by the Japanese during World War II in the middle of Kowloon (*The Economist* 1994), which is today the venue of some of Asia's most valuable real estate (*Vancouver Sun* 1995). More than 350,000 people are seriously impacted by the noise of the existing facility (Darmody 1993). The airport is about 15 miles from central Hong Kong, as the crow flies (*Aviation Week & Space Technology* 1994b).

New Seoul International Airport

The New Seoul International Airport (NSIA) will be built on reclaimed land between Joungjong and Yongyu Islands near Ichon, 52 kilometers (about 30 miles) west of Seoul (Shin 1993). A new airport is needed at Seoul because the existing facility at Kimpo Airport (which opened in 1958) will be saturated by 1997 (*World Airport Week* 1995a). Kimpo handles 91 percent of Korea's international traffic and 40 percent of its domestic traffic (Shin 1993).

A feasibility study for a new airport was conducted in 1989 and 1990. The site at Yougjongdo was selected in 1990. A master plan was finalized in 1992, and groundbreaking occurred in November 1992 (Shin 1993). The initial phase of the new airport is anticipated to be completed in 1999, and the airport will be operational in the year 2000 (*World Airport Week* 1995b). The first phase of the airport's construction began in 1992 and will be completed around the turn of the century, at a cost of $7 billion (Shin 1993).

Denver International Airport

In comparison to other airports, Denver International's location at 24 miles northeast of the Denver central business district (CBD) ranks it as one of the more remote airports in the world. To be sure, it is not as far removed as Washington Dulles (Fig. 6-1), Montreal Mirabel, Tokyo Narita, or the new Seoul International airports, but it represents a significant increase from old Stapleton's convenient 7-mile distance from the Denver CBD. Furthermore, with Stapleton's closing, DIA is the only commercial airport operating in Denver, unlike the situations in Washington, Montreal, Tokyo, or Seoul, where closer-in airports are still operating. Thus, Denver has the distinction of having to use the most remote airport in the world. To provide further insight into how the location of DIA was determined, the following section discusses the site selection process.

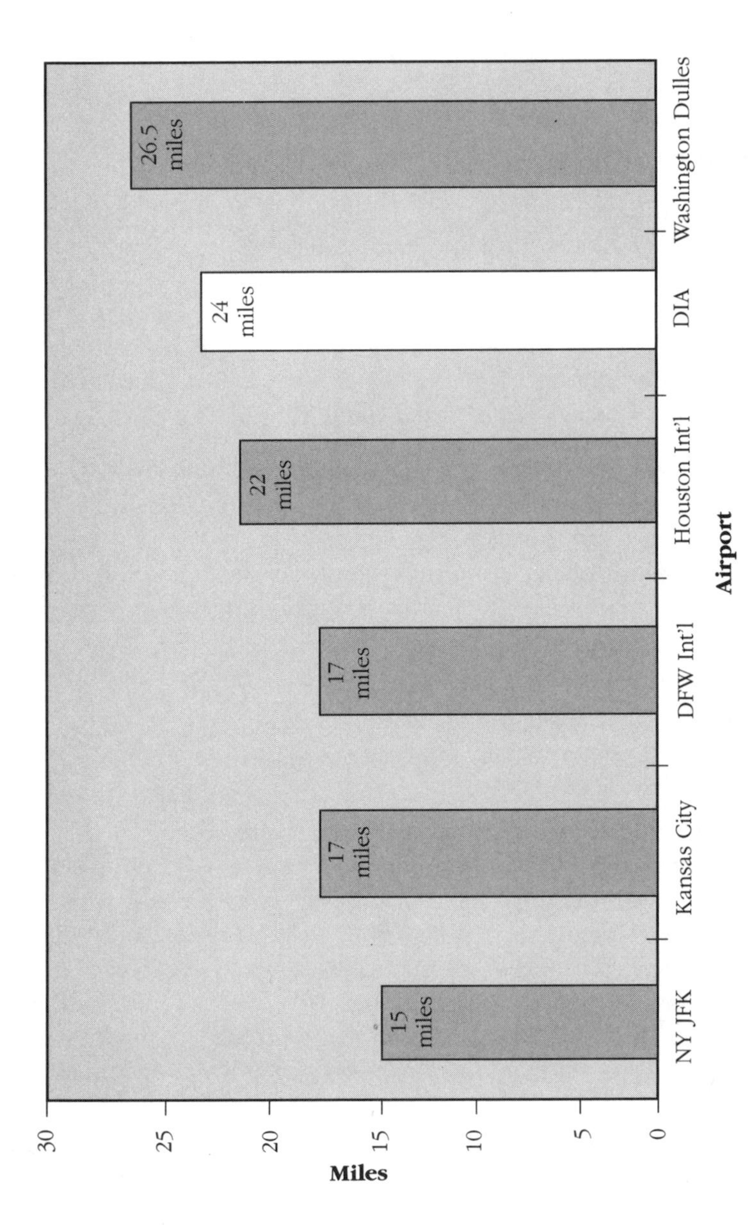

Fig. 6-1. *This chart depicts the relative distances from major U.S. airports to the downtown areas of the cities which they serve. Dulles International Airport is 26 miles from downtown Washington, D.C., but most major U.S. airports are closer to their downtowns than Denver's new airport.*

From Stapleton to DIA: Finding the best location for Denver's aviation activities

In the late 1920s, Denver mayor Benjamin Stapleton led the effort to build a new aviation facility on open farmland near Sand Creek, located 7 miles to the northeast of Denver's downtown. The local media and numerous critics decried the plan and charged that Stapleton was rewarding his political crony H. Brown Cannon by buying his dairy farm for the airport site (Leonard and Noel 1990, p. 433).

Whatever the motivations for the eventual site location of what was to become Stapleton International Airport, the decision to build the new airport in the northeastern sector of the city significantly influenced future urban development in the Denver metropolitan area (Fig. 6-2). In addition to the South Platte River valley, the north side of Denver had become the industrial part of the city as railroads, warehouses, mining smelters, and manufacturing plants came to dominate the landscape. From this initial northern orientation, the location of the airport contributed to a northeastern movement of the industrial sector, as new industries, refineries, trucking and rail terminals, and other transportation corridor activities came to be associated with areas near Stapleton and led to the development of industrial suburbs such as Commerce City. The siting of the Rocky Mountain Arsenal to the north of Stapleton in the 1940s also acted to stunt urban residential growth to the northeast. As a result, the principal axis of higher-income residential growth continued to move in a southeastern direction. The west side of Denver became dominated by moderate-income residential development, as growth pushed out toward the foothills of the Front Range of the Rocky Mountains.

These urban development patterns and the obvious physical constraints of the nearby mountains dictated certain functional realities regarding the potential sites for a new airport in the Denver metropolitan area. In the Denver Regional Council of Governments (DRCOG) Metro Airport Study Phase I in 1980, an airport study area was identified as the north-to-east quadrant from Denver's city center, extending 30 miles to the north, 40 miles to the east, and encompassing 1230 square miles. This quadrant was selected after considering the centroid of the Colorado/Wyoming Front Range population and the flatter topography of this area to minimize construction costs. Among the factors considered in the DRCOG site evaluation process were topography, airspace, noise, land use,

Fig. 6-2. *This satellite photograph reveals the location of DIA, the Rocky Mountain Arsenal, and Stapleton International Airport vis-à-vis the Denver metropolitan area and its major highway corridors.*

wildlife, recreation/scenic values, hydrology, agriculture, airport access, energy/resources, historic/archaeological conditions, and vegetation. Land areas were classified in one of three categories:

excluded, questionable or acceptable. Areas excluded from further consideration were populated areas; river valleys, lakes, reservoirs, and other water resource areas; areas deemed too distant from the Denver metropolitan area; and subsidence-hazard-potential areas where subsidence (settling) of 5 to 10 feet or more could occur instantaneously. Areas identified as questionable included those that had potential mineral (such as oil) resource importance; agricultural significance; important hydrologic features; somewhat steeper and less desirable topography; historical/archaeological significance; important vegetation and wildlife habitat; or scenic and recreational considerations (DRCOG 1983).

This elimination left six large sites that were deemed acceptable for airport utilization (Fig. 6-3). After further screening, Sites 1 and 2—located to the north of Denver near the towns of Brighton and Fort Lupton—were ruled out due to potential air operation difficulties, limited capacity for east-west runways, and adverse citizen opposition, among other factors (DRCOG 1983). The executive committee of DRCOG chose Site 3 (encompassing the Rocky Mountain Arsenal), Sites 4 and 5 (located to the east of the Arsenal and perceived as one site by most observers), and Site 6 (east of Box Elder Creek, north of the towns of Watkins and Bennett) to be studied further.

In Phase II of the Metro Airport Study, completed in September 1982, DRCOG more intensively analyzed each of these three locations to determine the optimal airport site. At the same time, the City of Denver was conducting its own study on the feasibility of expanding Stapleton onto the Rocky Mountain Arsenal. Eventually, DRCOG decided to include this Denver alternative in their final Phase II analysis. DRCOG's Aviation Technical Advisory Committee (ATAC) decided that Sites 3 and 4/5 should be eliminated from further consideration and that the Stapleton expansion alternative and the Site 6 alternative should be evaluated on an equal and parallel basis in Phase III of the Metro Airport Study (DRCOG 1983). Site 3, which equated to building a new airport using Arsenal land, was put aside in deference to the Stapleton expansion plan. Even though Site 3 was deemed to be the most economically desirable site largely because of lower access costs, the potential problems with decontamination of the Arsenal cast a questionable cloud over this site. DRCOG's concern over the Arsenal was equally applicable to Denver's Stapleton expansion plan, but DRCOG nevertheless decided to include the expansion alternative in its final decision. Meanwhile,

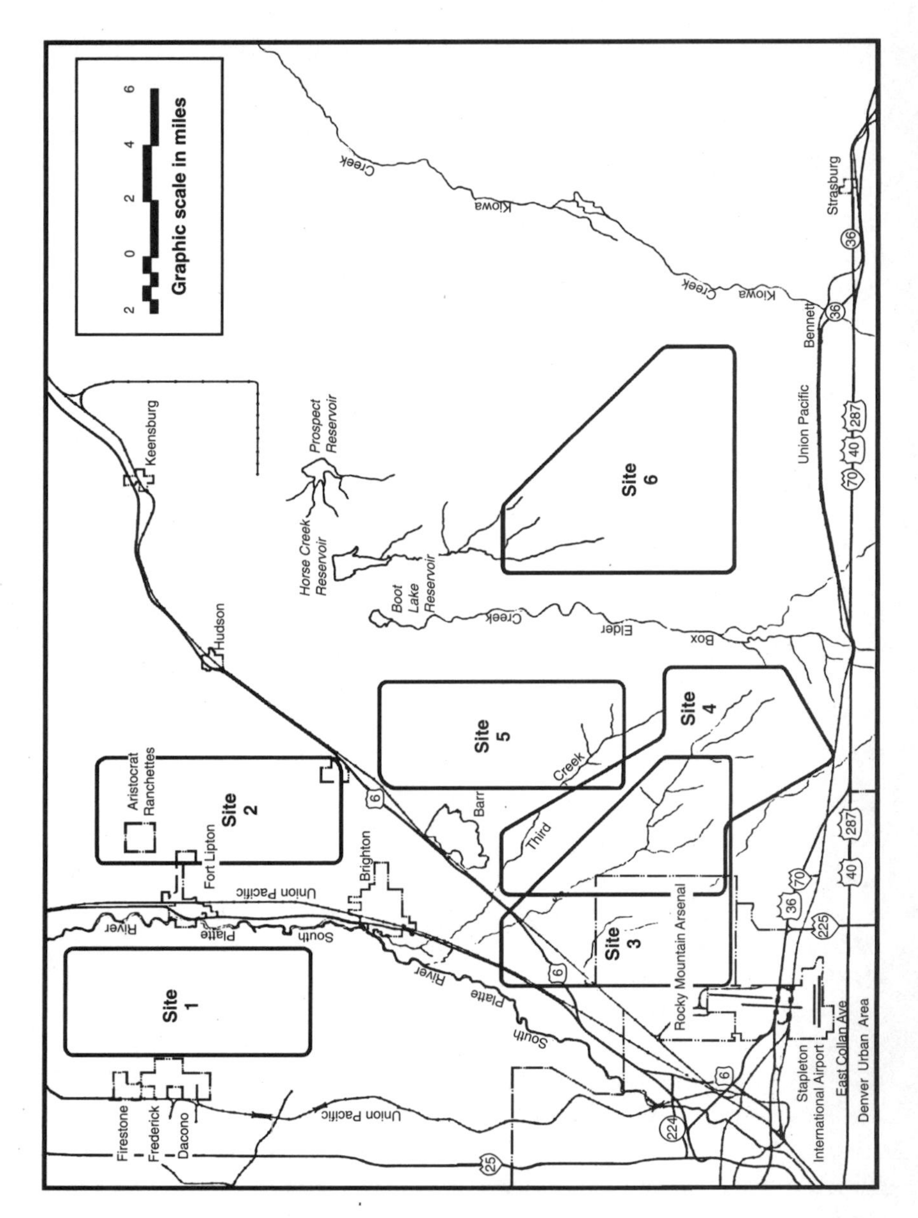

Fig. 6-3. *Locations of initial sites considered for a new airport in the Denver Regional Council of Governments (DRCOG) Metro Airport Study from 1979 to 1983.*

Sites 4 and 5, which most closely correspond to the eventual location of DIA, were rejected because they were deemed inferior to Site 3 based on ground access and economic cost considerations and inferior to Site 6 in terms of aircraft operations, airspace interactions, air quality, noise effects, ecological considerations, land-use development and control, and social effects (DRCOG 1983). Concerns over noise impacts on passive recreation activities at nearby Barr Lake State Park were also cited.

The original purpose of DRCOG's Metro Airport Study Phase III was to address the financing and implementation of the alternative selected in Phase II. Since the Stapleton expansion alternative and the Site 6 (Watkins-Bennett) new airport alternative were both selected in Phase II, the purpose of Phase III was expanded to include final selection of one of the alternatives, as well as financing and implementation. Extensive analyses were conducted for both sites that considered many factors, including the cleanup costs and availability of the Rocky Mountain Arsenal, airport access, noise and other environmental impacts, social impacts, and costs of construction.

The key cost findings that emerged from the Phase III analysis were

1. Capital costs for Stapleton expansion were slightly lower than for Site 6
2. Access costs were much lower for Stapleton expansion
3. Adding in Arsenal decontamination costs to the Stapleton expansion alternative might negate the other cost savings (DRCOG 1983)

It was estimated that decontamination costs of more than $1.45 billion would favor Site 6 over Stapleton expansion. It was unclear how much decontamination would actually cost, although the Army had mentioned figures as high as $2 to $6 billion (Wiley and Rhodes 1987). Possible additional costs for Stapleton expansion included acquisition of the Arsenal property ($33 million) and possible noise mitigation measures for 23,000 residents who would be affected by aircraft noise. Based on costs, DRCOG felt that limited expansion of Stapleton with transfer to Site 6 in the year 2000 would be preferable to early transfer to Site 6 (DRCOG 1983).

When assessing environmental impact, Site 6 was deemed to be preferable to Stapleton expansion because its more remote location meant that fewer people would be affected by noise, and air quality would not be as greatly degraded within the immediate Denver metropolitan area. DRCOG based its evaluations, however, on a Site 6 alternative that included building an interim east-west runway on the Arsenal by 1990. Thus, noise impacts between the two alternatives were assessed to be similar, at least until the new airport opened. Impact on wildlife, vegetation, and natural resources did not appear to be significant factors.

The final part of Phase III involved the implementation of either alternative. Based on an institutional analysis conducted in Phase I, it was assumed that the owners and operators of the airport in the Stapleton expansion alternative would be the City of Denver, while for the Site 6 alternative it would be a regional airport authority. Thus, selection of the Site 6 alternative would require the creation of a new regional airport authority by the State of Colorado. The City of Denver, already the owner and operator of Stapleton, would need no new airport authority but would have to address the Arsenal decontamination issue together with the army, the Department of Defense, Shell Oil Company, and the Environmental Protection Agency (EPA). But Denver would still maintain control of the airport if the Stapleton expansion alternative were selected.

On the basis of the technical assessments and ensuing discussions and public meetings, on July 20, 1983, DRCOG's ATC and its board of directors finally selected "the Rocky Mountain Arsenal alternative as the solution for the Denver region's long-term airport development needs and that the City and County of Denver be asked to take all necessary steps to implement the Board's decision, and that close cooperation with Adams County would be part of the working arrangements for expansion onto the Arsenal" (DRCOG 1983). While the City of Denver was pleased with DRCOG's selection, Adams County, as was discussed in Chapter 3, voiced immediate opposition and promised to forestall the implementation of this option through the environmental impact statement process and whatever other legal means it had available. Its two key environmental concerns were increased noise impact on nearby residents and contamination at the Arsenal. Adams County threatened to tie up the process in court for years to come.

It was at this point, from July 1983 through 1984, that the recently elected mayor, Federico Peña, and his airport project advisors reassessed the City of Denver's commitment to the Stapleton/Arsenal expansion plan and began seriously considering a new alternative. In early 1985, Denver and Adams County officials agreed to the Memorandum of Understanding (MOU) that laid out the principles of the intergovernmental agreement for the purpose of constructing a new airport on land that Adams County would allow Denver to annex, subject to referendum. Based on the 1985 Alternative Configurations Study conducted by consultants Howard, Needles, Tammen, and Bergendoff (HNTB), the MOU identified a site immediately east of the Rocky Mountain Arsenal for the new airport, thus discounting DRCOG's site recommendation (Site 6—between Watkins and Bennett) from its Metro Airport Study. The parties agreed, however, that the location, configuration, and use of the airport terminal and runways "should be as close as possible" to that recommended in the MOU, thus providing for some further adjustments if needed (Denver 1985).

Following the MOU, the City and County of Denver initiated a new airport master plan and hired HNTB to conduct a site selection study. The HNTB study analyzed eight alternatives: Base Case 1, Existing Stapleton; Base Case 2, Stapleton with Improvements; Stapleton Expansion into the Rocky Mountain Arsenal; A.1 Second Creek site; B.1 Box Elder Creek site B.2 Front Range site; B.3 Box Elder Creek North site; and C.1 Lost Creek site. It excluded the two base cases because noise impacts were high, future aviation demand would not be accommodated, and vehicular air pollution emissions would be generated in nonattainment areas. It also excluded the Arsenal expansion option because of noise impacts, increased aircraft delay from its constrained and inefficient airfield configurations, Adams County's staunch opposition, and potential major problems with Arsenal decontamination. The Front Range and Lost Creek sites were excluded because of negative factors emanating from their farther distances from the Denver metropolitan area. Thus, the HNTB study recommended parts of the Second Creek, Box Elder Creek, and the Box Elder Creek North sites as the final site area (HNTB 1986).

This site was much larger than the original site described in the MOU and also located farther to the north and east. Several issues emerged after the MOU had been signed that affected both the size and the location of the site. First, additional studies showed that a larger airport was needed because forecasted increases in passengers would be higher than originally thought, and a larger site area would be necessary to accommodate the enlarged airport. Specifically, the new airport was projected to eventually require 12 runways instead of the eight runways originally planned (given then-current aviation activity forecasts for Denver), so more land area was needed. Second, Air Force representatives at nearby Buckley Air National

Guard base began to voice concerns about the airport location because the proposed north-south runway alignments at the original MOU site would have interfered with security and defense radar systems at the base (HNTB 1986). Moving the airport to the north and east seemed to ameliorate the Air Force's concerns. Third, representatives from Adams County later expressed dissatisfaction with the location because economic development opportunities within Adams County's jurisdiction would have been limited to just the northern and eastern edges of the airport, unlike other locations that possessed greater potential for development. Thus, the City of Denver decided to open up an economic development "corridor" near the airport where the proposed E-470 highway could be routed and within which development could occur on both Denver and Adams County land (Colorado 1991).

Even though the City of Denver's airport master plan identified a general site area, the final legal boundaries of the new airport would be determined by political officials in Denver and Adams County, to be incorporated into the final airport layout plan. And of course, landowners in the general airport area were very interested in the actual location because there was much to be gained from ownership of particularly attractive properties.

Thus, the actual airport site was moved several times after the MOU signing in 1985. The original MOU location extended onto some of the Arsenal property on its western edge. The first shift moved the site farther to the east so that the western boundary would be contiguous with the Arsenal. According to Mayor Peña, this shift occurred because the initial configuration would have impinged on bald eagle habitat (Colorado 1991). Two more site changes occurred thereafter. The first major relocation to the north and east increased the size of the site to approximately 47 square miles. It occurred because of the increased size needs, additional eagle habitat avoidance, the radar conflicts, and the enhanced opportunities for Adams County and Denver economic development. One final shift occurred in January 1988 that moved the airport 1 mile farther north and 0.5 mile farther east to enlarge the site to 53 square miles. The size and economic development arguments have also been utilized to explain this move (Colorado 1991), as well as continued concerns by Adams County with noise impacts (Spensley 1995).

Still, as we saw in Chapter 3 and discuss again in Chapter 9, the original location decision and these final moves may well have been politically engineered to benefit specific individuals who owned land in the "corridor" leading to the airport. It is, however, appropriate to mention here that the Colorado Attorney General's Office did investigate these allegations, particularly "whether any property owners or others used, or attempted to use, illegal methods to influence City or County officials in this [site selection] process" (Colorado 1991). It concluded that "The investigation has not uncovered any evidence that any City or County officials were illegally influenced in their decisions concerning the selection of airport sites." Nevertheless, controversy continues to swirl around the site selection process and whether those who benefit from the decision may have had some influence on political or planning officials.

Environmental conditions at the new site

Once the final site was determined, additional questions were raised specifically about meteorological and soil conditions. Concerns over adverse weather included greater thunderstorm and snowstorm frequency, more microbursts or windshears, increased effects of mountain "air waves," and higher average wind speeds at the new site (GAO 1991). It was also suggested that new airport construction at the chosen site would be hampered by the presence of clay soils, such as bentonite, which expand when saturated with water (Booth 1990). Still other reports indicated that the airport was located near the Rocky Mountain Arsenal Fault, where more than 1500 earthquakes were recorded between 1962 and 1980 (Chandler 1993).

In early 1991, U.S. representatives Frank Wolf (Virginia) and Bob Carr (Michigan) asked the U.S. General Accounting Office (GAO) to investigate these and other concerns about the construction of DIA. After months of investigation, including discussions with scientists at the National Oceanic and Atmospheric Administration (NOAA) and the National Center for Atmospheric Research (NCAR), and analyses of available weather data, the GAO concluded that there were "no significant differences in weather between Stapleton and the new site" (GAO 1991). Potentially dangerous storms were not shown to be more frequent at the new site than elsewhere in the Denver area, and although average wind speeds were slightly higher, discussions with scientists and pilots suggested that overall weather differences were not significant and that higher wind speeds would in fact enhance flight safety (GAO 1991; Rhodes 1992). With regard to soils, the GAO, after consulting with the U.S. Geological Survey (USGS) and the Colorado Geological Survey, acknowledged that expansive clay soils were found at the new site, but they are also found extensively throughout much of the Denver region where much construction, including Stapleton and other airports, has already occurred. The GAO did recommend, however, that designers and engineers pay special attention to the expansive soil problem and implement measures to mitigate negative effects.

Despite these warnings, expanding soils have already affected the integrity of the airport's buildings. Soon after being occupied in April 1993, the FAA's TRACON building—the command and air traffic control center—developed cracks in the walls and foundations (Chandler 1993). Expansive soils also caused concrete floor slabs in the building to heave upward, while cracks in the runways and in the concourses were occurring (Kowalski 1993). It appears that the expanding soil problems may not have been adequately addressed during construction.

Environmental impacts

Environmental factors have an enormous bearing on where airports are sited, and in fact, whether they can be built at all. Many airports and other large-scale projects have been changed, slowed, or stopped due to findings of significant environmental impact through regulatory and planning processes designed to determine the nature and extent of impacts caused by the proposed project. Given the global sensitivity to environmental factors, new airports must be planned carefully so as to minimize negative effects on the natural environment as much as possible (DOT 1985).

Much of the current emphasis and concern about the natural environment stems from a reawakening of environmental awareness in the 1960s and early 1970s. It was during this time that influential books, such as Rachel Carson's *Silent Spring* and Ernest Schumacher's *Small Is Beautiful*, and the photographs of the earth taken by the Apollo astronauts from space underscored the importance of protecting our planet's environment. As a result, increased environmental consciousness led to growing political movements in the United States and elsewhere calling for major changes in attitudes regarding the primacy of urban and industrial development and for more regulations that would seek to preserve, defend, and protect the natural environment.

Environmental impact assessment process

In the United States the growing environmental movement led to promulgation of the National Environmental Policy Act (NEPA) of 1969, which still remains one of the most sweeping and powerful pieces of environmental legislation. A number of other acts regulating clean air, clean water, coastal zones, and other natural resources were also promulgated during the 1970's "federal environmental decade." NEPA created the Environmental Protection Agency (EPA) and further mandated that all projects receiving any federal funding or involving federal land be subject to a process involving submission and review of an environmental assessment (EA) and/or an environmental impact statement (EIS). Any major federal action significantly affecting the quality of the human environment requires the preparation of an EIS. The purpose is to ensure that any and all possible environmental impacts are identified, publicized, and assessed to determine the nature and extent of the threat to the relevant human and biotic communities. It is largely through this federally mandated process that public projects are legally open to scrutiny by environmental and other concerned groups or ordinary citizens.

The environmental impact statement is drafted by the agency responsible for the project and is circulated for official and public comment. After comments are received, the lead agency produces a final draft, receives more comments, and then renders a "record of decision" regarding the proposed project, explaining its rationale for the decision. The most prevalent benefit of this process is the identification and mitigation of environmental impacts for approved projects; in some cases, such as the proposed Two Forks Dam in Colorado, the lead agency may decide to abandon the project because of its excessive negative impacts. The EIS process also permits groups or individuals to take legal action to slow or stop the project through claims of procedural or other flaws in the impact assessment study. Several large projects, such as the proposed Tellico Dam in Tennessee (Reisner 1993, p. 325) or New York City's Westway Highway, as well as numerous nuclear power facilities, were stopped due to environmental concerns (Levy 1994, p. 248). Above all, the EIS process is one of the few specifically organized methods by which dubious or controversial projects can be scrutinized and possibly derailed by the public or by well-organized interest groups.

Thus, recent environmental legislation has done much to open up planning to increased citizen participation. It has adopted the view that "open air and sunlight are the best disinfectants" for rooting out bad projects and combating corruption and infestations occurring at the public trough. Today, of course, citizen participation is widely recognized by most local and regional planning agencies. Citizen advisory groups and formalized public participation are incorporated as part of the planning efforts, but it was not always that way. The entire NEPA process has done much to enlarge the scope of public participation and has been effective in allowing citizens to actually stop projects that are deemed to have significant negative impacts, environmental or otherwise. The specific public participation mechanisms and procedures that were implemented in the case of DIA, and their effectiveness, were discussed in Chapter 3.

Such participation was explicitly mandated for airport projects by the Airport and Airway Development Act of 1970, which called for the U.S. Federal Aviation Administration (FAA) and the U.S. Department of Transportation (DOT) to embrace the policies inherent in NEPA and to organize a process whereby any new or expanded airport project has to comply with environmental, ecological, and social impact requirements. The U.S. DOT has subsequently identified policy guidelines for considering environmental impacts, as follows (FAA 1986):

1. Avoid or minimize adverse environmental effects wherever possible.
2. Restore or enhance environmental quality to the fullest extent practicable.
3. Preserve the natural beauty of the countryside, public park, recreational lands, wildlife and waterfowl refuges, and historic sites.
4. Preserve, restore, and improve wetlands.
5. Improve the urban physical, social, and economic environment.
6. Increase access to opportunities for disadvantaged persons.
7. Utilize a systematic, interdisciplinary approach in planning and decision-making that may have an impact on the environment.

One of three types of environmental studies, depending on the extent of the proposed project, must be completed prior to final project authorization. If the proposed action is not expected to have a major impact, then a finding of no significant impact (FONSI) must be completed and made available to the public on request. If the project is expected to have some impacts, then an environmental assessment is undertaken that may serve as the basis for decisions regarding compliance with environmental requirements. For larger projects with significant impacts, an environmental impact statement must be approved that addresses:

1. The purpose of the project
2. Probable impacts on the environment
3. Mitigation for adverse impacts
4. Identification and evaluation of alternatives to the project (including the "do-nothing" alternative)
5. Identification and consideration of the loss of irretrievable resources (Ashford and Wright 1992, pp. 479–483).

In airport planning, the FAA is the lead agency which is responsible for producing the environmental impact statement, including incorporating public commentary and rendering final approvals.

The new DIA project, which relied on federal funding, involved federal agencies, and used federal government land, was most certainly subject to approval through the NEPA process. In fact, Denver was the first new airport to be built in the United States that was entirely subject to the NEPA guidelines (the Dallas/Fort Worth Airport, completed in 1974, was largely planned prior to NEPA, and thus was not subject to the same process). Once the City of Denver had determined its plans to go ahead and build a new airport, it hired James (Skip) Spensley, an environmental attorney, to head its New Airport Development Office. His consulting firm, Spensley and Associates, was contracted to oversee environmental impact assessment related to the airport. Early in the planning stages, Spensley negotiated with the FAA to organize the environmental approval process whereby the City and County of Denver would conduct an extensive environmental assessment to serve as a precursor to the FAA's official environmental impact statement. In this way, the city could conduct its own assessment in the same manner as an EIS and provide much of the background findings to the FAA for the actual EIS it was required to complete. To carry out the environmental assessment, the city hired several additional consulting firms to handle the more specialized impact studies, for example, Harris, Miller, Miller, and Hanson for noise monitoring.

Environmental impacts of airports

Because of the land they consume and the activities they support, airports are major contributors to human-induced impacts on the natural environment. Chief among these are impacts from aircraft and airport noise, as well as land-use development, air quality, water quality, hydrology, and ecology, including impacts on wildlife, waterfowl, flora, fauna, endangered species, and wetland and coastal zones. Related social impacts include displacement and relocation of households, businesses, and community facilities, and impacts on parks, recreational areas, historical places, archeological resources, and natural and scenic beauty (Ashford and Wright 1992, pp. 484–508; Horonjeff and McKelvey 1994, pp. 722–767). It is up to the airport's environmental planners to identify, measure, and assess each of these impacts for each potential airport site, select sites that have the least impacts, and develop strategies to mitigate whatever impacts will occur if a project is in fact pursued. The following subsections address in more detail some of the most critical environmental impacts of airports, especially those that are most relevant to the case of DIA.

Noise One of the most important and well-publicized environmental impacts of airports is noise (Stevenson 1992). As airports and aircraft have grown larger, through much of the history of aviation they have become noisier, and an ever-more sensitive public has grown more irritated as a result. Today, many older airports find themselves encroached upon by urban development and thus are experiencing an increasing problem regarding noise impacts on the surrounding community. Noise has become an important reason why new airports, generally located farther away from urban development, become a more attractive alternative instead of expanding in situ. Certainly this was the case in Denver (as discussed in Chapter 3), where residents from the nearby Park Hill neighborhood and surrounding municipalities vociferously opposed continued use and expansion of Stapleton and forced the city to consider more seriously the new airport option.

Excessive noise is disturbing to most people and can cause physical harm in extreme cases. Both behavioral and health/physiological

Mike Keefe, *Denver Post*. Reprinted by permission.

effects have been documented.[2] Behavioral effects include annoyance and interference with communication, mental activity, rest, and sleep. Health effects include loss of hearing and possible nonauditory impacts, such as cardiovascular disease and stress. Whereas it is known that prolonged exposure to excessive noise causes hearing impairment, the research evidence so far on nonauditory health is inconclusive, although concerns about this linkage have been raised. In one study, for example, Cornell University professor Gary Evans found that children living near Los Angeles International Airport had higher blood pressure readings than a similar socioeconomic group of children living in a quieter neighborhood (Evans and Lepore 1993). More recently, Evans concluded a before-and-after study around the old Munich airport and found that when children were exposed to more noise, they registered higher levels of the hormones that increased blood pressure (Evans and Lepore 1993; Kowalski 1995b).

The FAA and the International Civil Aviation Organization (ICAO), as well as the Environmental Protection Agency (EPA) and the Department of Housing and Urban Development (HUD), have all acknowledged the harmful effects of noise on people and have issued

2. Much of the next several paragraphs is based on the noise-impact discussion in Horonjeff and McKelvey (1994, pp. 724–762).

regulations and guidelines dealing with noise exposure. There are many ways to measure noise impact, but these agencies rely on a quantitative measure known as "day-night average sound level," commonly referred to as DNL or Ldn. This represents a 24-hour average sound level measured in decibels (dB), weighted an extra 10 dB for noise occurring between 10 P.M. and 7 A.M. The weighting tries to account for the increased human sensitivity to noise during nighttime hours. The FAA has used this measure in issuing guidelines specifying what land uses are compatible with certain noise exposure categories. For example, residential land use is not compatible in areas that have an Ldn of 65 dB or greater, and thus should be prohibited in those areas. Commercial, manufacturing, and recreational activity guidelines are not as stringent as residential, but have maximum Ldn limits ranging from 70 to 85 dB. Almost all land-use types, with the exception of transportation, agriculture, forestry, and mining, are prohibited in areas with Ldn greater than 85 dB. The FAA uses computer-based simulations from its integrated noise model (INM), which generates maps showing noise contours around airports, indicating areas of similar noise exposure (Fig. 6-4). Thus, local land-use planning jurisdictions might require that no new residential uses be built within the 65-Ldn contour specified by the computer simulation model, even though the FAA has not enforced this at the local level.

Even though the FAA, ICAO, and other agencies have chosen Ldn as the most appropriate measure to be used in establishing noise guidelines and regulations, there are some problems with its use. First, it must be recognized that human perception of noise can be very different from calculated noise, whether using the Ldn or any other measure. People can respond differently to the same measured sound; what is tolerable to some can be excruciatingly annoying to others. Confounding influences, such as the level of surrounding background noise, further complicate the issue. If background noise is high, people may not notice the aircraft noise as much. Likewise, when background noise is low, even relatively low levels of noise may be perceived as excessive. This is especially true when new flight paths or a new airport begin to affect areas that previously had little or no noise impacts. As we shall see below, this is precisely what happened in the case of DIA.

Another problematic aspect of using Ldn to measure noise is that it is an average sound level, and thus may not properly convey the severity of individual noise events. An Ldn is also a cumulative measure

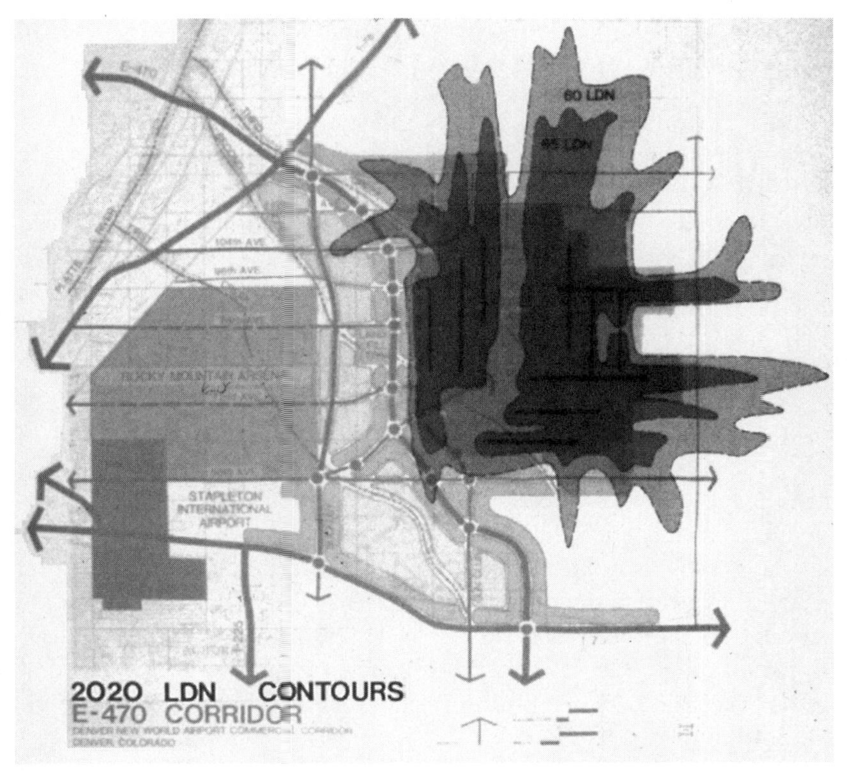

Fig. 6-4. *Computer-generated noise contours were developed and plotted according to the various alternative configurations before deciding on a final runway configuration for the new airport.*

over the course of a day, so that different mixtures of loud and quiet aircraft can produce the same Ldn, even though the timing and pattern of events can be radically different. Thus, the Ldn noise contours measured around airports may only provide a rough approximation of the noise impacts that residents actually experience.

These problems with measuring, tracking, and regulating noise impacts suggest some of the difficulties associated with planning and mitigation efforts aimed at avoiding impacts due to aircraft and airport noise. The most obvious way to avoid noise impacts is to locate airports in areas that are so remote from population centers that no one or very few will be affected. This approach must be balanced, however, with the economic locational imperatives of airports, which suggest convenient accessibility to local users and potential development opportunities near airports. Inevitably, some balance between the two must be struck, and it probably will involve *some* inconvenience in accessibility and *some* impact from noise on *some* local residents.

Because airports have had to contend with increasing public sensitivity to noise, a whole range of mitigation strategies has been developed, which include day and/or night limitations on noisy aircraft, alterations in flight paths and runway utilization, construction of noise barriers, and public involvement in helping to solve noise problems. More explicitly, the FAA has enacted a number of noise-specific Federal Aviation Regulations (FARs) that address several mitigation strategies. For example, FAR Part 36 sets noise standards for aircraft certification purposes, while FAR Part 91 uses those standards to regulate aircraft operation within the United States. In 1977, the standards were upgraded, and aircraft were categorized as either Stage 1 (older aircraft built when no standards existed; these aircraft have all been retired in the United States), Stage 2 (aircraft that meet 1969 standards but not the more stringent 1977 measures), or Stage 3 (newer aircraft that meet or exceed 1977 standards). The Airport Noise and Capacity Act of 1990 amends FAR Part 91 by calling for all Stage 2 aircraft to be replaced by quieter Stage 3 aircraft by the year 2000. This ongoing process should help to mitigate aircraft and airport noise impacts.

But airports in the United States and around the world continue to experience increasing volumes of traffic and increasing numbers of complaints due to noise. Residents living near the John Wayne (Orange County, California) Airport, for example, have forced substantial changes in aircraft and airport operations for the purpose of reducing noise impacts. Pilots have to perform unusual takeoff and landing procedures that involve steep ascents and descents when using the Orange County airport. Serious noise restriction measures are also in place at Washington National, Chicago O'Hare, and New York LaGuardia Airports. After only 20 years since its construction, residents now living near Dallas/Fort Worth International Airport have started to complain about excessive noise impacts from increased air traffic. In Sydney, Australia, increased noise from more air traffic and the expansion of Kingsford-Smith Airport have led to protests in which the airport was boycotted and the runways were taken over and rendered inoperative by a militant segment of the public. Even though these are some of the more extreme examples, virtually every city in the world with a major airport has a "noise problem" that must be confronted on an almost daily basis.

By comparison, the situation in Denver seems almost tame Nevertheless, as was discussed in Chapter 3, a lawsuit filed in 1981 by residents of the Park Hill neighborhood of Denver (Fig. 6-5) identified a noise problem at Stapleton Airport that eventually became frequently cited as one of the reasons why a new airport needed to be built. The lawsuit charged that operations from Stapleton regularly exceeded the maximum decibel levels under a Colorado noise abatement statute (*Denver Post* 1981). Partly as a result of a series of legal victories for the plaintiffs, the City of Denver eventually settled the lawsuit in May 1985 by agreeing to reduce takeoffs over Park Hill, reduce all evening flights over the neighborhood, require planes to climb higher before turning, ban certain types of aircraft from using the airport, and hire an airport "noise officer," among other measures (*Denver Post* 1985). But most importantly, the settlement included a provision for Denver to close Stapleton to all aviation activity once a new airport (agreed upon in the Memorandum of Understanding signed in January 1985) was built, a provision that the city deliberately included to ensure that Stapleton would remain closed and never endanger the financial viability of DIA (see Chapter 3).

Residents in neighboring Adams County felt similarly. Time and again, the threat of environmental lawsuits due to excessive noise impacts was used as the principal weapon by commissioners from Adams County seeking to prevent Denver from expanding Stapleton. It was largely the credibility of these threats that forced Denver to take Adams County's concerns seriously and led to the decision to pursue more vigorously a new airport alternative.

As part of the new airport environmental assessment process, expected operations at Stapleton as well as each new potential airport site were evaluated in terms of anticipated noise impacts. Based on 1987 activity levels, it was estimated that nearly 14,700 people resided within the 65 Ldn contour at Stapleton. This figure actually represented an improvement from the approximately 34,000 people exposed to Ldn 65 or greater prior to 1982, largely due to the noise mitigation efforts at Stapleton. The 1995 impacted population was expected to be roughly equivalent to the 1987 figures. The Stapleton expansion into the Rocky Mountain Arsenal alternative was projected to affect more than 46,000 residents within the 65 Ldn contour by 1995 (Denver 1988).

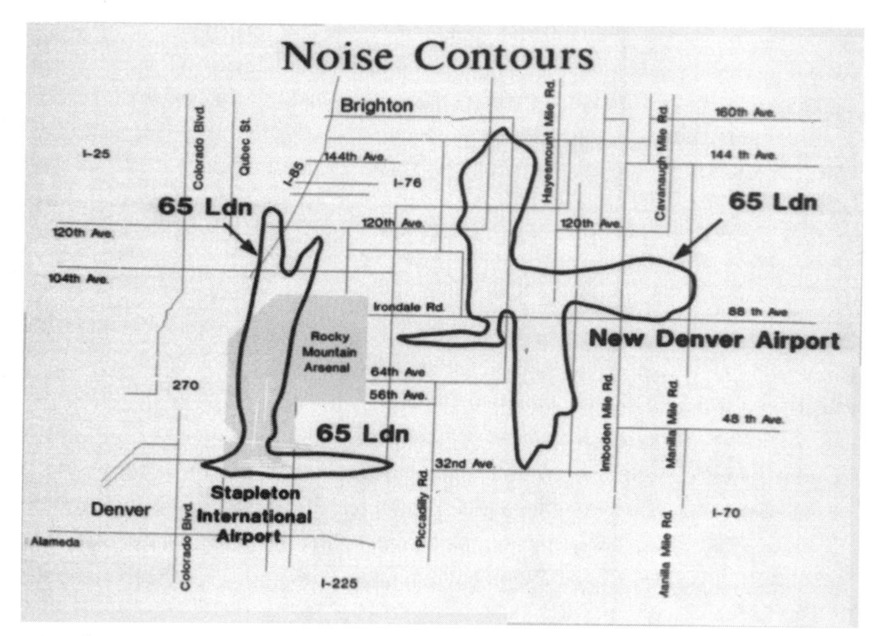

Fig. 6-5. *While aircraft noise generated by Stapleton International Airport affected a significant number of suburban residents (about 15,000 people resided in the 65 Ldn area), the noise impact from DIA was expected to be virtually nil. The expectation was false, of course, since thousands of suburban residents have complained about DIA's noise.*

Alternatively, each of the completely new sites under consideration was shown to reduce substantially the number of people subjected to high noise levels. From 629 people within the 65 Ldn contour at the B.2 Front Range site to only 48 at the C.1 Lost Creek site, it was clear that any of the new sites would be vastly preferable to the existing Stapleton or Arsenal expansion alternatives in terms of these impacts. Eventually, once the actual site of DIA was fixed, it was estimated that 559 people would be residing within the 65 Ldn contour, but the City of Denver would purchase these residences and relocate the occupants (Denver 1988). Thus, Denver appeared to solve its noise problem with the move to a new airport. But this was not to be the case. Opening the new airport represented the beginning of a new round of noise problems.

On April 21, 1988, less than a month before the Adams County airport site referendum, the City of Denver and Adams County conducted an intergovernmental agreement (IGA) that included provisions to safeguard Adams County neighborhoods against noise impacts from DIA. The two main elements of this agreement were:

1. To monitor the location of the actual 65 Ldn contour to ensure it does not expand beyond the originally projected contour.
2. To enact noise exposure performance standards (NEPS) at 101 sites in three noise-sensitive areas of Adams County to ensure that actual noise levels do not exceed projected levels.

The projected 65 Ldn contour and the NEPS were derived from a simulation using the FAA's integrated noise model (INM). The NEPS at each of the 101 sites were established using the Leq measure of noise, which is the same as Ldn except that it does not include the additional 10 dB penalty for nighttime noise.

Whereas monitoring the 65 Ldn contour is necessary to comply with the FAA's final environmental impact statement, the specified NEPS were different. To compel Denver to abide by the NEPS, the intergovernmental agreement established a schedule of fines for violations, amounting to $500,000 for each site found to be more than 2 dB over projected levels measured over an annual period. The NEPS that were established and agreed upon in April 1988 have Leq levels ranging from a high of 51.7 dB for a site near Barr Lake down to only 31.4 dB at a site near Buckley Air National Guard base (DIA 1995a). The City of Denver agreed to these noise levels even though some technical experts acknowledged that compliance would be difficult to maintain. Ostensibly the city felt compelled to agree to these standards to mollify the concerns of Adams County residents on the eve of the important new airport referendum. Many Adams County residents, however, feel that the process still allows too many loopholes in proving a noise violation and that Denver will be able ultimately to circumvent any punitive actions (Kowalski 1995a).

Nevertheless, with these agreements in place, the new airport was approved in Adams County, and, after all of the implementation difficulties documented elsewhere in this book, DIA finally opened on February 28, 1995. Right from the beginning, however, noise complaints from residents started to pour in. In the first week of operation, 277 noise complaints were registered, which represented the lowest weekly total during the new airport's first year. The number of complaints grew exponentially, reaching a high of 3840 during the week of September 10, 1995, before stabilizing between 2000 and 3000 per week. By December 2, 1995, DIA had totaled an astounding 66,742 noise complaints, compared with 431 at Stapleton during all of 1994 (*Airports* 1995; DIA 1995a)!

Why has there been such an outcry from residents concerning noise at DIA when the airport's remote site was so carefully selected to minimize noise impacts? Were the 65 Ldn noise contours incorrectly simulated, or is Ldn not capturing the full extent of the noise impacts upon the community? Could the outcry represent more than just dissatisfaction with aircraft noise?

Noise technicians were stunned at the public backlash concerning DIA. Actual monitoring of the 65 Ldn contour around DIA illustrates a relatively close conformity with what was projected, with two notable deviations. First, some residential areas to the northwest of the airport, south of Barr Lake (Lake Estates and the Van Aire subdivision) were actually within the 65 Ldn even though original projections indicated they would not be. Second, two residential areas to the northeast (Vantage Estates and the Great Rock Estates) were also receiving noise in excess of 65 Ldn. This occurred because aircraft taking off to the north were starting their turns to the west or east earlier than the original FAA-approved flight paths required. To remain in compliance with the EIS and with Adams County's IGA, the FAA and the city need to make sure the actual departures conform to the original flight paths called for in the EIS.[3]

Even though many protests have been received from these residential communities that have unduly suffered from aircraft noise greater than 65 Ldn, the sheer volume of complaints suggest that they are not the only areas that have been affected. Indeed, many have come from areas that are 5 to 30 miles away from DIA, in communities such as Aurora (southeast of Denver, 10 to 15 miles southwest of DIA), Westminster and Northglenn (north of Denver, 20 miles west of DIA), Parker and Elizabeth (southeast of Denver, 20 miles south of DIA), Nederland (northwest of Denver, just west of Boulder, more than 30 miles northwest of DIA), and other areas in the foothills west of Boulder. In these cases, new flight paths into or out of DIA are creating aircraft noise where none or little existed previously. Noise monitoring operations in these areas indicate that

3. It should also be noted that the noise contours developed in the final EIS were based on operations using six runways, not five, as is presently the case. Without the sixth runway, the takeoff and landing patterns are different, resulting in different noise impacts.

individual noise episodes may be quite loud, but 24-hour averages are still well below the 65 Ldn threshold. Apparently, levels of background noise in these areas are so low that even moderate levels of aircraft noise irritated their inhabitants.

Not all of the people living in these areas appear to be equally affected by noise impacts, because the total number of inhabitants is far greater than the number of households filing complaints. The first 58,000 complaints were registered by just over 3000 households, thus each of these households is averaging nearly 20 calls. Noise-complaint tracking also indicates that a small number of people have been engaged in an organized calling campaign and are responsible for the majority of the calls (*Rocky Mountain News* 1995b,c). Ten families account for one-fourth of all complaints, and one household in Elbert County has accounted for 11 percent of all complaints received (*Rocky Mountain News* 1995a).

Still another major headache that the City of Denver faces regarding noise impacts is compliance with the IGA with Adams County, in particular, the NEPS. During the airport's first year, noise monitoring operations at DIA revealed that roughly 75 percent of the 101 NEPS sites were not in compliance with the IGA (DIA 1995a). The yearly average noise level at these sites did not achieve compliance by DIA's first anniversary, February 28, 1996. The city, however, has an additional year to bring these levels into compliance. If it is unable to comply, the City of Denver could be facing fines ranging between $35 and 40 million, payable to Adams County.

In August 1995, Denver mayor Wellington Webb and Adams County commissioner Elaine Valente wrote a joint letter to Department of Transportation Secretary Peña asking him to force the FAA to adopt more vigorous noise-abatement procedures. They reminded Peña of his pledge when he was mayor of Denver that "no existing residential areas in Denver, Adams County, or Aurora will be adversely affected by noise" (Kowalski 1995c). Peña responded by meeting with local and FAA officials in Denver in late August 1995.

To address the expansion of the 65 Ldn contour, the noncompliance with the NEPS sites, and the huge number of noise complaints, the City of Denver previously came up with nine specific noise-abatement recommendations in July 1995. Subsequently, the FAA decided to implement the agreed-upon measures. They include rerouting departure and arrival paths to minimize noise impacts on the most affected communities, reducing the number of nighttime flights, maximizing the use of certain runways (especially the runway heading east), rerouting 107 Stage II aircraft departures to the east instead of the west, maintaining stricter adherence to flight paths originally outlined in the final EIS, defining a standard departure profile that requires aircraft to gain altitude more quickly, and providing strong support for construction of a sixth runway, which is expected to distribute noise impacts in a manner more consistent with the final EIS (DIA 1995b). As of December 1995, roughly half of these nine measures had been implemented, while the others were being phased in (McKee 1995). As these mitigation efforts were fully implemented, noise impacts were reduced dramatically. As older and noisier Stage Two aircraft flights were rerouted away from populated areas in March 1996, complaints dropped sharply. The number of violations carrying fines dropped from 69 to 2, and potential fines fell from $34.5 million to less than $1 million (Flynn 1996). Still, noise impacts from DIA continued to be a political and public relations issue for the airport.

Air quality Another major environmental concern related to airports is air quality. Airports increase urban air pollution due to aircraft engine emissions, aircraft fuel venting, and motor vehicles carrying passengers, visitors, and employees to and from the airport. Aircraft fueling systems, ground service equipment, airport heating operations, and construction operations can also contribute to air pollution problems near airports (Ashford and Wright 1992, pp. 502–503).

The 1977 Clean Air Act Amendments established national ambient air quality standards (NAAQS) for six major air pollutant categories (carbon monoxide, hydrocarbons, nitrogen oxides, sulfur dioxide, suspended particulates, and ozone and other photochemical oxidants), and required states to formulate state implementation plans (SIP) to meet those standards (Ashford and Wright 1992, p. 478; Horonjeff and McKelvey 1994, p. 722). Planning for any federally supported project, including airports, must now fully document in an environmental impact statement how the proposed project will impact air quality and how states or metropolitan areas will be able to comply with the air quality standards. Although much progress has been made in reducing emissions of air pollutants since the 1970s, several U.S. metropolitan areas have had difficulties in meeting these standards and thus have been categorized as nonattainment areas.

This designation places these metropolitan areas at risk of sanctions, and, as a result of the recent 1990 Clean Air Act Amendments and the 1991 Intermodal Surface Transportation Efficiency Act (ISTEA), nonattainment can mean ineligibility to receive federal funds for new transportation projects such as highways or airports.

A key aspect of the ISTEA legislation is its emphasis on linking transportation planning with environmental protection. In this regard, it complements the 1990 Clean Air Act Amendments (CAAA), which identified the automobile as a major polluter and explicitly holds transportation planning liable for meeting air quality goals. Federal matching funds for transportation projects can be withheld from metropolitan areas that do not attain air quality standards. The ISTEA legislation provides considerable funding for transportation projects but requires planners to construe transportation planning more broadly and to explicitly consider air quality and land-use planning as part of the process for project approval (Hanson 1995, p. 22). Lisa Wormser of the Surface Transportation Policy Project, which lobbied for many of ISTEA's provisions, has referred to ISTEA as the "carrot," and CAAA as the "stick" (Plous 1993). These two pieces of legislation represent a major shift in American urban transportation planning and will make it more difficult to build new highways or expand existing ones in years to come.

The Denver metropolitan area has experienced severe air quality problems over the last several decades, and has been a nonattainment area for carbon monoxide and particulates. As population, economic activity, and motor vehicle traffic have all increased rapidly in the region, problems with air pollution have grown much worse. Because of topography and climate, Denver suffers from the infamous "brown cloud" problem, a layer of particulates and other pollutants that forms over the CBD, the industrial north side, and the South Platte River valley. The problem becomes worse during the winter months, when temperature inversions are more frequent. Cold air in the valleys is sealed off by warmer air at higher elevations, which acts to trap automobile exhausts and other pollutants near the ground. After several days of wind-free conditions, air quality in Denver can deteriorate seriously, causing a soupy, brown cloud to hang over the city. Some progress has been made in cleaning up Denver's air, but the region has had difficulty in maintaining NAAQS, especially for carbon monoxide, ozone, and particulate matter (FAA 1986). "Red air pollution alert" days, when driving is discouraged and wood-burning bans are in effect, still occur quite frequently.

Given the concerns over air quality in Denver, a plan that proposed to build the largest airport in the world at a distance of 24 miles from the CBD was bound to become highly controversial. Denver area residents have become very sensitive with regard to air quality and, at various times in the recent past, have ranked it as the most serious problem facing the region. Furthermore, environmental protection has been a consistently strong philosophical and political theme throughout Colorado. In 1972, Denver and Colorado became the first, and only, host community in history to refuse the Olympic Games (the 1976 Winter Olympics), largely inspired by concerns over growth and environmental degradation in the nearby Rocky Mountains. The leader of that movement, former Governor Richard Lamm, also was able to stop or slow other growth-oriented projects, such as the proposed I-470 circumferential beltway around Denver (Fig. 6-6). The new airport, with all the new growth it promised to bring, was anathema to those who had supported Lamm and who feared further degradation of the region's natural environment. But by the mid-1980s, the region's economy had already hit the skids, and politicians such as Governor Roy Romer and Mayor Federico Peña ushered in a new mindset that emphasized economic development and job creation through major public works projects. To be sure, environmental issues still mattered, but the political and business elite were determined that environmental issues would not impede or slow down the drive to build a new airport.

Yet preliminary studies showed that a new airport would become a giant new pollution source because of the increased aviation activity it would bring and the extra distance that cars and other motor vehicles would have to travel to reach the facility. Early estimates indicated six to eight times as much pollution as at Stapleton, with the new airport accounting for as much as 10 percent of the carbon monoxide pollution in the entire metro area by 2020, compared to 1 to 2 percent by Stapleton in 1985. These findings were, however, contested by other experts who indicated that the new airport's remote location would actually improve downtown air quality by dispersing air pollution over a wider region (Kowalski 1988). The new airport site was believed to be out of the downtown Denver air drainage basin, so that the extra emissions might not contribute to the "brown cloud." Perhaps more importantly, the new airport site

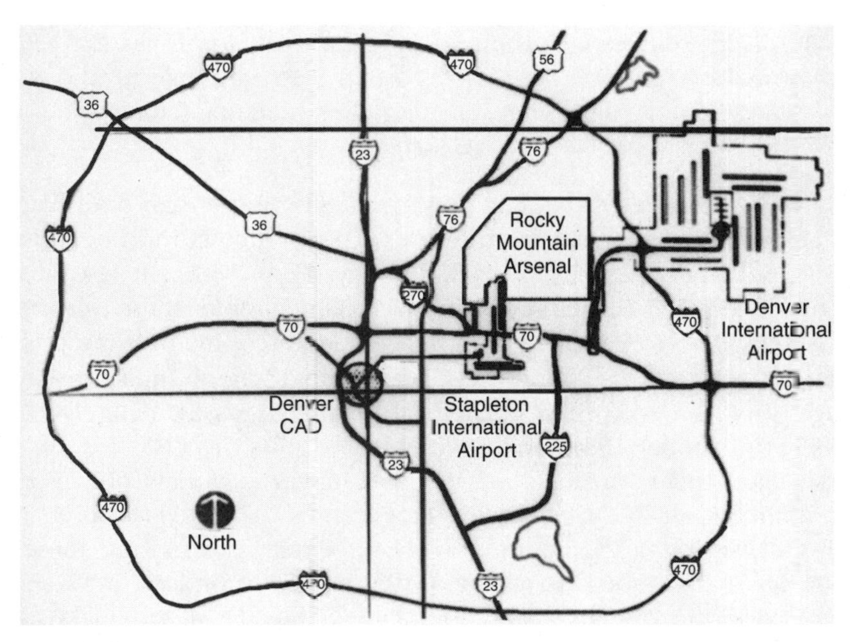

Fig. 6-6. *Eventually, a major beltway, E-470, is planned for Denver. It will provide better access to DIA from the northern and southern suburban communities. However, political opposition to its construction suggests that its completion is in jeopardy.*

was geographically located outside the area designated as nonattainment by the NAAQS.

Shortly after this fundamental contradiction between experts surfaced, the City of Denver's New Airport Development Office, in April 1988, unveiled parts of its environmental assessment. Preliminary results indicated that the new airport would be less of a polluter than Stapleton. Mayor Federico Peña announced that "if we continue to operate Stapleton . . . air pollution will be a much more serious problem than at a new airport" (*Denver Post* 1988a). Soon thereafter, however, several critical mistakes were uncovered in the air quality analysis that called into question the city's findings. These included:

1. Using a site closer to downtown Denver than the actual site of DIA's terminal thus understating total emissions from automobiles and trucks traveling the extra distance to DIA
2. Counting ground traffic only from the northeastern quadrant of the Denver metropolitan area to calculate total emissions

3. Using a different, and more stringent, study to predict air quality impacts for continued use of Stapleton than for the new airport, thus overestimating continued Stapleton's pollution (*Denver Post* 1988a).

The City of Denver responded in May 1988 by more than doubling its air pollution estimates for the new airport, from 115 to 274 tons of air pollutants per day by 2020 (*Denver Post* 1988b). It also acknowledged that this figure might pose major problems for Denver in meeting the NAAQS.[4] In June 1988, a federal scoping meeting was held in which the FAA and the City of Denver agreed that a new study was needed for the purpose of air quality assessment (DOT 1989). In October 1988, these two agencies plus the EPA, the Colorado Department of Health (CDH), the Denver Regional Council of Governments (DRCOG), and the Metropolitan Air Quality Council (MAQC) signed a Memorandum of Agreement committing these agencies to a comprehensive air quality modeling project "in order to provide the primary basis upon which the [FAA] can make a sound judgment about the impact of the proposed project on regional air quality" (DOT 1989). This study was to include emissions inventories for carbon monoxide (CO), nitrogen oxides (NOx), hydrocarbons (HC), and particulate matter (PM10), regional dispersion modeling for CO, PM10, and ozone (O3), and terminal area dispersion modeling for CO. In November 1988, the results of the study were released, indicating "that there would be no significant degradation of regional air quality directly or indirectly resulting from construction or operation of the proposed new airport, either during Phase I (1995), or in the year 2010." The study also found that "the construction and operation of the relocated airport will not delay the date of attainment or maintenance of the EPA's National Ambient Air Quality Standards (NAAQS) in the Denver Metro area" (DOT 1989).

But this air quality study too made some assumptions, most importantly that numerous proposed ground transportation projects would be completed when the new airport opened (Obmascik 1988). The projects included several lines of a new light rail system by 1995, the E-470 and W-470 beltways by 1995, and six new rail transit lines by

4. It should be noted, however, that the NAAQS are not based on measures of gross emissions. They are measures of concentration (parts per million) averaged over a specific time period (two-hour, four-hour, or eight-hour average periods). So, it was not necessarily evident that the city's increased emission estimate would result in nonattainment of NAAQS.

2010, including one out to the new airport. These assumptions allowed the study to conclude that traffic would continue to move as rapidly in 1995 and 2010 as it did in 1988, and that more people would be using transit, thereby significantly reducing estimated pollution emissions. Even so, the administrators of the study were uncertain whether these projects could actually produce the promised reductions in air pollution, because they said funding was not available to write new computer programs that would have determined these effects. This concern was somewhat moot anyway, because each of the projected transportation projects was extremely expensive (the total cost was estimated to be $11 billion), making it highly dubious that each one of them could actually be built in the assumed time frame (Obmascik 1988).

By 1995, a short light rail segment through downtown Denver and small parts of the E-470 beltway had been built. But the other rail lines and beltways assumed in the air quality study had not been built. The E-470 beltway may be finished by the early 2000s, but it is highly unlikely that the W-470 beltway or the rail transit lines will be built by 2010, if ever (see the section "Ground access" in this chapter). Essentially, this 1988 air quality study used a "wish list" of metropolitan area transportation projects to estimate future impacts. In fairness, many of these projects had to be included in the air quality analysis because many of the related agencies (for example, Federal Highway Administration) insisted on their inclusion. But the 1995 assumptions have proven to be flawed, and the 2010 assumptions are likely to be even more inaccurate.

Nevertheless, the FAA accepted the study's results and concluded in its draft EIS released in February 1989 that the new airport would not violate federal air quality standards (Kowalski 1989). The EPA did not immediately, however, endorse the findings mainly because of the nitrogen oxide emissions estimates. Bob Yuhnke, an attorney with the Environmental Defense Fund, also questioned the air quality analyses from the city and the FAA, claiming that "They . . . played a funny game with nitrogen oxide." Skip Spensley countered that Denver made the best determination it could on nitrogen oxides in noting, "We don't have any way of adequately modeling (nitrogen oxide). We have nothing that indicates that there's a violation of any standard" (Kowalski 1989).

Because of the "unique air quality situation in the Denver metropolitan region," the FAA and the EPA developed a special grant condition, which was agreed on by the City of Denver. It stipulated that continued federal assistance was contingent upon meeting mitigation responsibilities contained in an approved state implementation plan (SIP) for the purpose of addressing existing or predicted violations of air quality standards due to the new airport (DOT 1989). But the final EIS reiterated the FAA's earlier conclusion that the new airport would not significantly degrade regional air quality and that the project could go forward on this basis. But it still remains to be seen whether DIA will contribute significantly to the degradation of the region's air quality.

Other environmental impacts The most controversial environmental impacts from DIA were noise and air quality. Nevertheless, the final EIS had to address other important potential impacts, including water quality; land use; social impacts; surface transportation; induced socioeconomic impacts; historic/archaeological; biotic communities; endangered species; wetlands; floodplains and hydrology; coastal zone management, coastal barriers, and wild and scenic rivers; farmlands; energy supply and natural resources; solid waste impact; light emissions; construction impacts; and design, art, and architecture application (DOT 1989).

Some water quality problems existed at Stapleton due to an aging wastewater treatment system that was built to meet earlier, less stringent water quality standards. Increasing incidents of fuel line and storage tank leaks also compromised nearby water quality at Stapleton. The new airport was built to new water quality standards, thus representing an improvement over existing conditions. Discharges from the facility are controlled in receiving ponds, especially the capture and treatment of stormwater contaminated with de-icing agents (ethylene glycol) [Denver 1988].

Land-use impacts focused on the transformation of agricultural land to commercial, industrial, and residential land uses. Chief among concerns here was compatibility of airport vicinity land use with airport operations, particularly noise impacts. Effective land-use planning among the relevant jurisdictions was identified as critical to prevent noise-sensitive encroachment upon the airport (see Chapter 4 for more on land-use impacts).

Social impacts involved the acquisition of 186 parcels of land and the relocation of occupants of 152 residential units. Because of the limited population previously residing at the new airport site, significant problems were not anticipated and no significant problems emerged.

Two important induced socioeconomic impacts were identified as the redevelopment of Stapleton and secondary development around the new airport, thus creating new spatial patterns of development within the Denver metro area. Whether this was a concern or not depended on one's long-term vision for the region. Some viewed the airport, together with the E-470 beltway, as catalysts for vastly expanded decentralized development in the eastern part of the Denver metropolitan area. These assertions recalled the debates of the 1970s over growth, sprawl, and the "Los Angelization" of Denver. Concerns over air quality, water resources, and sustainability of growth in this region were the major themes. Others felt that growth was desirable and inevitable and that movement to the east would benefit the economic development of the metro area, particularly in the city of Aurora and Adams County. Furthermore, it was reasoned that air quality in Denver might be improved by dispersing air pollutants over a larger region. Much of this debate, therefore, centered on alternative views of decentralized growth and its costs versus benefits.

Other induced socioeconomic impacts mentioned were the operation of nearby Front Range Airport and power transmission line relocation. It was acknowledged that Front Range Airport would be severely impacted and would have to reconfigure its operations to be compatible with DIA. Some difficulties were also encountered with relocating a 230 KV electric transmission line that ran right through the middle of the new site. Denver, Adams County, and Aurora went through a messy negotiation to reroute the line.

Regarding endangered species and other wildlife, there was some concern about effects on bald eagles and their nesting areas at nearby Barr Lake State Park. The U.S. Fish and Wildlife Service found that "development of the Airport will not jeopardize the continued existence of the Bald Eagle. However, primary and secondary construction and general urbanization in and around the project area will affect Bald Eagles and their habitats" (DOT 1989). Agreements were reached among the Wildlife Service, the FAA, and the City and County of Denver on conservation actions to minimize the worst effects on bald eagle habitats. No negative effects were identified for the black-footed ferret or whooping cranes. Some impacts on the habitats of sandhill cranes and white pelicans were identified, but the Fish and Wildlife Service felt that these species would not be adversely affected to any significant degree.

The other impacts detailed loss of acreage of biotic habitat, wetlands, floodplains, creek channels, and farmlands. Some oil and gas wells were located on the site. There were only a few historical sites and archaeological properties identified by the State Historic Preservation Officer (SHPO) at the new airport site. Based on conditions of possible future recovery and other recommendations, the SHPO and the Council on Historic Preservation agreed to a finding of no adverse effect. Native American groups also approved the new airport site in terms of lack of burial grounds or significant artifacts.

Ground access

Ground accessibility to and from airports is an extremely important aspect of urban transportation planning that deserves more attention. As it has become faster to travel from city to city by air, the ground transportation portion of the journey has become a larger part of the total time expended in getting to the final destination (Ashford and Wright 1992, p. 418). Thus, savings from reductions in air travel time may mean little if ground travel time increases substantially. Of course, this is a function of the entire transportation network of a metropolitan area or a large region, but ground access must be considered in any holistic assessment of air transport and airport benefits.

In this regard, recent policy initiatives toward promotion of intermodal transportation are quite relevant. The 1991 Intermodal Surface Transportation Efficiency Act (ISTEA) provided $151 billion in transportation system improvements with an emphasis on more "flexible" spending for different modes, including transit, bicycle paths, and pedestrian walkways (Plous 1993). Intermodalism also means improving connectivity and efficiency of the linkages between and among different modes. Intermodal projects include such things as park-and-ride facilities, transit links to airports, and multimodal stations that allow passengers to transfer easily from one mode to another (Plous 1993). Classic examples include many European cities like London, Amsterdam, and Frankfurt, which provide good-quality, multimodal access to their airports. U.S. cities are notoriously poor when it comes to intermodalism, since the overwhelming dominance of private automobiles and interstate highways has stunted the development of a more balanced transportation system.

Today, despite the ISTEA legislation and the Clean Air Act Amendments of 1990, most U.S. cities still consider the problem of ground access to airports as one of providing enough highway or roadway capacity to the site so that private automobiles can be readily accommodated. Of course, there are a range of other modes that can be used to improve ground access to airports, including conventional railway (intercity systems), urban rail transit (heavy rail or light rail), urban bus transit, bus charters, special buses or van service, limousines, taxicabs, and even helicopters or waterborne modes (Ashford and Wright 1992, pp. 423–431). Likewise, not all ground traffic to an airport is attributed to airline passengers. A considerable number of travelers to an airport can be senders and greeters, airport employees, air cargo access personnel, persons who supply services to the airport, and other visitors (Ashford and Wright 1992, p. 419). In this way, airports should be viewed by urban transportation planners as major activity centers. Considering airport employment alone (London Heathrow at 48,000, Frankfurt at 41,000, and Los Angeles at 35,000), airports constitute major employment centers that are larger than some central business districts (Ashford and Wright 1992, pp. 420–421). Thus, ground access to airports has to be planned and built in coordination with the entire urban transportation system.

Recent international airports: Ground access considerations

Before discussing the ground access situation at DIA in detail, it is appropriate to summarize the existing situation for some other recent airport projects around the world.

Munich's Franz Josef Strauss Airport

Munich had the foresight to link the airport to its center city, not only with its ubiquitous autobahn system, but with direct subway routes. Connections to the central city railroad station take 38 minutes, with trains departing every 20 minutes, and the passenger paying only about $5 one-way (Upton 1992). In contrast, the new airport is a $60 taxi ride from downtown Munich (Mecham 1992). Planners anticipated that 40 percent of the airport's passengers would arrive by train, while 60 percent would take buses, automobiles, or taxis. Train and bus passengers are able to check their bags in the railway station and proceed directly to their gates (Mordoff 1988). However, unlike Frankfurt, Munich's new airport does not have access to Europe's intercity rail system.

Osaka's Kansai Airport

Kansai International Airport has highway, light rail (linking to a bullet train), and express ferry connections (*Kyodo News International* 1994). The train runs directly into the terminal, allowing passenger movements from the airport to Osaka city center in 29 minutes. Kansai actually serves Osaka, Kobe, and Kyoto, which are western Japan's major distribution center, Japan's largest port, and one of the nation's leading tourist destinations, respectively (*Aviation Week & Space Technology* 1994b). A marine terminal will allow high-speed passenger boat service between the airport and Kobe in 28 minutes. Air terminals will be established in Kyoto, Osaka, and Kobe to allow passengers to check in before reaching the airport (Ogawa 1993).

Macau International Airport

A new four-lane bridge links the airport island of Taipa to the Macau peninsula, making the airport about a 10-minute drive from downtown Macau (*Travel Weekly* 1995). A second inter-island bridge is under construction to handle airport traffic (*Asian Wall Street Journal* 1994). The north side of the airline terminal will include a ferry terminal for transfer to Hong Kong, about 40 miles away, or mainland China at Guangzhou, about 70 miles away (Donoghue 1995). Highway and railway lines are anticipated to link the airport to the Shuhai special economic zone and Guangzhou (Paisley 1995). The railway station, located west of the air terminal, will block expansion in that direction. Dramatic improvements in rail and road infrastructure in the Pearl River basin are anticipated, which should feed traffic into the Macau airport (*World Airport Week* 1995d).

Hong Kong's Chek Lap Kok Airport

Hong Kong's new airport, built on an island, will be linked to the city by tunnel, as well as a 4475-foot-long suspension bridge (*Orlando*

Sentinel 1995). It will be tied to the city by railroads, subways, highway tunnels, and container ship terminals (Mufson 1994). More than 40% of the airport's passengers are expected to arrive and depart by rail (Darmody 1993). The road and bridge infrastructure are being built by the Mass Transit Railway Corporation; a private sector franchisee will build and run the Western Harbour Crossing (Mok 1993).

Adjacent to the terminal building will be a transportation center, which will contain arrival and departure platforms for the high-speed rail to the city's urban and business districts. It will also serve as a bus depot, and taxi and rental car facility. A ferry terminal will provide sea access to Hong Kong, Macau, and points along the coast of southern China and the Pearl River estuary (Mok 1993).

New Seoul International Airport

A new 54.5-kilometer expressway, a high-speed railway, and a suspension bridge are being built to link the New Seoul International Airport with Seoul's industrial area and its seaport (*World Airport Week* 1995a). The upper level of the 4.4-kilometer suspension bridge will be used for automobiles and trucks, while the lower level will be used for rail traffic (Shin 1993). The rail line will link with the Seoul-Pusan high-speed railway, which is under construction. Seaport facilities will be constructed to allow intermodal cargo movements (*World Airport Week* 1995a).

Bangkok Nong Kgu Hao International Airport

The new Bangkok Nong Kgu Hao International Airport is being built 18.6 miles east of central Bangkok, near the Bangkok-Pattaya highway (*World Airport Week* 1995c). Bangkok's existing airport has extremely poor surface access. As John Meredith observes:

> *In Bangkok, heavy road congestion often makes journeys to and from the city longer than the flight time to adjacent Asian centers. Surface access is often one of the most neglected aspects of airport development.*
>
> *For the passenger journey does not begin or end at the airport. At both ends of the journey passengers have to go through an often lengthy and arduous process involving travel to and from the airport. If the passenger's journey is to be rapid, efficient and comfortable, it is essential that*

> *ground links to and from airports are improved to match the increasing capacity planned for airways and airports. Failure to make these improvements will simply transfer congestion in the skies, on runways, or in terminals to the road and rail links that serve airports (Meredith 1995).*

The government is planning a road to link Bangkok's two airports, since Don Muang will likely become the nation's primary domestic airport, while SBIA will be the international airport (*Travel Trade Gazette* 1993).

Denver International Airport

The principal path of access to Denver International Airport is the 12-mile (19.3-kilometer) Peña Boulevard, a limited access, four-lane highway that connects the airport with Interstate 70 leading to downtown Denver (24 miles away) and the rest of the metropolitan area (Figs. 6-7 and 6-8). The overwhelming majority of ground travel to the new airport will be via the private automobile. More than 14,000 parking spaces are available on the site, and more are being built due to increased parking demand. The Regional Transportation District (RTD) operates public buses along four different routes, including service to downtown Denver, Boulder, and the Denver Tech Center (DIA 1995c).[5] The one-way fare is $6 to downtown Denver and $8 to suburban locations. Charter buses, commuter shuttles, and mountain carriers (to ski areas) operate services to and from the airport, while hotel, motel, and off-airport parking companies provide courtesy shuttle service. Rental cars are available from a remote lot accessible by company shuttles. Limousines and taxicabs are also available on demand. A one-way taxi ride to downtown Denver is approximately $40. There is no rail service.

Getting from here to there: The story of ground access at DIA

Throughout much of the period when the new Denver airport was being planned, little serious consideration was given to the problems of ground access. At first, much of the focus was on the airport it-

5. Much of the following information is obtained from *Denver International Airport Visitor's Guide* 1995.

Fig. 6-7. *Peña Boulevard connects DIA with Interstate 70.*

self, considering that there was still great uncertainty whether it would be built or not. But even after the two referenda, the EIS approval, and groundbreaking in 1989, it took several years for ground transportation planners from the City of Denver, Adams County, DRCOG, the Regional Transportation District (RTD), and Colorado Department of Transportation (CDOT), in addition to local politicians, the media, and the public at large, to realize that a new airport 24 miles away from downtown needed a coordinated and comprehensive plan to address ground access issues. Because no one agency had dominant responsibility for ground access to the airport, the issue was neglected and it turned into a complex interjurisdictional interface problem.

Aside from planning for the 12-mile access road (Peña Boulevard), providing enough parking for cars, and making transit right-of-way accommodations in the highway design, airport planners did not view ground access as their concern. Original designs for Peña Boulevard (originally named Airport Boulevard) called for six to 10 lanes of divided highway with a transit right-of-way in the median. Today, only two lanes are available in each direction, and no rail transit line serves DIA. The airport planners saw their immediate problem as the airport's costs, and they viewed ground access as a local, regional, or state transportation planning matter. At the same

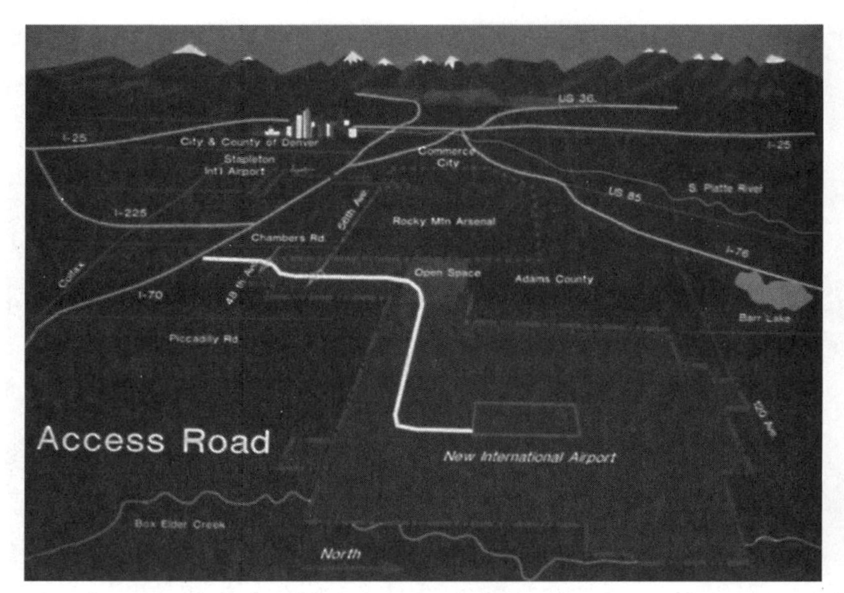

Fig. 6-8. *DIA is served by a single four-lane highway (Peña Boulevard) linking it with Interstate 70. This graphic depicts the major highways in the area.*

time, ground transportation planners from numerous agencies viewed access into DIA as an airport or Denver issue for which they preferred not to assume primary responsibility. Much of this uncertainty and abdication can be attributed to funding questions, especially which government entities would have to pay for the ground transportation projects. Adams County and Denver bickered over improving local access roads. Club 20, a Colorado Western Slope promotional organization, lobbied the Colorado Transportation Commission to refuse provision of state funds to Denver or Adams County for roads to the new airport. Denver promised that no state tax funds would be used for any part of the airport (Miller 1992). Unfortunately, even the federal government, despite DOT's strong support of the airport project, stepped away from the ground access problems, leaving the locals to "work out the kinks" (George 1992). Ground transportation to and from the airport was a responsibility that no one wanted to assume, and thus it fell between the cracks throughout the planning process.

Eventually, however, some ground transportation planning activity involving DIA began to occur. The DRCOG released a study in September 1992 analyzing the problems of access and mobility at the new airport (DRCOG 1992).[6] Because of the lack of transportation capacity in the new airport region, DRCOG envisioned increased traffic congestion along roads leading to Peña Boulevard, such as I-70, I-225, and segments of Tower Road and 56th Avenue To reduce congestion, they recommended several short-term strategies, including new traffic management measures (improved traffic signal synchronization, intersection capacity improvements, etc.) and provision of good bus service, as well as implementation of the strategies contained in its 2010 long-range plan, such as construction of beltway segments E-470 (around the eastern half of the metro area) and W-470 (in the northwest quadrant of the metro area), and increased provision of transit service (including perhaps some form of rail transit). Of course, DRCOG also noted the obvious constraint to these plans—obtaining the necessary financial resources, subject to the same impediments previously mentioned.

That the situation will not improve quickly is attested to by the checkered history of the E-470 and W-470 beltway projects. They were initially a part of the proposed I-470 circumferential beltway to be built around the entire Denver metropolitan area, a project which then-Governor Lamm vowed to kill in 1976 by "driving a silver spike through its heart" due to concerns over highway-induced pollution, sprawl, and the "Californication" of Colorado (Sutton 1995, p. 18).[7] But part of the proposed project later sprung back to life in the form of C-470, a limited access state-funded beltway in the southwestern quadrant of the area. While C-470 was being built from 1982 to 1990, additional attention focused on completing the beltway's E-470 and W-470 segments. Several local and state referenda that contained initiatives to publicly fund these projects had already been voted down. But in 1985, the E-470 Authority, a private group, was created and developed plans to complete a 50-mile, self-sustaining toll road on the eastern half of the metro area. Financing this plan is proving difficult, but the E-470 Authority was able to start a 5.5-mile stretch between I-25 and Parker Road in southeastern Denver. Current plans call for construction of additional segments of E-470 leading to the DIA area to be completed by mid-1999. The fate of W-470 is still in limbo, as opposition to it has remained steadfast. But it is included in DRCOG's 2010 plan and has been identified as an important link

6. The rest of this paragraph references DRCOG (1992).

7. Much of the following beltway discussion is based on Sutton (1995).

to improve access to the new airport. Both E-470 and W-470 continue to face significant funding and public opposition hurdles, especially in light of the current environment for building new highways locally and nationally as reflected in, for example, the 1991 ISTEA legislation.

By the early 1990s, once it became apparent that the new airport was going to be built, an increasing chorus of local media and citizen groups began to call for some type of rail access to DIA. It was also at this time that the local Regional Transportation District (RTD) was trying to push ahead with its plan to build a light rail line from downtown Denver to Stapleton Airport. Even though Stapleton was closing, one of the arguments for building this line was that it could eventually extend out to the new airport. RTD, however, encountered stiff opposition to this plan, especially from residents along the proposed route to Stapleton. Still, RTD was determined to build a light rail line and finally settled on a short, 6-mile line from Broadway and I-25 through downtown to the nearby Five Points neighborhood, nowhere near DIA. RTD intended this Metro Area Connection (MAC) line to be the central artery of a much larger proposed system. Current plans call for an expansion of light rail along Santa Fe Drive (U.S. Highway 85) in the region's southwest corridor. It is highly unlikely, however, that any light rail line will be built to the new airport in the foreseeable future.

A somewhat more promising rail alternative is the "Air Train," a proposed commuter rail link from downtown Denver Union Station to DIA using existing Union Pacific Railroad lines and new tracks into the airport area. It would represent a relatively unique public-private partnership involving the City of Denver and Union Pacific, a freight rail carrier that owns the railroad rights-of-way in the corridor from downtown toward DIA. The final 12 miles of rail alignment would be in the median strip of Peña Boulevard. The Air Train's cost is projected at $140 million, with operation and maintenance at $14.4 million per year in 1993 dollars (Kelly 1995). Prices would be $6 one-way and $10 round-trip from Union Station to DIA. Somewhat lower fares would be available from access at the former Stapleton airport as well as a stop in Aurora. The City of Denver has estimated 5840 daily riders in the train's first year of operation. That would be a very large percentage of the average of 30,000 vehicles that operate on Peña Boulevard every day.

Much has been said and written about the Air Train, but little has been done. As always, financing is the main stumbling block, and the city has been preoccupied with getting DIA to function effectively. Still, the Air Train represents the best hope of a rail alternative to the airport in the near future. Of course, without a comprehensive rail system in the metro area or Front Range region, the Air Train line would have very limited connectivity. This shortcoming could, of course, be alleviated if it were included as an essential part of an overall strategy that attempts to achieve intermodal objectives for the entire region by linking the Front Range communities by rail. To maximize its utility, any rail system connecting the Denver metropolitan area to its airport should first attempt to identify the residential areas for most of its patrons, and build rail lines to serve them. Identification could be done by correlating automobile license plates, which pass through DIA's toll booths, with zip codes on automobile registrations. Moreover, Air Train has some potential to help mitigate air pollution impacts from the increased traffic forced to cover the extra distance to the new airport. Whether enough people would use the Air Train to help justify its environmental benefits, however, is still questionable. Overall, it represents the right direction for ground transportation in the region, particularly to the airport, but its construction, implementation, financially responsible operation, and integration into an entire system remain problematic.

Any form of rapid transit would have serious financial implications for DIA. Since it is so difficult to get to DIA any other way for most of the residents and businesspeople in the Denver metro area, the private automobile is the overwhelmingly dominant mode of travel to the new airport. And with more than 14,000 spaces available for parking, with more on the way, the parking garage is one of the major financial success stories since DIA's opening.

Every car that enters DIA must pass through a toll plaza located several miles from the terminal to pick up a ticket indicating time of ground arrival. Whether parking or not, every vehicle is subject to paying $2 per hour for access to the airport after the first 70 minutes. This means that cars just dropping off or picking up passengers are subject to paying fees unless the vehicle can depart the toll plaza within 70 minutes of original arrival. This arrangement represents an improvement over the original rate structure, which called for fees after only a 30-minute grace period. Given the distances, it was virtually impossible for cars to get in, pick up or drop off passengers, and get out in less than 30 minutes. This was especially true because queues at the toll plaza were exceedingly long during the first year of operation, and many irate travelers were forced to wait in line and pay the access fees.

Recent changes in parking and access rate have lowered costs for the traveler, but DIA is still producing healthy total revenues from its parking and access operations. In its June 1995 bond prospectus, the city estimated a $19-million net profit for the airport in 1995, due largely to parking and concession revenues (Leib 1995). It is thus not surprising that the City of Denver and DIA officials are not too concerned about intermodalism or ground transportation alternatives to the new airport.

In retrospect, and given ISTEA's intermodal mandate, the first thing built at the DIA site should have been a standard gauge rail spur from the Union Pacific's main line, only a few miles away. The spur could have been used to reduce transportation cost and expedite delivery of the vast quantities of heavy steel and concrete used for DIA's construction. The main terminal building could have been built around such a spur, so that when funding ultimately was secured for rolling stock, passengers could have arrived by commuter train in the basement of DIA, as they do at many international airports, such as Frankfurt-am-Main and Amsterdam Schiphol. Amtrak's Zephyr, which provides daily service between Chicago and Salt Lake City (and beyond), could also have made a scheduled stop at DIA. Unfortunately, intermodal transportation issues were largely ignored by DIA's planners. Highway congestion and environmental pollution are the consequences of such myopia.

That airport planners think only about airports is perhaps understandable. They want their finite capital pouring runway concrete rather than highway asphalt. But that the U.S. Department of Transportation (DOT) and its subsidiary organs (Federal Aviation Administration, Federal Highway Administration, and Urban Mass Transit Administration), which provide essential financial support for airports, highways, and transit authorities, respectively, did not insist on intermodal coordination is unforgivable. Since its inception in 1967, DOT has been given the mandate to foster seamless intermodal integration, a responsibility it often tends to forget. Indeed, an essential purpose for DOT's creation was to coordinate the national transportation system. With the lever of withholding financial support, DOT could have insisted that the local airport, highway, and transit authorities come together to provide a comprehensive transportation plan for access to the new airport. No new airport should be built without DOT's insistence on a coordinated approach to intermodal transportation planning, or better yet, a consolidation of regional highway, transit, and airport authorities under a single roof.

Lessons learned

- *Locate the airport to minimize adverse noise impacts.* Several Asian airports are being built on the ocean, not only because of the dearth of suitable level land, but because of the noise impact on dense population clusters. Landfill on seabed is among the world's most expensive and complicated engineering feats. But once constructed, 24-hour-a-day takeoffs and landings may be possible.
- *Develop and implement noise mitigation procedures in consultation with affected publics.*
- *Monitor the validity of the noise impact forecasts.*
- *Incorporate multimodal ground access into the terminal airport.* Most of the world's major new airports (DIA is the obvious exception) incorporate rail directly into the airport terminal, allowing efficient, high-speed, environmentally sound ground access.
- *More effective institutional mechanisms must be developed to address complex interjurisdictional issues such as ground access to airports.* Given the cross-sectional character of civil aviation, many problems are encountered because of jurisdictional uncertainty. Ground access to airports is a classic case in point.
- *Develop a comprehensive intermodal and multimodal transportation hub.* At Seoul, Hong Kong, Macau, and Osaka, we see efforts to link all modes of transport together and a key desire to accommodate the rapidly growing cargo sector. At Denver, with the exception of one four-lane highway, intermodal transportation was treated as "somebody else's responsibility" by DIA's planners.
- *Air quality impact must be considered for an entire region.* The affected area should not be defined in a manner that automatically assumes that airport-generated pollution will

have a minimal impact on the nearby metropolitan area. A more holistic approach to air quality impact analysis for airports should be used.

- *Site selection processes must be open to the public.* When selecting sites for public planning and other purposes, many factors are assessed in the process of a technical analysis. This often involves the identification and relative weighting of key locational criteria (e.g., site must be on relatively flat terrain) that are then used to develop scores for potential sites based on how well they address each of the stated criteria. From this, several "good" potential sites are chosen and then an optimal site is selected. But what may start out to be a very objective, technical procedure inevitably becomes subject to political forces that might sway decision-making toward other directions and sites. Typically, in fact, once general areas are identified that meet the key criteria from the technical analysis, the specific site selection tends to become more a consequence of the political decision-making process. In any site selection analysis, both the technical and political dimensions of the process must be understood to explain the final locational decisions. The public must be informed and made aware of how and why these decisions are made.

References

Airports, vol. 12, no. 13. 1995. Denver receiving record airport noise complaints. March 28.

Allen, Roy. 1979. *Major Airports of the World.* New York: Charles Scribner's Sons.

Ashford, Norman and Paul H. Wright. 1992. *Airport Engineering*, 3d ed. New York: John Wiley & Sons.

Asian Wall Street Journal. 1994. Airport to be Macao's gate to region. October 31.

Aviation Week & Space Technology. 1994a. U.S. funds feasibility study for second stage of Macau airport. June 6:36.

————. 1994b. Growth outpaces Asian airports. August 29:58.

Black, Alexandra. 1994. Trade and tourist boom from new airport. *Inter Press Service Network.* September 9.

Booth, Michael. 1990. Expanding soil poses costly threat to new airport. *Denver Post.* January 29.

Brown, David. 1987. Japanese building international offshore airport to serve Osaka. *Aviation Week & Space Technology*. July 13:38.

Chandler, David. 1993. Terminal condition. *Westword*. July 21.

City and County of Denver (Denver). 1985. Memorandum of Understanding. Signed with Adams County, January 28.

————. 1988. New Denver Airport Draft Environmental Assessment, Executive Summary.

Colorado National Banks. 1989. *Ready for Takeoff: The Business Impact of Three Recent Airport Developments in the U.S.* Prepared by Coley/Forrest, Inc.

Darmody, Thomas. 1993. The design and development of world class airports. Paper presented to the IBC International Conference on Airport Development & Expansion, Hong Kong, October 28.

Deem, Warren and John Reed. 1966. *Airport Land Needs*. Cambridge, MA: A.D. Little.

Denver International Airport (DIA). 1995a. Aircraft Noise Abatement Office Quarterly Report. 2d Quarter.

————. 1995c. *Denver International Airport Visitor's Guide*.

DIA Noise Abatement Office. 1995b. DIA Recommendations for Current Noise Impact Issues (revised 7/18/95).

Denver Post. 1981. Residents near Stapleton sue city for noise levels. March 28.

————. 1983. Metro Airport Study: Final Report.

————. 1985. Denver agrees to cut Stapleton noise. May 29.

————. 1988a. City goof understates pollution at new airport. April 27.

————. 1988b. City officials revise air pollution estimates for new airport. May 7.

Denver Regional Council of Government (DRCOG). 1992. The New Airport Area—Access and Mobility.

Donoghue, J. A. 1995. The Pearl-Y gateways. *Air Transport World*. February 1:75.

The Economist. 1994. Fasten your seat belts. November 12:42.

————. 1995. How to avoid that sinking feeling. February 4:73.

Evans, Gary W. and Stephen J. Lepore. 1993. Nonauditory effects of noise on children: A critical review. *Children's Environments Quarterly* 10:31.

Federal Aviation Administration (FAA). 1986. Policies and Procedures for Considering Environmental Impacts. Order 1050.1D. Washington, D.C.: Federal Aviation Administration.

Feldman, Elliot J. and Jerome Milch. 1982. *Technocracy Versus Democracy: The Comparative Politics of International Airports*. Boston: Auburn House Publishing Company.

Flynn, Kevin. 1996. Rerouting of jets cuts risk of fine. *Rocky Mountain News.* June 7.

George, Mary. 1992. Feds support airport: Ground transport left up to locals. *Denver Post.* July 24.

Hanson, Susan. 1995. Getting there: Urban transportation in context. *The Geography of Urban Transportation*, 2d ed. Susan Hanson, ed. New York: The Guilford Press.

Hill, Leonard. 1991. Countdown on Munich 11. *Air Transport World.* June 1:70.

Horonjeff, Robert and Francis X. McKelvey. 1994. *Planning and Design of Airports, 4th ed.* New York: McGraw-Hill.

Howard, Needles, Tammen & Bergendoff (HNTB). 1986. New Airport Master Plan, Section 4: Staff Recommendation Final Site Area. Prepared for the City and County of Denver.

Kelly, Gus. 1995. DIA may land a train, too. *Rocky Mountain News.* May 23.

Kowalski, Robert. 1993. Expanding soil plagues airport. *Denver Post.* November 18.

————. 1995a. DIA noise pact raises the roof. April 5.

————. 1995b. Peña focuses on DIA noise. August 11.

————. 1995c. In an increasingly loud world, how much noise is too much? December 7.

————. 1988. $3 billion project to pollute more than Stapleton does. *Denver Post.* January 31.

————. 1989. FAA's airport report questioned. *Denver Post.* February 2.

Kyodo News International. 1994. Kansai Int'l Airport inaugurated. September 5.

Leib, Jeffrey. 1995. DIA projections healthy, woes also cited in bond report. *Denver Post.* June 6.

Leonard, Stephen J. and Thomas J. Noel. 1990. *Denver: Mining Camp to Metropolis.* Niwot, CO: University Press of Colorado.

Levy, John K. 1994. *Contemporary Urban Planning, 3rd Edition.* Englewood Cliffs, NJ: Prentice Hall.

McKee, Mike. 1995. Interview. DIA Noise Abatement Office. December 11.

Mecham, Michael. 1992. Munich turns out lights on Riem. *Aviation Week & Space Technology.* May 25:20.

Meredith, John. 1995. Room to boom. *Airline Business.* January:38.

Miller, Ellen. 1992. Club 20 asks state to bar funds for new airport roads. *Denver Post.* May 29.

Mok, John. 1993. The development of Hong Kong's new international airport. Paper delivered at International Conference on Aviation & Airport Infrastructure, Denver, Colorado, December 8.

Moorman, Robert. 1994. Osaka to me. *Air Transport World.* October 1:62.

Mordoff, Keith. 1988. Air Transport Munich's new international airport expected to begin operations in 1991. *Aviation Week & Space Technology.* February 22:92.

Mufson, Steven. 1994. Accord boosts Hong Kong airport. *Washington Post.* November 5.

Obmascik, Mark. 1988. Airport study assumes huge transit outlay. *Denver Post.* December 18.

Ogawa, Zenjiro. 1993. Kansai International Airport Projects. Paper delivered at International Conference on Aviation & Airport Infrastructure, Denver, Colorado, December 8.

Orlando Sentinel. 1995. Hong Kong builds $20.3 billion airport. April 9.

Paisley, Ed. 1995. On a wing and a prayer. *Far Eastern Economic Review.* May 5:76.

Plous, F. K., Jr. 1993. Refreshing ISTEA. *Planning.* February:9–10.

Reisner, Marc. 1993. *Cadillac Desert: The American West and its Disappearing Water.* New York: Penguin Books.

Rhodes, S. L. 1992. Meso-scale weather and aviation safety: The case of Denver International Airport. *Bulletin of the American Meteorological Society* 73(4):441–447.

Rocky Mountain News. 1995a. DIA drowns out noise complaints. August 20.

————. 1995b. DIA reroutes planes to cut noise. September 29.

————. 1995c. Din grows at DIA along with risk of fines. October 10.

Shin, Jong-Heui. 1993. Airport developments in Korea. Paper delivered at International Conference on Aviation & Airport Infrastructure, Denver, Colorado, December 8.

Spensley, Skip. 1995. Interview. December 20.

State of Colorado (Colorado). 1991. Report of the Attorney General: Airport Land Acquisition Investigation.

Stevenson, Gordon McKay Jr. 1992. *The Politics of Airport Noise.* Belmont, CA: Duxbury Press.

Sutton, Christopher J. 1995. The Socio-economic, Land Use, and Land Value Impacts of Beltways in the Denver Metropolitan Area. Ph.D. Dissertation. Department of Geography, University of Denver.

Travel Trade Gazette Europa. 1993. Bangkok to add gateway. April 8:20.

Travel Weekly. 1995. Far East facility slates December opening. July 3:29.

U.S. Department of Transportation (DOT). 1985. Airport Environmental Handbook. FAA Advisory Circular 150/5050.4A. November.

————. Federal Aviation Administration. 1989. Final Environmental Impact Statement, New Denver Airport. Volume 1.

U.S. General Accounting Office (GAO). 1991. New Denver Airport: Safety, Construction, Capacity, and Financing Considerations. GAO/RCED-91-240.

Upton, Kim. 1992. New Munich airport stresses ease of travel. *Los Angeles Times.* May 17.

Vancouver Sun. 1995. Hong Kong airport construction site strictly high-tech. June 3.

Wiley, Karen and Steven L. Rhodes. 1987. Decontaminating federal facilities: The case of the Rocky Mountain Arsenal. *Environment.* April:16–20, 29–33.

World Airport Week. 1995a. Air traffic growth forces expansion of Seoul airport construction. May 30.

————. 1995b. NSIA privatization opportunities. June 20.

————. 1995c. NSIA considers opening with extra runway. July 1.

————. 1995d. Macau's presence in Pearl River Delta will be as reliever airport. July 18.

7

Airport layout, design, and technologies

"Denver International Airport is 'the world's safest and most efficient airport.'"—George Brewer, Federal Aviation Administration[1]

The master plan and modular design

In this chapter, we review the planning process at DIA, evaluate the airfield and terminal layout (with an examination of the various airport design types that have emerged since the dawn of commercial aviation), and discuss the major technologies. After evaluating the specifics of the DIA case, we examine how the new airports that are springing up around the globe deal with each of these issues.

An airport's master plan ordinarily is developed through three phases:

1. Determination of airport requirements
2. Site selection
3. Airport layout and land-use planning

The third phase is the subject of this chapter. This phase typically includes four components (Wells 1992, pp. 108–109):

1. The airport layout (i.e., the actual configuration of runways, taxiways, and aprons)
2. Land use (i.e., designation of areas for terminals, maintenance facilities, commercial buildings, industrial sites, ground access, and noise buffer zones)
3. The terminal area (landside and airside)
4. Airport access (primarily automobile and rail).

1. Imse 1995.

Deregulation and airport design

The role of the airport (i.e., whether it is to serve as a hub connecting facility or an origin-and-destination facility) must be determined prior to undertaking its physical layout and design (Gesell 1981). The fundamental change in major air carrier distribution systems that emerged after the Airline Deregulation Act of 1978—the hub-and-spoke phenomenon—created new demands in airport design in the United States. Most of America's airports were built for the pre-deregulation linear-route system, oriented toward satisfying the needs of the local origin-and-destination (O&D) passenger (Fig. 7-1). In the early 1980s, Denver had become a city serving as a hub for three major airlines, a phenomenon that existed at no other airport in the world. But the Darwinist competition unleashed by deregulation made that a short-term phenomenon—an aberration, really. With Continental's acquisition of Frontier in 1986, there were two hub carriers operating at Stapleton. Most hub airports were effective monopolies. In the late 1980s, Denver (with United and Continental), Chicago (with United and American), Dallas (with American and Delta), and Atlanta (with Delta and Eastern) were the only hub duopolies in the United States. Today, only Chicago is a hub duopoly. All other hubs are effective monopolies.

Fig. 7-1. *A 1970's airport design, excellent for origin-and-destination traffic because of the convenience of parking proximity to departing gates. Examples include Dallas/Fort Worth International Airport.*

Prior to DIA, the last major new airport built from scratch in the United States was Dallas/Fort Worth International Airport (DFW), completed several years before deregulation. But DFW was designed to serve O&D traffic and was poorly laid out for the hub operations of American and Delta that came to use it as a transfer point. Only Atlanta reconfigured its airport (Hartsfield International) in 1980 into an efficient design for a hub airport—a main terminal connected to three parallel concourses via an underground train, surrounded by four parallel runways (Fig. 7-2). DIA is Hartsfield perfected (Thurston 1993).

Fig. 7-2. *Atlanta's Hartsfield International Airport was the revolutionary design that DIA perfected, with a separate landside terminal accessible on both sides by automobiles, and underground access to three parallel concourses.*

The planning, design, and construction process

Design of DIA's airfield and airspace was undertaken by an Airline Airport Affairs Committee composed of several technical subcommittees, each focused on separate pieces of the puzzle—airfield, terminal, baggage, fueling, and de-icing. The city, the airlines, the FAA, the Air Transport Association and the Air Line Pilots Association were represented on salient subcommittees. The airlines were lukewarm about a new airport, but if the city was determined to proceed, the airlines wanted some input in the planning process. Their participation on the airfield subcommittee worked well. But the carriers, especially United Airlines, attempted to derail the terminal subcommittee's work in a multitude of ways.

The facilities design phase of airport master planning includes development of an airport layout plan (ALP) to address airfield configuration, terminal design, land use, and surface transportation. Objectives include optimization of efficient aircraft operations, enhancing passenger flow with minimum walking distances, accommodating intermodal surface transportation connections, and avoiding environmental degradation (Spensley 1991).

Denver's airport planner, Dick Veazey, masterminded the planning, producing a master plan in 1986. Veazey laid out the design of the runways, concourses, fire stations, and access roads. He has been described as "the man responsible for most of the DIA features that aviation experts say will make the new airport function as one of the best airfields in the world" (Booth 1995). Veazey headed the city's planning of the technical side of the DIA project. In 1985, the city signed contracts with six consulting companies to assist the planning process. These consultants focused on

1. Noise
2. Terminal and airfield configuration
3. On-airport ground access
4. Off-airport ground access
5. Cost estimating and engineering oversight regarding constructability
6. Forecasting and financial analysis

Some 105 technical tasks were parceled among them, with one consultant having primary responsibility, and coordinating consultants participating as appropriate (Veazey 1995).

Skip Spensley, who was Peña's head of the new airport office, chaired the public affairs side of the process, with jurisdiction over site selection, public participation, environmental impacts, and several ad hoc studies. These tasks, and Spensley's work on the airport, are discussed in other chapters.

Of course, airport planning and design in the United States is definitively governed and guided by several FAA Advisory Circulars and the aircraft manufacturers' design and performance criteria. These provided the template for runway design, layout, and engineering specifications at DIA (ATA 1977; DOT 1976, 1985a, and 1985b).

As the planning process began in 1986, the FAA took the unusual step of assigning officials to the airport's planning team. Each major division of the FAA (e.g., flight standards, airway facilities, and airports) was assigned to the "DVX" team, so named as not to be confused with Denver's DEN airport code—the three-letter code that survived the transition from Stapleton International Airport to Denver International Airport. The FAA's George Brewer set up shop as principal liaison in a suite of offices in the city's Aviation Division in the Department of Public Works at Stapleton International Airport with orders to "make it happen" (Brewer 1995). Brewer's constant presence on the property allowed the new airport's planners to interact over such features as runways, airfield lighting, and the control tower, and to resolve problems expeditiously (Green 1992). Each piece of the airport could be planned with a view to satisfying FAA rules and regulations at the outset. Each FAA division developed a checklist to facilitate an orderly airport review process. The FAA set up a complete mockup of the air traffic control tower in cardboard to ensure that everything needed would fit (Veazey 1995).

Because of the scale of the new airport project, the city hired nine firms to refine various aspects of the master plan. In 1987, the city established a program management team (PMT) consisting of airport staff and consultants under engineer Bill Smith, associate director of aviation for Denver. PMTs are increasingly common for large and complex projects. The city awarded the $34 million PMT contract to a joint team of Greiner Engineering, Inc. (an engineering, architectural, and airport planning firm), and Morrison-Knudsen Engineering (a design and construction firm), known at DIA as Greiner/MKE (Green 1992). The PMT was responsible for coordinating the schedule, cost control, information management, and administration of approximately 100 contracts, 160 general contractors, and 2000 subcontractors (Russell 1994). Greiner's Dick Haury said:

> *We looked at how we would build the airport before we designed it. Early in the process we agreed on the length of time*

for building each component. Too much construction time results in too much for interest. Too short a period, and there's not enough labor and materials available (Green 1992).

But the process has not been without its critics. Two principal charges have been levied:

- Responsibilities were not clearly defined. The muddled chain of command was named by all the designers as the most frustrating aspect of the project. Architect James Bradburn faulted the city, which he said "infiltrated" the PMT with its own staff in such a way as to inappropriately undercut the authority of Greiner/MKE personnel nominally in positions of higher responsibility.
- The process took over. The project became beholden to the needs of negotiators and bond-rating agencies, to the detriment of the completed facility. Those negotiating for the city seemed not to grasp the construction consequences of changes they permitted the airlines to exact. The consequences of redesigning the terminals and concourses so late in the process will be felt for the life of the facility (Russell 1994).

Indecision in the mayor's office also became a problem after Wellington Webb replaced Federico Peña. Aviation Director George Doughty found that the expeditious decisions and affirmations on technical issues he enjoyed under Mayor Peña deteriorated under Webb so that meetings ended with neither a final decision nor authority given to Doughty to resolve the problem. Doughty left a year after Webb took office because he felt he could not work effectively in such a political and indecisive environment (Doughty 1995).

The PMT organizational structure was awkward. It might have been better to have a PMT that reported to the city, rather than one that was a quasi-part of city organization (DeLong 1995).

DIA master plan

DIA's master plan called for 12 runways surrounding four (or five) center-field 60-gate airside terminal buildings (concourses) linked to a landside terminal building (with five modules at full build-out) by a 5100-foot underground train. Ten runways were to be 12,000 feet long, and two were to be 16,000 feet long. The runways were laid out in a four-quadrant symmetrical configuration, allowing arrivals and departures from four separate directions (Rue 1993).

DIA opened in 1995 with five 12,000-foot runways, three concourses (with 22, 44, and 20 gates, on Concourses A, B, and C, respectively), and a terminal building with three modules. The repetitive modular design of both the landside (Figs. 7-3 and 7-4) and airside terminals and runways allows future expansion at reasonable cost while preserving their functional and architectural integrity (McCagg 1993). One source noted, "The key DIA feature is its modular design, meaning that the airfield is laid out in such a way that runways, concourses, and terminal space can be added in pieces in the future if traffic growth demands it" (Booth 1995).

Everything in the design process was done with a view to maximizing expandability and flexibility of the airport over time (Veazey 1995). Virtually everywhere, ample space was left for future expansion (Fig. 7-5). As we shall see, the mid-field location of the terminals, although adding distance from the city's center, allows reduced aircraft taxi time and enhanced operational efficiency (Rue 1993). DIA's main terminal designer, Curt Fentress, observed:

> *Commissions who generate airports echo [efficiency] concerns and add the need for memorable imagery and the desire to create landmarks unique to their location. Encapsulated in all of their basic requirements is the need for expandability. It has become imperative to create new designs which are inherently expandable and flexible. By not taking these into consideration, the project will be doomed to eventual failure and a waste of public funding (Fentress 1995).*

When ultimately built out (sometime after the year 2020), DIA's capacity would be 110 million passengers a year (compared with about 31 million actual passengers in 1995) and an astounding 1.23 million takeoffs and landings (O'Driscoll 1994a). This is more capacity than any other airport in the world.

DIA is linked to Interstate 70 by a 12-mile, four-lane boulevard, with sufficient space in the median for future construction of a rail line. Rental car facilities are located along the north side of that boulevard, while air cargo facilities are on its south side. Peña Boulevard can be

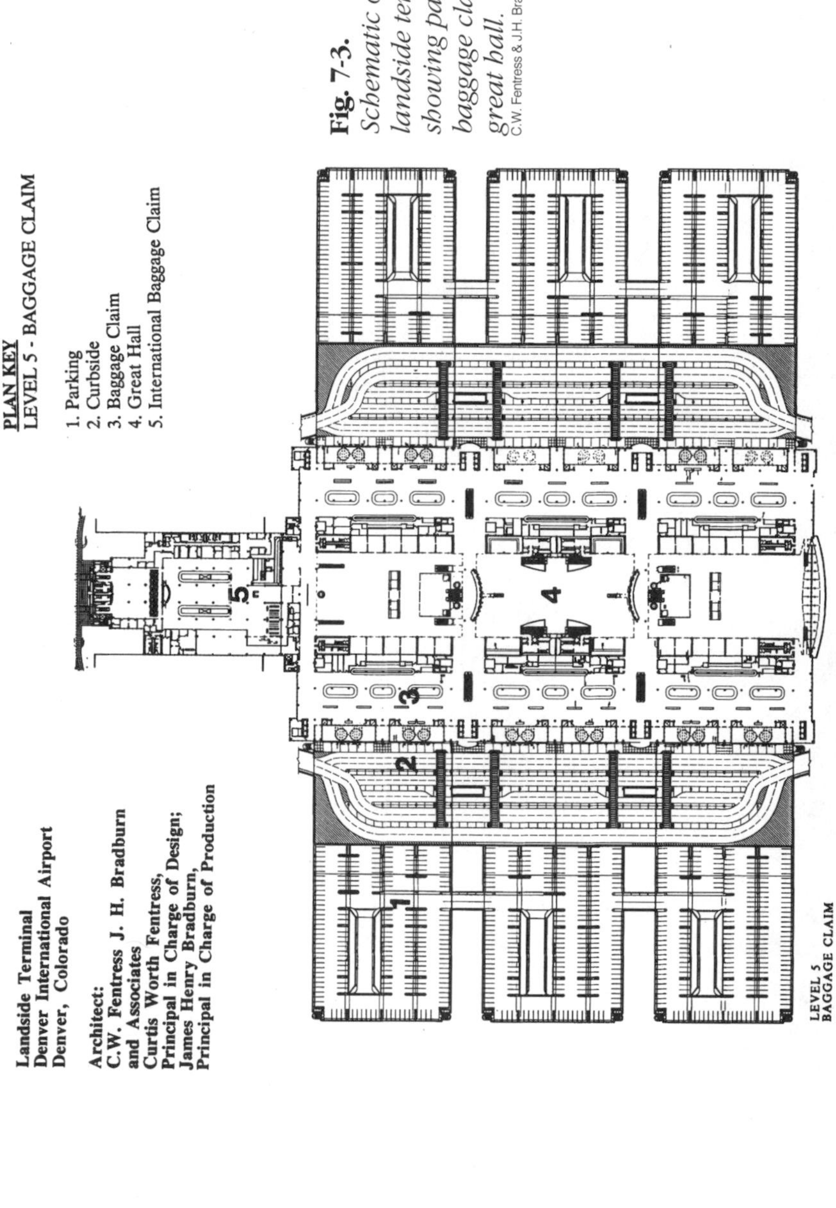

Landside Terminal
Denver International Airport
Denver, Colorado

Architect:
C.W. Fentress J. H. Bradburn
and Associates
Curtis Worth Fentress,
Principal in Charge of Design;
James Henry Bradburn,
Principal in Charge of Production

PLAN KEY
LEVEL 5 - BAGGAGE CLAIM

1. Parking
2. Curbside
3. Baggage Claim
4. Great Hall
5. International Baggage Claim

LEVEL 5
BAGGAGE CLAIM

Fig. 7-3. *Schematic of level 5 of the landside terminal building, showing parking, curbside, baggage claim, and the great hall.* C.W. Fentress & J.H. Bradburn.

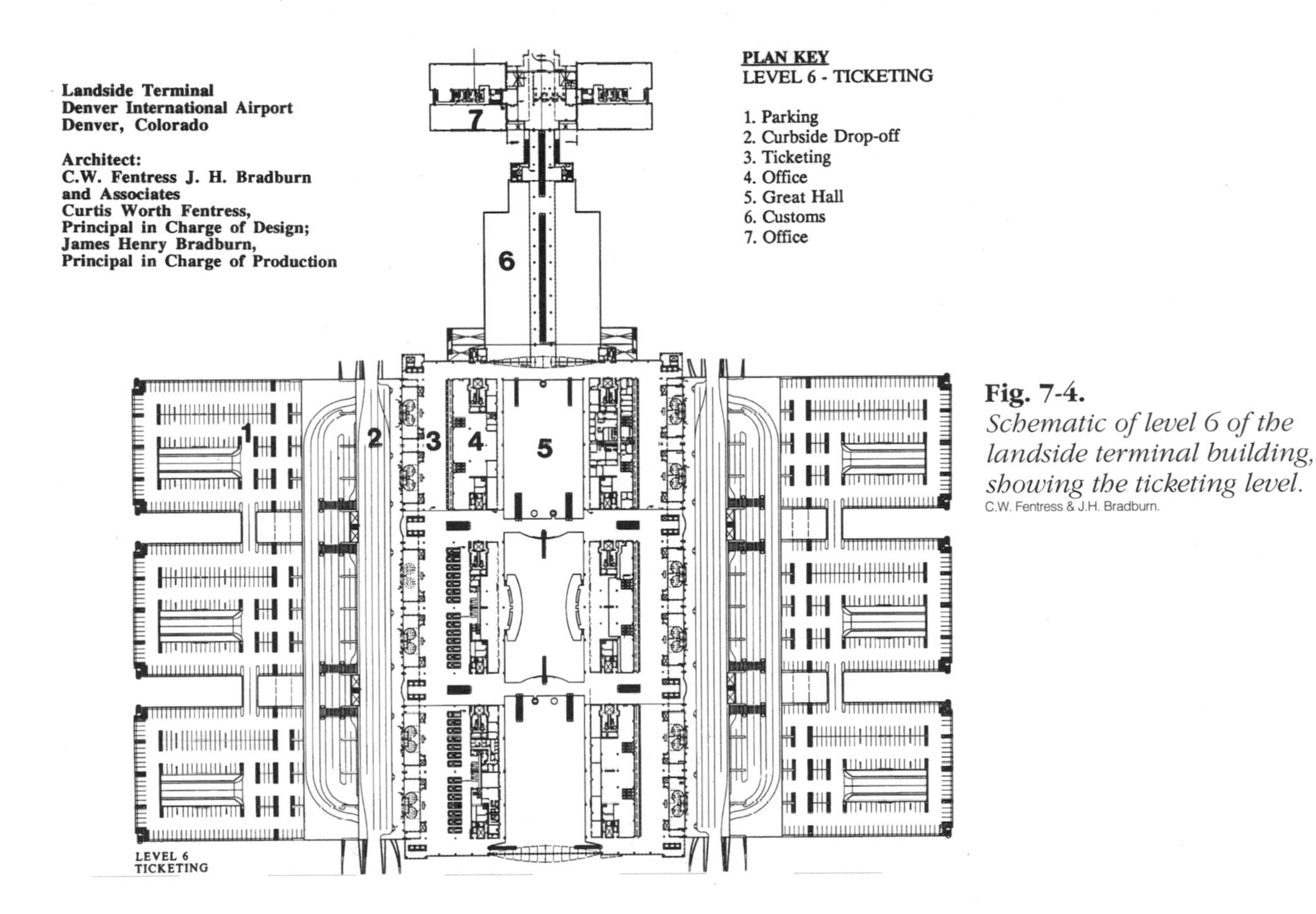

Fig. 7-4.
Schematic of level 6 of the landside terminal building, showing the ticketing level.
C.W. Fentress & J.H. Bradburn.

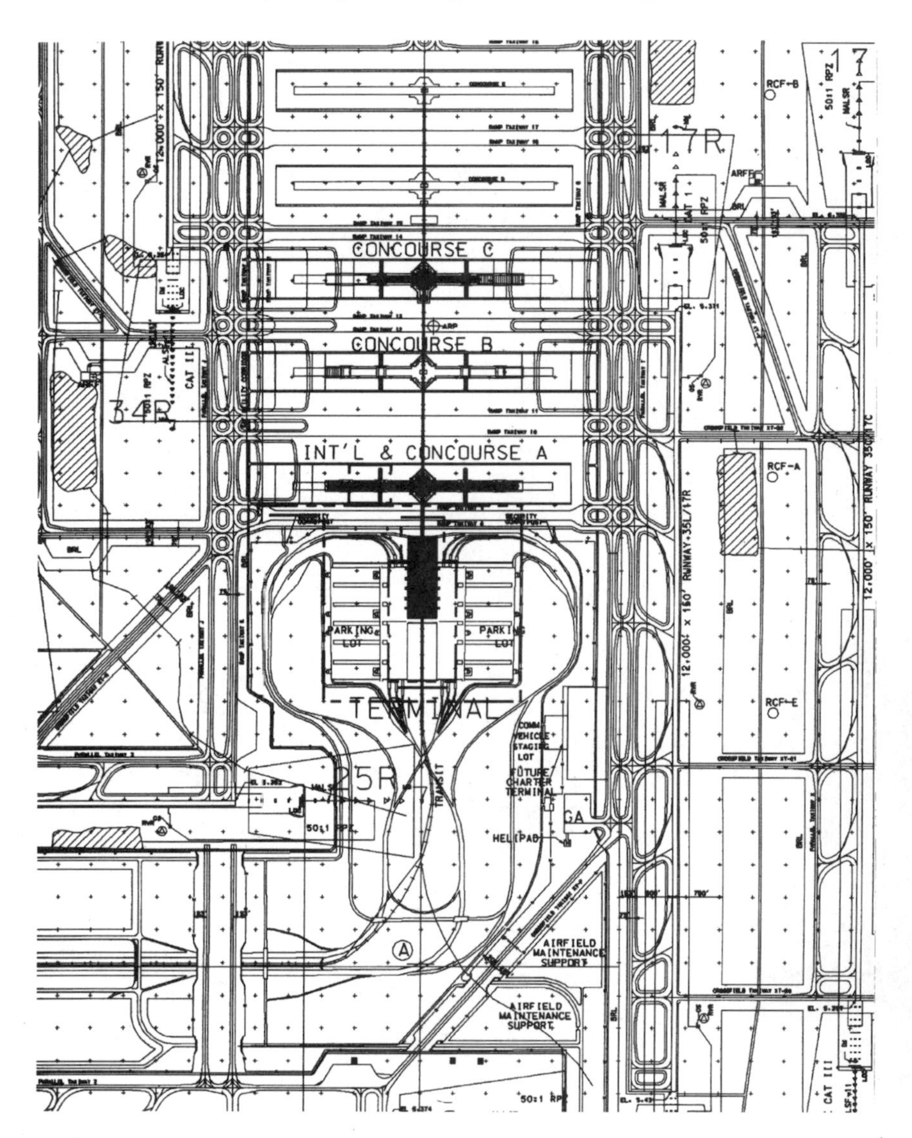

Fig. 7-5. *DIA plan for concourses and passenger terminal. DIA left ample land for the construction of two additional concourses and doubling the size of the main terminal building.* C.W. Fentress & J.H. Bradburn.

widened to 12 lanes and still leave land for a rail line in the median (O'Driscoll 1994a), although the bridges are only two lanes, and entry into the terminal area by a rail line may be impeded by highway viaducts. Four lanes to DIA contrasts sharply with the 14 that connected the smaller Stapleton Airport to Denver.

Airfield design

The principal concerns driving airfield design are safety, efficiency, noise, and flexibility for future growth (de Neufville 1976; Horonjeff 1975). At DIA, first came the land—53 square miles of it, more than could be consumed by an airport in the lifetime of its managers. Assuming continuing growth of air traffic, future expansion would be unimpeded until well into the mid-21st century. Then came runway configuration. Terminal location and design came later.

One objective was to ensure that the flight paths did not blast residential neighbors under the aircraft arrival and departure streams with undue noise. Denver had been taught by the residents of Park Hill and Aurora (both under Stapleton Airport's flight path) what a political headache residential neighbors can be. The agreement between Denver and Adams County ensures that no homes will be built within 5 miles of the end of any runway or within 2 miles of the airport boundary (Nordwall 1993).

Denver was also painfully aware of the debilitating impact of inclement weather on airport operations. Stapleton Airport's disastrous runway configuration (two were separated by 900 feet and two by 1600 feet (Fig. 7-6)—insufficient space between parallel runways) often made it a one- or two-runway airport when the snow fell, as it has a propensity to do in Colorado (Eddy 1995). Fog, thunderstorms, windshear, and heavy crosswinds also froze Stapleton on occasion, causing a ripple effect throughout the national air transport system (Broderick 1995a). In fact, Stapleton's primary inadequacy was its runway configuration, and that, more than anything else, was the dominant air transport motivation for a new airfield (Fig. 7-7). Stapleton's terminal, although congested on the airside, was ample for Denver's needs, particularly with the recent addition of Concourse E (Fig. 7-8). Of course, as we have seen, economic considerations (jobs, growth, and prosperity) were also important motivations for building a new airport.

From the outset of the DIA design process, the goal was to come up with a runway configuration that allowed at least two incoming air lanes to be open at all times (Schwartzkopf 1986). Only after the airfield was configured would a terminal be designed.

The city originally chose an L-shaped configuration for runways to carry noise away from populated areas. But the FAA and the city's

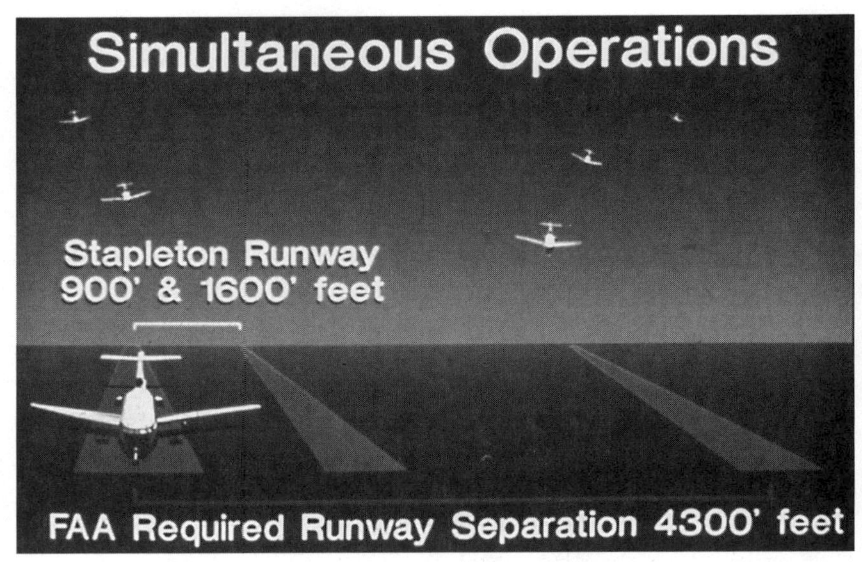

Fig. 7-6. *The principal deficiency of Stapleton's airfield was the close proximity of parallel runways, which forced runways to close down during periods of inclement weather.*

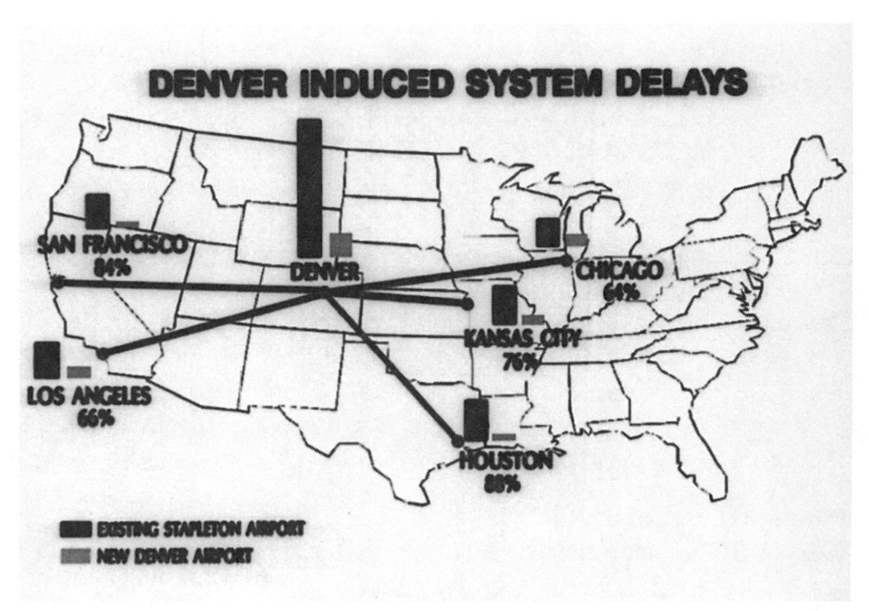

Fig. 7-7. *Inclement weather at Stapleton International Airport had a ripple effect on air traffic across the United States. This chart reveals its impact on delays at Denver and several other major U.S. airports.*

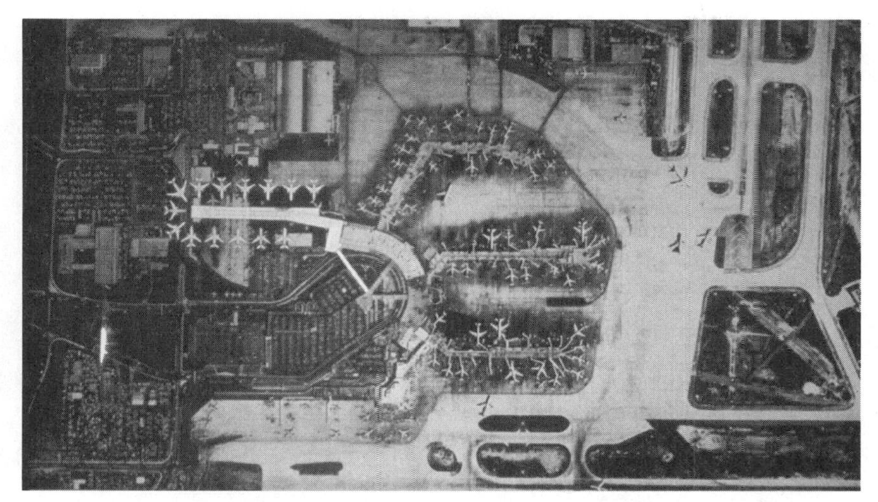

Fig. 7-8. *An aerial view of Stapleton International Airport, showing the 1980's expansion of Concourse E at the far left.*

aviation consultants promoted the efficiency of a pinwheel configuration, and that design won out, with the terminal and concourses placed in the center of the pinwheel (Green 1992). The final configuration is shown in Fig. 7-9. The FAA was also aware of the bottleneck Stapleton Airport posed to the national aviation system (particularly the huge networks of United and Continental Airlines) during inclement weather and was determined that the new facility would help relieve gridlock (Nordwall 1993). The FAA's George Brewer observed, "[Denver] has been a choke point in the whole U.S. air traffic system. The hope is to minimize those delays" (Goldstein 1993).

DIA's airfield planning subcommittee identified 20 designs, ultimately reducing the alternatives to four, on which it performed extensive computer simulations. The simulations began with maximum long-term capacity (1,230,000 flights and 110 million passengers), and worked backward to closer-in-time projected capacity (640,000 flights and 50 million passengers) to select an appropriate airfield configuration for DIA. Ultimately, the FAA reconfigured 240,000 square miles of navigable airspace to accommodate the flight paths of the new airport (Veazey 1995).

Although the site on which DIA sits is rolling semi-desert farmland, it needed some significant leveling. To flatten the site, 110 million cubic yards of earth were moved, about one-third the amount of earth moved to build the Panama Canal. If dumped in one place, it

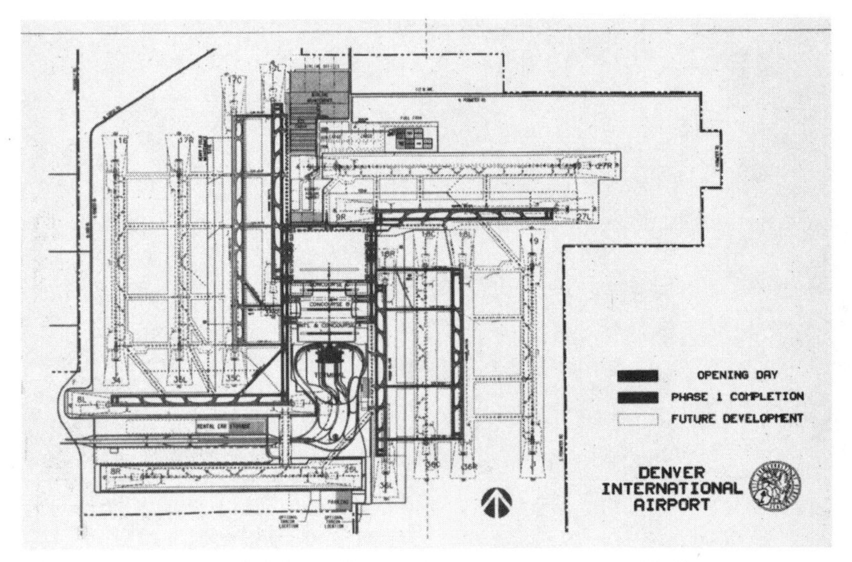

Fig. 7-9. *DIA airfield configuration at opening, with five runways, and a sixth originally planned to be opened one year later.*

would cover 32 city blocks in a quarter mile of dirt (Hoversten 1995). It would be enough to fill a 10-by-30-foot ditch stretching from Denver to New York City (Searles 1992).

DIA's runways are designed and sized to satisfy the FAA's "Group 6" category aircraft, the 600-seat planes that some aircraft manufacturers eventually would like to build (O'Driscoll 1994a). DIA opened with three north-south and two east-west runways, all surrounding the terminal and concourse complex. Each runway is 12,000 feet (3658 meters) long and 150 feet (46 meters) wide. In contrast, Stapleton International Airport had only one 12,000-foot runway (Nordwall 1993). Specifications for the runways called for a 40-year concrete life instead of the conventional 20-year standard (Fig. 7-10).

Construction began on a sixth runway but was canceled in May 1994 as the airport faced delays and cost overruns, and Congress became concerned over various allegations of illegality (see Chapter 9). In October 1995, the city awarded a contract to complete initial grading of the sixth runway, which will be located parallel to the north-south runway in DIA's northwest quadrant (Flynn 1995b). In 1996, Congressional funding was derailed when Mayor Webb failed to submit a letter pledging that all future airport contracts would be issued on the basis of the lowest qualified bid. Figure 7-11 shows the location of the planned runway.

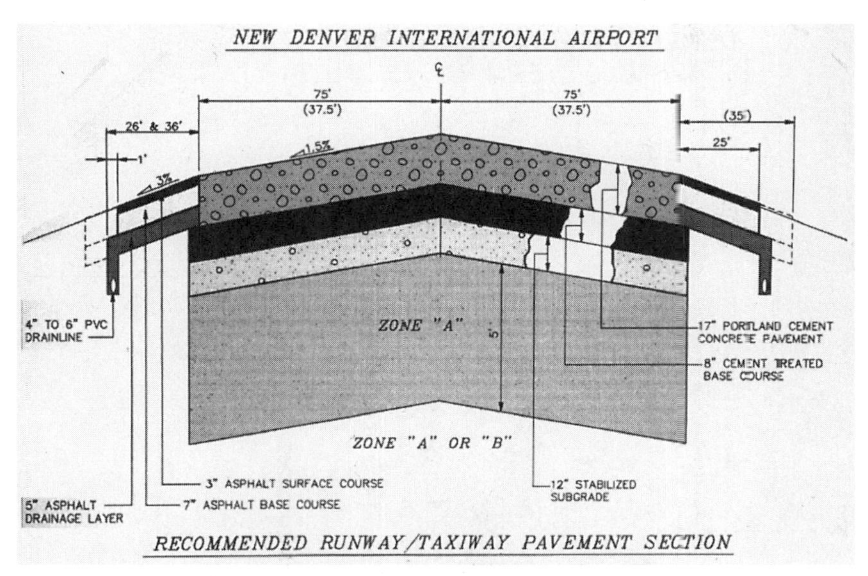

Fig. 7-10. *Advanced runway and taxiway pavement technologies were embraced at DIA to give them longer life.*

The sixth runway would be 16,000 feet (more than 3 miles or 4800 meters) long, allegedly sufficient for a fully loaded wide-bodied aircraft to take off for Asia in Denver's thin air and sizzling summer temperatures (Fohn 1994; O'Driscoll 1994a; *Denver Post* 1995a). Like Johannesburg, South Africa, Denver sits at an elevation (5200 feet or 1 mile) that, because of its thin, dry air, makes takeoffs of heavy aircraft difficult during hot summer days. Boeing and McDonnell-Douglas studied Denver's problem and concluded that Asia is within Denver's reach. At 87 degrees Fahrenheit, a loaded Boeing 747-400 can reach Seoul, Korea, Beijing, China, Europe and all of South America. At 95 degrees Fahrenheit, that plane can still reach Tokyo, most of western Europe, and all of South America from Denver, at a takeoff speed below 235 miles per hour, which is the temperature breakpoint for aircraft tires (Veazey 1995). Apparently, 3 miles of runway will do the trick.

The spacing between the runways (at least 4300 feet, as required by the FAA to accommodate simultaneous instrument approaches) gives air traffic controllers flexibility to move traffic from one runway to another because of changes in wind direction or inclement weather. In fact, the north-south runways east of the terminal are 5700 feet apart; the north-south runways west of the terminal are 7600 feet apart, and the east-west runways are 13,500 feet apart (Searles 1992). Each of

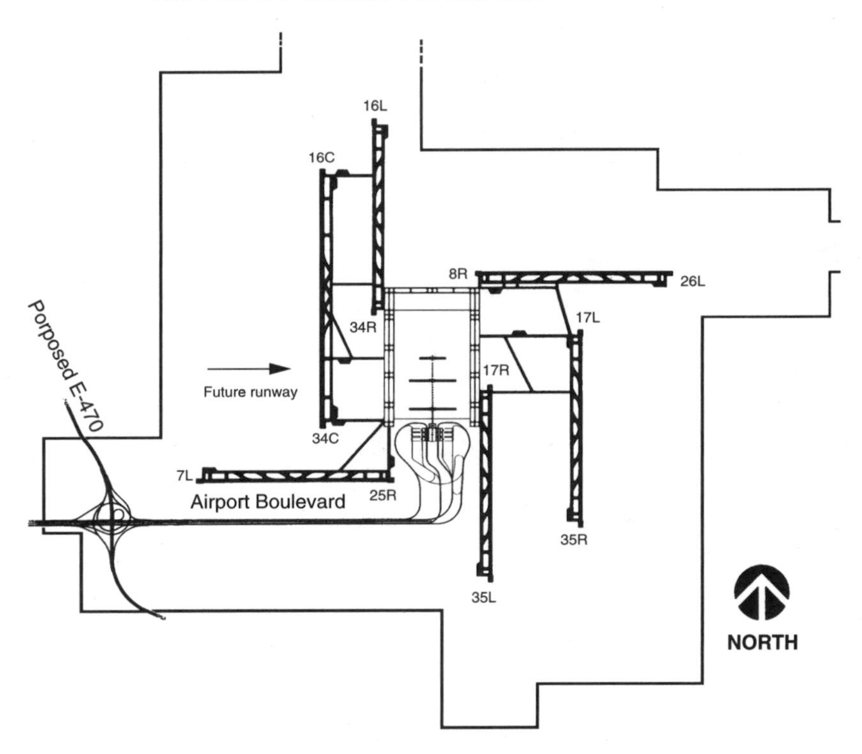

Fig. 7-11. *DIA airfield layout with six runways.*

the quadrants is separated by 3 nautical miles and are therefore operable as separate and independent airfields (Veazey 1995).

The land mass is so vast that one quadrant of DIA can have debilitating inclement weather, while another can be free of it. With such an enormous airfield, you need an air traffic tower that scrapes the sky. The 327-foot air traffic control tower is the tallest in North America (Fig. 7-12). It is designed to sway no more than ½ inch in an 86-mile-per-hour wind. Each panel of distortion-free glass in the tower weighs 11,000 pounds. On a clear day, you can see Wyoming and New Mexico (Lopez 1995).

No runway crosses another, thereby reducing the possibility of delays and collisions. The runways and taxiways are also embedded with 18,000 white, green, yellow, red, and blue lights and sensors, designed to lead the aircraft around the airfield without collision

Fig. 7-12. *DIA's 33-story air traffic control tower, the highest in the world, essential to view the world's largest airfield.*

(Thomas 1994). These lights and sensors help guide pilots during periods of low visibility. DIA is one of only four airports in the United States with an advanced airfield lighting system (Nordwall 1993).

The runways are configured around a central terminal core with widely spaced airside concourses to minimize taxi distance and time and thereby enhance operational efficiency (Fig. 7-13). As one source noted:

> *Each runway has high-speed exit taxiways from the runway to the parallel taxiway. This allows landing aircraft to exit the runway quickly, reduce runway occupancy time, and maximize runway capacity. Parallel taxiways for all runways are located between the runway and the concourses, permitting the aircraft to exit or enter the runway directly from the concourses without crossing a runway. There are also a number of crossfield taxiways that allow more efficient taxiing from the runways to the concourses. A few of the crossfield taxiways are angled to provide for a much shorter taxi distance (Rue 1993).*

Because the runways are staggered, with runways originating or terminating near the concourses, an aircraft can land on one runway and taxi directly to its gate without stopping (Nordwall 1993). Spacing

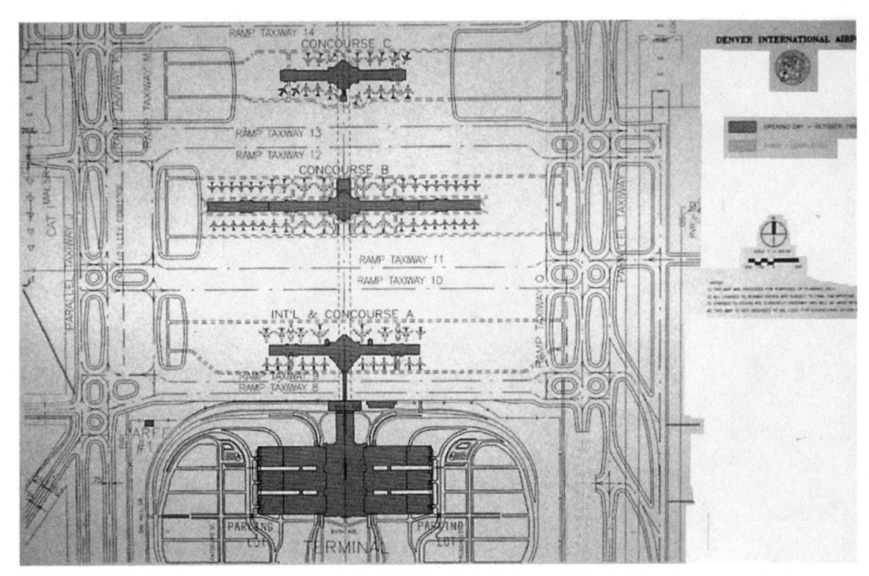

Fig. 7-13. *Adequate spacing between DIA's concourses was essential to allow the unimpeded flow of aircraft between them and to and from their gates.*

between concourses also permits aircraft to taxi in both directions and to back aircraft away from gates without encountering another aircraft (Del Rosso 1994). After the aircraft deplanes and enplanes passengers, it can back away from its gate and taxi directly to the end of a runway, pursuing a takeoff in the opposite direction from which it landed.

As a consequence, DIA may be the most efficient large-hub airport in the world. As explained in Chapter 2, hubbing involves flying large numbers of aircraft into an airport, transferring passengers and baggage between them, and flying them all out again. Because of DIA's airfield design and concourse spacing, the effects of weather and congestion will be minimized as impediments to the free flow of aircraft into and out of the airport. Other major hubs are plagued with delays as aircraft fly in holding patterns waiting for space to land or sit in "penalty boxes" on the ground waiting for gates and access taxiways to become clear. Despite its marketing and yield advantages, hubbing is a high-cost distribution system that sacrifices aircraft and flight crew utilization and fuel consumption, while creating enormous peaks and valleys in gate and corresponding ground crew utilization. After deregulation, all major airlines, with the sole exception of Southwest, adopted hub-and-spoke systems. After deregulation, only Southwest (which flies a pre-deregulation linear-route system) experienced consistent profitability.

The principal benefit to the hubbing airlines that use DIA is its relative efficiency vis-à-vis other major airports in the world, which enhances equipment and labor utilization and fuel consumption by reducing flight and ground delay and compressing the hub rotation—the window of time in which hub aircraft are landed, passengers and baggage are transferred, and aircraft take off again.

But one source pointed out that DIA's efficiency will, on occasion, be less than optimal:

> *Myth of the month: The airport's layout allows planes always to land toward the concourses, park, then depart away from them, saving taxi time and fuel.*
>
> *The facts: While this is true most of the time, it won't always happen that way. In certain weather conditions and in peak periods, some planes will land to the north on the main departure runway, necessitating a 180-degree turn and a 2- to 3-mile taxi back to the concourses. Similarly, departures to the north sometimes will occur on the two southernmost north-south runways, requiring a long taxi to the southern ends of those strips (Flynn 1993).*

Furthermore, building an airport geared to meeting the needs of hub airlines and connecting passengers is not necessarily the most convenient or efficient design for local origin-and-destination passengers.

DIA technologies

In 1991, the FAA designated four airports, including Denver, as demonstration projects on the ground movement of aircraft. Throughout the design process, DIA planners focused on safety and efficiency. Each runway has a 400-foot asphalt blast pad at each end. Advanced lighting, signage, and control were all embraced to avoid runway collision. Additionally, the following technologies were adopted:

- Runway centerline, edge, and touchdown zone lights for all five runways

- Centerline lights for all taxiways and taxi lanes
- Taxiway centerline lights into the runway for all exit taxiways
- Taxiway edge lights for all taxiways and apron edges
- Runway and taxiway signage
- Clearance bar lights at critical taxiway intersections, ILS critical areas, and on taxiways throughout the runway approach
- Circuit selector switch gear on the airfield to control stop bars and lead-on/lead-off lights (Rue 1993).
- Advanced weather tracking (including 29 wind speed and direction sensors scattered around the airfield to gather windshear information)
- Navigational aids (including a microwave landing system), 152 miles of fiberoptic communications
- A radar center nearly 10 times the size of Stapleton's (Broderick 1995b)

The key elements of the major technologies are described in the following sections.

Radars

The Terminal Radar Approach Facility (TRACON) is not a new technology, but it is being used innovatively to permit three planes to land simultaneously on three separate runways, even in zero-visibility conditions. The 68 TRACON controllers and seven supervisors (as compared to 43 and seven at Stapleton) work in a building far from the terminal and the concourses. There they monitor the flights on high-resolution screens. The system is driven by a new technology, Final Monitor Aid (FMA), which is designed to provide a better, faster, and more accurate picture of aircraft location than traditional radar (Thomas 1994). It enables the controllers to adjust the view of the airspace from a high-level perspective to one that shows only a section of a runway. The controllers watch the planes as they enter the three runways, and, if one leaves the "safety zone," voice and display alarms flash, and the controller can use a frequency shared with the control tower to warn the pilot. The new system doubles the number of inbound and outbound routes from four to eight

(Kowalski 1994; Monroe 1994; Park 1994). Its technical features have been summarized as follows:

> *The premise behind the FMA is to give controllers additional time to perceive a potential problem developing when three airplanes are approaching the airport on parallel ILS approach courses [that] are about a mile apart. One component of the system is an ASR-9 radar unit, which senses Mode S transponder returns. This signal from the system is seven times more accurate than current air traffic control radar beacon systems. Additionally, the system incorporates very high resolution controller monitors and a 4.8 second refresh rate for the precise control of approaching aircraft (Sellars 1993).*

The FAA is also testing a new tool that alerts controllers three hours in advance of a plane's estimated arrival time (ETA). The Terminal Traffic Management Advisor (TTMA) uses an algorithm to analyze data gathered by FAA radars and presents the ETAs next to a radar display of the actual airspace. By allowing controllers to make early decisions, it is hoped that more planes will be able to land on schedule (Monroe 1994).

Surface detection radars (ASDE-3) have also been deployed at DIA. These radars are designed to prevent runway accidents by providing the controllers with an accurate picture of the position and movement of all aircraft, vehicles, and baggage carts in all kinds of weather (Phillips 1995). As we shall see, performance of these radar systems has been less than perfect.

Airport lighting

DIA features an advanced airport lighting system, consisting of 18,000 lights, as well as a low-visibility Surface Movement Guidance and Control System (SMGCS) plan. The latter ensures that planes will be guided across the long taxiways with a steady stream of guiding lights (Park 1994). The advanced airfield lighting technologies are controlled by a unique touch-screen computer system, which, because of its ease of operation, contributes greatly to enhanced safety and efficiency. Located in the control tower, it permits the air traffic controllers to communicate quickly and accurately with the pilots in any weather. It consists of four touch-screen monitors, each of which can handle all the airfield operations. The operator obtains a representation of a runway or a section of a runway and selects the illumination (in sequence and at optimal intensity for the weather conditions) to guide the pilot to or from a taxiway. Other lights that might mislead the pilot are automatically switched off.

Communications system

The data and voice communications at DIA (including the "smart" runway system) are handled through a $40 million communications network, based on SONET technology, that is one of the most advanced to be found at any airport in the world. DIA is the only airport that uses SONET technology on such a scale. Designed as an integral part of the airport, the system is totally self-contained and controlled by the city. It is the largest privately owned network of its type in the world. Marshall Parsons, an expert who worked on the project, described the challenge and achievement:

> *Airports have extensive communication requirements, from voice to security and network management, making them among the most challenging of all communications projects. . . . However, Denver has established a new standard for airport communications systems . . . . There's no question that DIA is going to serve as a communications model for multitenant campus environments for years to come (Kissler 1994).*

The system (multiplexed transmission through fiber) provides both voice and data services to all the tenants, whether they be coffee shops or car rental agencies, and accommodates the network protocols and architecture requirements of the airlines in regards to their reservation systems and the airport's weather, baggage, and flight information systems. It also provides video security and an 80-channel TV network. In addition, it tracks the thousands of commercial vehicles using the airport daily—each is fitted with bar-coded tags that are scanned and read when they enter and exit the facility.

The network consists of 5300 miles of fiberoptic cable. Approximately 9000 "universal" voice-data outlets are located throughout the airport. The network incorporates SONET transmission systems and Fujitsu network-management software. The SONET technology permits easy expansion because it uses rings and sub-rings, the transport speeds of which can be enhanced without new fiber installations and which can be extended to new facilities. The system can handle technological developments of all types, such as broadband services, switched multimegabit data service, frame rely, and asynchronous transfer mode (ATM), and is designed to handle a 400 percent increase in capacity. The technology also ensures survivability for its ring architecture, which incorporates total redundancy and diversity routing, making it relatively invulnerable to the lightning storms that are commonplace at its location (USWest).

People movers

Given the distances between the terminal and the concourses, an underground train system was required (Fig. 7-14). Accordingly, AEG Westinghouse designed and built the Automated Guideway Transit System (AGTS), which operates through two parallel tunnels running north and south from the middle of the terminal to the concourses. There are five two-car trains plus six spare cars. The trains travel at 30 miles per hour and arrive at a station every 90 seconds. They can move almost 6000 passengers per hour in each direction and make the round-trip in 10 minutes. Track crossovers, placed in strategic locations, provide enhanced reliability, since a breakdown will not prevent the other trains from operating normally, at least in theory. A spare train is always at the ready. The AGTS is controlled by a

Fig. 7-14. *The underground people-mover train at DIA extends from the main terminal to all three concourses.*

sophisticated computer system with large-screen, high-resolution monitors with touch-screen capabilities. The system underwent extensive testing before the airport opened to ensure that the downtime would not exceed 2 percent (Lawrence n.d.). However, passenger loads exceeded estimates, and additional cars and entry and exit doors had to be added. Furthermore, as we shall see, unexpected problems have arisen with respect to the people-mover system.

Baggage system

Although important advances have been made in baggage-handling systems technology, the traditional "tug and cart" system remains the most common, although its speed and efficiency have been improved through such innovations as sorting at each concourse. Conveyor belts are also commonly used for short distances, and that technology has advanced greatly in recent years. Programmable logic controllers, bar-code scanners, diverters, and tracking software have enhanced its speed, flexibility, and reliability (Knill 1994b). Fully automated systems that take a single bag in a destination-coded vehicle (DCV) have been developed and installed at various airports. But typically, these systems are small, used by a single airline, and loaded manually. San Francisco's automated system has only 300 carts, loads on a static spur and uses simple tracking (Knill 1994a), a far simpler and smaller system than the one at Denver. Frankfurt's

Mike Keefe, *Denver Post.* Reprinted by permission.

airport has an integrated system, but it uses conveyor belts rather than individual carts. Furthermore, operational and maintenance problems drive the costs up, and some airports (e.g., DFW and JFK) have reverted to more conventional systems (Bozman 1994).

The DIA system was designed to be the largest and most sophisticated in the world. It was to be fully automated, but not with conveyor belts or in the concourse-specific way. On the contrary, the system was designed to serve the whole airport and to alter a bag's destination if a change was required (Rifkin 1994).[2]

The system is essentially a continuously operating miniature railroad (Fig. 7-15). Anyone who has experienced Disneyland's Space Mountain would have an idea of what the carts look like whizzing around corners on their rail tracks. Five loops, each with inbound and outbound sides, tie the terminal to the concourses. Concourse B has its own loop for transfer luggage. Small, free-rolling cars with trays that tilt to load, carry, and unload, run along tracks suspended from the terminal basement ceiling through tunnels linking the terminal to the three concourses. Linear induction motors mounted at various points along the track propel the cars with alternating magnetic forces. Overall, 20 miles of track and 6 miles of conveyor belts are involved. On these tracks run 3100 regular and 450 large cars (designed to handle skis and other oversize baggage) traveling at speeds up to 17 miles an hour—three times the speed of conveyor belts. The system is designed to move 700 bags per minute to their load or destination points in less than 10 minutes and in six minutes on United's kilometer-long Concourse B.

The system includes 5000 electric motors, 2700 photocells, 59 laser bar-code reader stations, 311 radio-frequency readers, more than 150 computers, workstations, and communication servers, and 14 million feet of wire (Rifkin 1994).

A software program sends empty carts to appropriate destinations as needed. For the system to work, the bags that are loaded onto the carts must be recognized when they are placed on the conveyor belt in the terminal and when they are dropped into the cart. Bar-coded tags are attached to each piece of luggage.

2. Rifkin (1994) contains a fine description of the baggage system and is the primary source for the following section.

Fig. 7-15. *The automated baggage system at DIA became among its most controversial and costly components.*

As the bag moves down the conveyor belt, it passes a series of laser scanners that sends the data to the computer, which, in turn, sends a message to the BAE sorting computer, which identifies the flight number, matches it with a gate, and transmits the data to a radio-frequency identification tag on the moving cart into which the bag is deposited. Throughout, the BAE computer tracks the cart with the specific bag via a series of radio transponders that read the labels at various points along the track as it is directed to its destination, where it unloads the bag (Rifkin 1994). The system operates 18 hours per day, every day.

Such a system "posed unprecedented performance requirements" because of the distances involved, the amount of baggage, the delivery times, and its complete integration (Di Fonso 1993). But the corollary, that meeting such requirements necessitated the use of technologies that had never been used in a project of this scale and complexity, remained unspoken. It soon became clear that the city had a disaster on its hands, especially since the city, driven by what proved to be unseemly pride and unrealistic expectations, invited the media, TV cameras and all, to witness the launch on a small loop of the system. Instead of smoothly running carts, mayhem ensued with misloaded bags, jammed carts, spilled luggage, and general chaos. How and why this situation arose is discussed in Chapter 9. Here we consider only the technical aspects of the problems.

That the system encountered major difficulties should not have been surprising and was, in fact, anticipated by various consultants and experts. The system was on the technological cutting edge and incorporated elements that themselves had never been tested on the scale and complexity of DIA. The successful functioning of the system depended on several elements working perfectly, individually and in integration with the other components. These elements included:

1. The mechanical structure with its complex carts and rails
2. The computer software system
3. The laser scanners

Each of these encountered significant problems (Knill 1994a).

The concept of automated cart systems dates back to the early 1970s. Such systems were installed at several airports, but, as noted, each proved troublesome or was far simpler and smaller than the Denver project. At DIA, for example, the linear induction-propelled carts had fins on the bottom that slid through motors mounted under the tracks. The state of this technology has been described by Knill as follows:

> *Fast straight-ahead travel is no problem. Things get sticky when stops, slowdowns, and diverts have to be built into a high-speed system that has plenty of curves, inclines, and declines. The use of vertical friction wheels and permanent magnets to fine-tune a linear induction motor system is more an art than a science in a system the size of DIA (Knill 1994a).*

The use of scanners to pass information from bar codes to a radio-frequency identification system is a proven technology for industrial material handling systems. Its use in a complex, high-speed baggage system represented a first (Knill 1994a). Hence, numerous failures occurred because the scanners were not always positioned perfectly or the bar codes were not readable because of dust and dirt.

Even more troublesome was the programming. Ginger Evans, the chief of construction for DIA, estimated in May 1994 that 65 percent of the problem lay with the software and 35 percent with the mechanical aspects (Knill 1994a). Several factors generated problems.

The system had to interact with each airline's computers, a difficult and error-prone undertaking. Furthermore, the software proved unable to send carts out as required so that jams became commonplace (Rifkin 1994). The complexity and scale of the system proved its undoing. The computer software was simply incapable of controlling the varied routings of so many cars and the location of so many bags throughout the airport (Auguston 1994; Flynn 1995a).

Could these technical problems have been foreseen? Almost certainly—if the planners had understood the difficulties in developing large-scale systems, especially those that involve large technological leaps. There is no doubt that the entire project involved high risks, risks that were never properly appreciated, and inadequate attention was paid to how these might be minimized. As Professor Richard de Neufville of MIT noted in 1994, the project had been doomed from the start because highly variable systems never function according to theoretical performance. In his words:

- *There was inadequate consideration of the potential for "car starving." Enough empty cars must be provided to fulfill peak capacity, which means that the effective capacity of an automated baggage system can be only half of its theoretical capacity.*
- *Despite the manifest risks, the city had developed no contingency plans for the system's failure. The decision for a backup baggage system was not made until 10 months after the originally scheduled opening date.*
- *Risks were not recognized, and contingency plans were not developed to deal with these risks (de Neufville 1995).*

Such plans could have involved a detailed simulation run to identify the breakpoints in the system, design for peak loads, prototyping new technologies, and hedging by building a backup system (Auguston 1994). The planners of Munich's airport, for example, tested all their systems for months before the scheduled opening—and they tested the relatively simple baggage system, based on conveyor belts, for a whole year (O'Driscoll 1994b).

Confronted with the need to open the airport, extensive discussions involving United, the city, BAE (the system designer), and external consultants, resulted in a decision in September 1994 to drastically simplify the system to serve, at first, only United's concourse and, simultaneously, to build an alternative system of traditional tugs and carts to serve the other concourses.

Since the load was too great for the computers, United reduced by half, to 30, the number of carts per second on each track. Furthermore, the amount of track serving its concourse was expanded; and the loop assigned to Concourse C was modified and linked to the Concourse B loops (Bozman 1994; O'Connor 1995). Together with other changes that affected practically every component—the motors, the track, the hardware, the software, and the supports to minimize the extensive vibration—the system became sufficiently operational for Concourse B so that DIA was finally able to open on February 28, 1995. By the summer of 1995 it handled more than 600 bags per minute, with a capacity of 1000. The inbound portion became operational shortly thereafter and the system was then to be expanded to Concourse A (O'Connor 1995).

The decision to limit the automated system to Concourse B led to major protests by the other airlines, especially American Airlines, located at Concourse C, the farthest from the terminal, which feared that United would gain a significant competitive advantage because its baggage would be delivered much more expeditiously (Leib 1995). Ironically, these concerns proved to be unfounded—the alternate system that used conventional technology—conveyor belts, baggage-claim units, baggage sorters—worked so well, even for the airlines located on Concourse C, that the tenant airlines opposed plans to expand the automated system there and, when the city insisted that it be expanded to Concourse A, two airlines announced that they would move to Concourse C rather than pay the additional costs imposed by the automated baggage system. Ultimately, however, the Concourse A carriers were billed for the system, although it has never been operational for them.

The backup system consists of outbound conveyor belts that move the bags from the ticket counters and curbside to airline bag stations that were built in one of the parking structures. Each bag is tagged with a bar code that is read by laser scanners, which relay the information to a controller that runs a sorter that funnels each bag to one of two carousels. From there the luggage is loaded manually on carts that are tugged to the appropriate concourse through the tunnels and loaded on the plane. Incoming luggage is carried by tugs and carts to the bag station, where it is transferred manually to the baggage-claim carousels. This system will eventually become the backup (O'Connor 1995).

Security

Aviation security has become a major problem for airport planners and administrators throughout the world, since airplanes and airports have come to represent tempting targets for terrorists. The 1980s was a disastrous decade for aviation and confirmed the existence of a dangerous trend toward greater violence against air transport. Overall, 26 planes suffered from explosions, causing 1207 casualties, as compared to 650 deaths caused by 44 explosions in the 1970s and 286 deaths from 15 explosions in the 1960s.

These developments revealed a troubling fact—the number of deaths was increasing because terrorists were changing their tactics and making use of new technologies. In the 1960s, the favorite terrorist tactic had been hijacking major airliners. This threat to airline travel was met in the United States and abroad by the development and installation of passenger screening devices and processes at all major airports. So successful were these measures that, within a few years, the number of hijackings had dropped to a handful per year.

Terrorists adapted quickly and switched their tactics. They now had access to more sophisticated and lethal technologies—automatic weapons and deadly plastic explosives, such as SEMTEX, that are easily molded and shaped. They attacked airports using pistols and bazookas (between 1975 and 1985, 30 incidents took 156 lives) and developed numerous ingenious ways to turn innocuous-looking suitcases and radios into lethal bombs. As a result, in 1992 there were 20 hijackings of non-U.S. air carriers and 800 bomb threats worldwide.

In response to the heightened level of security challenges to civil aviation, new measures were mandated by the Aviation Security Improvement Act of 1990. Among its provisions was a charge to the FAA to oversee security at major airports. Responsibility for supervising the security arrangements was assigned to FAA security managers. The task of security officials was expanded well beyond checking passengers as they boarded to safeguarding the entire airport from terrorist acts. In addition, the FAA now carries out periodic threat-and-vulnerability assessments and publishes guidelines on such topics as airport design, public notification of threats, security personnel investigation and training, cargo and mail screening, research and development activities, and international security negotiations. The Federal Aviation Regulations (FARs) lay out general requirements concerning systems, procedures, and technologies. Of particular relevance is FAR 107, which deals with airports (Laird 1994).[3]

These guidelines were taken into account by DIA's project management team, although no security committee as such was established. Rather, DIA's security manager and the FAA's federal security manager (a position created by the Aviation Security Improvement Act of 1990) worked with all the relevant committees, such as the Communications Committee and the Technical Committee, to ensure that security concerns were taken into account in every part of the project.

Essentially these concerns were divided into "airside" and "landside" elements. The former included access to the airport operations area (AOA), location of facilities, location of the security identification display area (SIDA), separation of general aviation and unscreened passenger flights from screened flights, doors and gates providing AOA access, airside roads adjacent to perimeter fences, perimeter and ramp lighting, isolated aircraft parking positions, a K-9 kennel facility, a magazine for explosive storage, bomb-search building, and a bomb disposal area. The "landside" considerations included: location of public parking, use of glass, screening checkpoints, sterile areas, location of public lockers, doors providing AOA access, observation decks, security support facilities, emergency command post, and protection of vulnerable areas. One important innovation was to have the predeparture gate screening conducted in a centralized location in the terminal prior to boarding the train rather than at the entrance to each concourse.

Security considerations, however, were often a secondary consideration when key decisions were made, so security was not as integrated as thoroughly as might have been anticipated and is not optimal from the FAA's perspective. Still, the end result is generally judged to be a good one. Furthermore, DIA features practically all the latest security technologies, including up-to-date passenger-screening devices, surveillance equipment, and related technologies.

3. For this and the following section, we have drawn on Laird (1994).

Especially noteworthy is its unique computerized access control system, which is the most advanced in the world today (Fig. 7-16).

Designing a system for a facility with a 30-mile perimeter and 6.5 million square feet of floor space was a complex undertaking.[4] It provides each employee with color-coded and magnetically coded badges that permit authorized personnel to access areas on a "per person/per portal basis." In other words, those who need access to certain locations are able to enter, but only at certain times and in certain areas. Cleared personnel cannot enter into areas they should not be in or enter authorized areas when they are not on duty. Another feature of the system is its "anti-tailgating" capability, which ensures that only one person or vehicle passes through any of the 25 different types of doors and vehicle gates per card.

The entire system is fully integrated with a state-of-the-art distributed computer system that permits information to be sent from the central database to field computers and control panels. Intelligent device controllers not only handle the transactions of up to 20,000 cardholders each but also constantly monitor all door alarm contacts, magnetic door locks, photoelectric beams, gate controllers, help

Fig. 7-16. *The command center at DIA allows centralized computer oversight of all major systems.*

4. The system is described in detail in Howard and Rochon (1993).

buttons, and enclosure tamper switches. The cardholder data can be changed instantly by the central computer and the individual transactions can be uploaded to the mainframe.

Because it is a distributed system, a failure is confined to the computer on the specific data line. In other words, the computer system is set up in a "fault-tolerant" configuration, an important consideration because security specialists now recognize the centrality of computers in security systems. Sabotaging the computers compromises the security system. The DIA system effectively safeguards against this eventuality.

Snow, wind, and technology

Because of its layout and technologies, DIA is allegedly "the world's first all-weather airport," capable of landing 99 aircraft per hour in a snowstorm, compared with Stapleton's 80 in perfect weather (Searles 1992). The four-quadrant runway design allows triple simultaneous Category III ILS (instrument) landings during inclement weather, something no other airport can do (Green 1992). This is possible not only because of the runway configuration, but because of the technologies previously discussed. Nevertheless, DIA will suffer from delays during heavy, blowing snows and freezing temperatures, because snow-removal equipment will still have to clear the runways (Broderick 1995a). At most airports, de-icing occurs at the gates. At DIA, large de-icing pads, each capable of simultaneously de-icing six aircraft, are located at the end of each concourse and near the runways, to minimize circuitry and the amount of time between de-icing and departure (Nordwall 1993).

The first Category III instrument landings at DIA were remarkably consistent, with each aircraft's tire tracks coming down precisely on top of the preceding one, just 6 inches to the left of the runway centerline. A subsequent adjustment moved them 6 inches right, to line up perfectly with the centerline (Brewer 1995). With the construction of additional runways, DIA will eventually be able to accommodate quadruple simultaneous independent parallel ILS approaches (Thomas 1994). According to FAA Administrator David Hinson, "because we have essentially the same runway capacity in either IMC or VMC [instrument or visual meteorological conditions], we will not have the ripple effects we had before. Delays in Denver . . . for the most part will disappear" (Scott 1995).

This prediction was borne out in its first months of operation, since DIA exhibited enhanced operational efficiency vis-à-vis the airport it replaced, in all kinds of weather. During DIA's first two hours of operation, controllers cleared aircraft to land in low visibility at the rate of 92 per hour, compared with Stapleton's maximum of 32 per hour under the same conditions. DIA's official maximum landing rate is 120 per hour. It handled 42,000 flight operations during its first month, only about one-half of one percent of which were delayed. At Stapleton, delays were 3.3 percent of flight operations during a like period the preceding year (*Rocky Mountain News* 1995). During DIA's first six months, it experienced 1030 flight delays (or 0.43 percent of operations), compared with 5550 (or 2.04 percent of operations) at Stapleton. That made DIA's performance more than twice as good as Minneapolis, three times as good as Detroit, four times as good as Boston, and seven times better than St. Louis.

Especially noteworthy is the fact that during snowstorms, DIA accommodated triple landings. But automobile access to and from the airport proved to be difficult, with multiple accidents, and no snow fences to block the blowing snow on Peña Boulevard (O'Driscoll 1995). Thus, local O&D passengers missed flights, while connecting passengers enjoyed timely connections.

But the first real test of how the "all-weather airport" performed in blizzard conditions was October 22 and 23, 1995, when everything that could go wrong, did go wrong:

1. Although only 5 inches of snow fell, winds reached 60 miles per hour, making it impossible for snow plows to keep up with drifts on runways and surrounding roads.
2. A United Airlines 727 aborted a landing to avoid a Chevy Suburban parked on the wrong runway (ground radar with cataracts did not pick up the misplaced vehicle), and hours later, a power failure shut down the ground radar altogether.
3. A United Express commuter aircraft made a wrong turn and got stuck in snow drifts on a taxiway, shutting down landings at the only open runway for 45 minutes.
4. More than a dozen United flights were diverted to Colorado Springs Airport, while 25 flights were canceled.
5. Delayed United Airlines aircraft had to wait on aprons because many gates were occupied because of flight delays (flights were delayed up to five hours).

6. The 327-foot FAA air traffic control tower's ceiling began to leak, requiring snow to be vacuumed off equipment and causing a ceiling tile to fall on a controller, with blowing snow entering a failed joint seal between an exterior wall panel and the metal flashing running around the parapet of the roof.
7. A baggage cart slid on the ice into a Mexicana jet, causing eight-hour departure delays.
8. The 56th Avenue access for employees was shut down, putting all automobiles on four-lane Peña Boulevard, which slowed to a crawl in bumper-to-bumper traffic, with ice and snow drifts amassing because snow fences had not yet been installed (Amole 1995; Booth 1995; Flynn 1995c; Meadow 1995). A few days later, winds reached 40 miles per hour, forcing the airport to use only its two east-west runways, shutting down the three north-south runways on three occasions (Gutierrez 1995).

DIA is located on the Denver Convergence Zone, an area where summer winds off the mountains meet opposing winds off the Palmer Divide, converging to generate thunderstorms, tornadoes, and brisk gusting. The National Center for Atmospheric Research found that "Wind speeds at the new airport are significantly higher than at Stapleton. Winds last three to four hours longer . . . . Incidence of storms is greater . . . . High winds are more common" (*High Country News* 1995). The flatness of the desolate land denies DIA any natural object to block high winds.

However, a report by the U.S. General Accounting Office concluded that "potentially dangerous storms do not appear to be more prevalent at the new airport site than elsewhere in Denver [and] current data suggest no discernible differences in the occurrence of severe storms between the two airport sites" (GAO 1991). Nevertheless, air traffic controllers and airport officials have complained that the new weather-reporting systems at DIA do not work as well as the human observation system they replaced, because the new systems take frequent readings but miss longer-range weather issues only human eyes can see (*Denver Post* 1995b).

Alternative terminal designs

Several general types of landside and airside terminal designs have been embraced by major U.S. airports. In this section, we examine several of the most prominent types.

Denver's Stapleton International Airport terminal building was much like the ad hoc pier finger designs of airport terminal buildings around the world, consisting of a ticketing (upstairs) and baggage claim (downstairs) complex tied to long concourses jutting awkwardly out from it, in what looks like sort of an E design from the air. Typically, such airport terminals originally were built in the piston era of commercial aviation, then expanded over the years in a clumsy metamorphosis reflecting Band-Aid–type solutions to immediate capacity demand constrained by available land (Wells 1992, p. 147). Passenger connections between concourses usually require a circuitous walk back to the main terminal area. Examples of such airports are abundant, with St. Louis Lambert International Airport, Minneapolis/St. Paul International Airport, and New York LaGuardia Airport as prominent examples.

In the 1950s, 1960s, and 1970s, airport planners became more creative about terminal design, attempting to accommodate the convenience of passengers with the needs of aircraft. New York Kennedy (JFK) Airport and Los Angeles International (LAX), both large international gateways, came up with an interesting design, consisting of several independent airline terminal buildings situated around a central highway core, shaped something like a U or an omega from the air. For example, LAX has one large international terminal at the bottom of the U, with four satellites on each side of the U. Buses carry passengers between domestic and international terminals for connections. While suitable for O&D traffic, and on-line connections, the distances between terminal buildings make interline connections (between different international and domestic aircraft) difficult (Wells 1992, p. 153).

Yet another type of terminal design is reflected by airports built at Kansas City and Dallas/Fort Worth (DFW), both of which were designed as magnificent O&D airports, with terminals shaped like strung-together Cs at DFW and isolated Cs at Kansas City. In the center of the C is ample parking so that passengers have a short walk to their gate. On the outside of the C are aircraft parked at gates. These types of terminals preceded airline deregulation and the massive shift of airline route systems to "hub-and-spoke" operations. While delightful from the perspective of the O&D passenger, such terminals could be major headaches for connecting passengers, since with gates on only one side of the terminal, the walk to the connecting aircraft could be some distance away from the arriving aircraft. DFW attempted to remedy this somewhat by building a train linking remote parts of its terminal facility.

Washington's Dulles International Airport inaugurated the mobile-lounge concept of transporting passengers from a main terminal directly to their aircraft. This has since been replaced with a central satellite airside concourse. The most awkward piece of the airport remains the large, sluggish, elevated bus-type vehicles used to transfer passengers from the aesthetically inspired landside terminal to the remote airside terminal.

Another design that emerged in this era was the mid-field airside satellite terminal, which allows aircraft to be parked at both sides, or around a circular central core, for more convenient connections. Until 1980, Atlanta's Hartsfield Airport had a terminal much like Stapleton's. But that year, it completed a wholly new terminal building with a revolutionary design developed by HOH Company and Stevens and Wilkinson Architects, embracing and expanding the mid-field airside terminal concept. It solved the Dulles passenger-transfer problem with an underground train. Atlanta's terminal is situated mid-field, enhancing efficiency of aircraft access to gates. It consists of a large landside terminal building sandwiched on two sides with viaducts and automobile parking and three parallel airside concourses that are shaped like I's when seen from the air. One source explained the key features of the design:

> *The key element of concern here was expandability and rapidity of passenger movement from A to B. It was thought that the best condition would be an [underground] automated people mover so that aircraft could traverse all sides of the concourse without impedance. A previous experiment in 1973 with a circular system at Seattle International Airport had already proved very confusing and very difficult, if not impossible. Therefore, it was proposed that the Atlanta concourses were to be developed as needed along with a perpendicular transportation axis sequentially away from the main terminal and accessed by an automated train traveling on a*

> *pinched-loop system. Such a concept would lend itself to expansion as long as space was available (Fentress 1995).*

As noted, Denver chose to perfect this design. More about that in a moment.

Several airports embraced the mid-field airside concourse accessible via underground transport. Seattle-Tacoma's remote concourse is accessible via underground train. The United Airlines Concourse B at Chicago O'Hare also reflects the mid-field airside concourse concept, although there the passenger connections are via moveable sidewalks through a futuristic Disneyland-type neon-and-tinkle-bell environment.

Two other designs are worthy of mention. One is Orlando's circular pod airside terminals linked to the landside terminal via aboveground monorail, whisking passengers over ponds and through palm trees—also appropriately Disney-like, given the location. Pittsburgh's recently reconfigured mid-field airside terminal is in the shape of a huge X, which allows efficient passenger transfers between aircraft (Fig. 7-17). The terminal is more like a shopping mall than a traditional airport, with BAA (the privatized British Airports Authority) leasing concessions and maximizing revenue like a shopping-mall landlord. Like Atlanta and DIA, Pittsburgh has a landside terminal accessible by automobile from two sides.

These, then, are several of the major terminal types that have been embraced in the United States. Professor Alexander Wells has observed:

> *Clearly, no single design is best for all circumstances. Traffic patterns, traffic volume, and flow characteristics (peaking), the policies of individual carriers using the airport, and local considerations (esthetics and civic pride) dictate different choices from airport to airport and from one time to another. The airport planner, who is required to anticipate conditions 10 to 15 years in the future, must often resort to guesswork. Even if the guess is correct initially, conditions change and result in a mismatch between terminal architecture and the traffic to be served. To guard against this, airport planners now tend to favor flexible designs that can be expanded modularly or offer the opportunity for low-cost, simple modifications as future circumstances might demand (Wells 1992, p. 153).*

Let us now examine the design chosen for DIA.

Fig. 7-17. *Pittsburgh reconfigured its airport with a mid-field terminal in the shape of an X.*

Landside terminal building

As noted above, at DIA, runway configuration preceded terminal design. As a hub airport, the emphasis was on the efficient movement of aircraft to gates, coupled with the ease of passenger transfer between aircraft, as well as fulfilling the needs of local O&D passengers.

The earlier thoughts about terminal design were something along the lines of a "combination of the long concourse airport, such as Dallas or Stapleton, and the 'hub' or 'pier' layout of Hartsfield International in Atlanta or Seattle-Tacoma International Airport" (Schwartzkopf 1986).

DIA's terminal building opened with 1.4 million square feet of floor space and 1200 linear feet of ticket counters (Fig. 7-18). At full build-out, it will have more than 2.2 million square feet and 2400 linear feet of ticket counters (*Aviation Week* 1991). In other words, the terminal could nearly be doubled in size with a carbon copy built to the south if passenger traffic exceeds 50 million passengers per year (O'Driscoll 1994a). If there is one obvious design flaw, it is that no landside train was incorporated into the basement to allow rail access. Given that Union Pacific Railroad trackage is located only a few miles away from DIA, a rail spur could have been built to accelerate

DIA's construction and used subsequently to ferry passengers to and from downtown Denver.

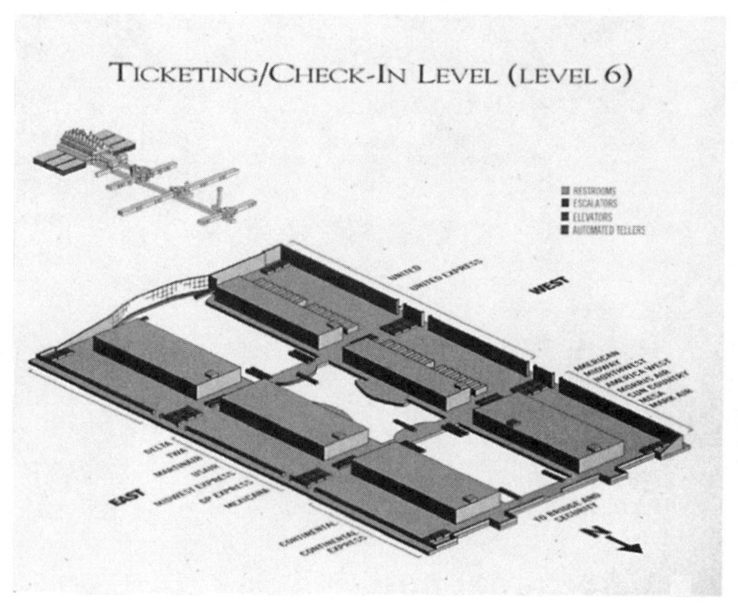

Fig. 7-18. *The ticketing level of DIA was designed for maximum ease of passenger flows.*

The main terminal building (Fig. 7-19) and each of the concourses were designed to incorporate ample retail stores, restaurants, pubs, and newsstands. The airport has more than 60 stores, 48 restaurants, and two information centers (Bidwell 1995).

Walking distances (between parking and ticketing, ticketing and departure gates, arrival gates and connecting gates or baggage claim, and baggage claim and ground transport) must be taken into account in terminal design. As Professor Wells notes:

> *One of the most important factors affecting the air traveler is walking distance. It begins when the passenger leaves the ground transportation vehicle and continues on to the ticket counter and to the point at which he or she boards the aircraft. Consequently, terminals are planned to minimize the walking distance by developing convenient auto parking facilities, convenient movements of passengers through the terminal complex, and conveyances that will permit fast and*

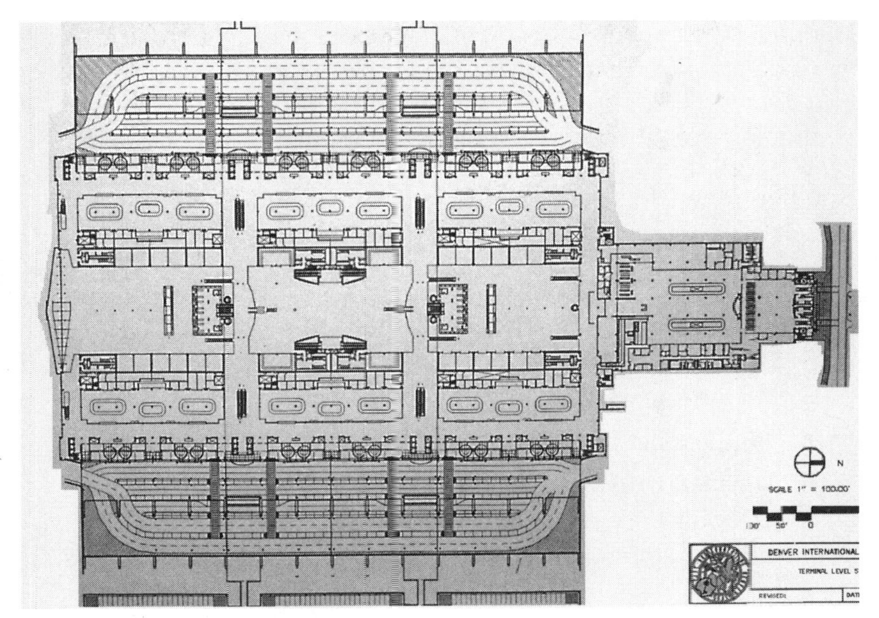

Fig. 7-19. *Schematic showing layout of the interior of the main terminal building at DIA.*

> *efficient handling of baggage. The planner normally establishes objectives for average walking distances from terminal points to parked aircraft. Conveyances for passengers such as moving walks and baggage systems are also considered (Wells 1992, p. 142).*

At DIA, 750 feet separates the most remote of the 13,000 parking spaces from the terminal (or 450 feet from the most remote space in the parking garages that sandwich the main terminal building). However, with the backup baggage system consuming a huge section of the parking decks, many passengers have complained about the insufficiency of parking spaces and the circuitous pedestrian traffic patterns to enter the main terminal from that level. (A new outdoor lot is being built to relieve the pressure.) Elsewhere, distances were considered carefully. The longest walk from the brushed chrome ticket counters to a gate is 450 yards—less if the moveable sidewalks are used. DIA's sidewalks are one and a half times wider than those at most airports and move one and a half times faster than at most airports (Hoversten 1995). Concourse B has four lanes of wide, parallel moving sidewalks to whisk passengers quickly to their gates. As noted by DIA architect Curt Fentress, modern travelers "require efficiency in movement from point-to-point—simple, easily understood circulation systems to help ease the stress of finding one's way in a strange new place with the pressure of a time clock continually weighing on each decision" (Fentress 1995).

Another thing DIA's planners did well was to limit the number of decisions a passenger had to make between ticketing to gate:

1. Check bags on upper level
2. Proceed down escalator to security
3. Proceed down another escalator to train
4. Get off train at concourse A, B, or C
5. Take escalator up two flights
6. Turn left or right and proceed to gate (usually by moving sidewalk).

Down or up, left or right—at any point in time, decision-making is limited to two simple choices. As a consequence, DIA is an easy airport for passengers to navigate.

However, several areas in the main terminal were designed with insufficient space:

1. Standing space in the ticketing and check-in area—35 feet of space was eliminated during the design process (Veazey 1995)
2. Standing area in the upper-level food court
3. Walking space and lack of direction in the train arrival area (where a tight bottleneck requires passengers departing from the train to make two sharp turns to go up the escalator on their way to the baggage claim area; at this point, passengers are disgorged into the cavernous Great Hall and can be seen rubbernecking their way to the hard-to-locate baggage claim areas).
4. The bottleneck of traffic passing through security and descending down escalators to the train level in the main terminal building
5. The bottleneck in escalators ascending from the train level, particularly in Concourse C

Some of the space problems might have been alleviated had the lower level of the Great Hall (which contains the security screening area leading to escalators that take passengers to trains departing for the concourses) not been designed to be so vast—aesthetics over efficiency.

As noted above, DIA is a splendid design for a connecting hub airport. But focusing on its role as a national hub necessarily caused trade-offs that sacrificed convenience for the O&D passenger. For example, locating the groundside terminal in the center of the airfield necessarily places it several miles further from downtown than locating it on a piece of airport property nearer the city, thereby increasing ground transit time, delay, and cost, as well as automobile emissions. Moreover, the primary access highway—Peña Boulevard—was laid out in a circuitous upside-down L-shape, adding unnecessary distance to the trip, despite the fact that the shortest distance between two points is a straight line (see Chapters 6 and 9). By late 1996, it was apparent that DIA's $6 million automobile toll plaza was inadequate.

Several parts of the terminal and landside access emphasize the focus of its designers on connecting traffic and have led to complaints by local O&D passengers. Beyond distant airport terminal location, these include:

1. The baggage claim level (level 5) is accessible only to public transport (taxis, shuttle vans, and buses) and not to private automobiles, which must pick up passengers on a lower level (level 4). Arriving passengers must haul their luggage down one level to be picked up by private automobile.
2. To get from the parking garage to the drop-off or pickup levels, or vice versa, automobiles must loop 3 miles around to return to the terminal area.
3. Indoor parking space is often inadequate, particularly since the backup baggage system consumes much of the level 3 parking garages.
4. Underground trains connecting the main terminal to the concourses are often packed.
5. Not until more than a year after DIA opened was a single gasoline station built along the 12-mile Peña Boulevard.
6. With only two lanes in each direction, Peña Boulevard can often be congested.
7. No snow fences were originally built to protect the boulevard from blowing snow.
8. Denver police turned the boulevard into a 55-mile-per-hour speed trap, which some complained was yet another brazen attempt by the city to extort more money from local DIA users (Allison 1995). After public outcry, the speed limit was raised to 65.

Design and planning lapses on the landside have created "crowded parking garages, poor signs and long lines at toll booths," prompting the city to hire even more consultants to recommend solutions (Kerwin 1995).

Out-of-town O&D passengers must share some of the inconveniences. A taxi ride from DIA to downtown Denver costs $40 one-way from a disgruntled taxi driver who has been sitting in a queue for three hours because the state liberalized taxicab regulation and flooded the market with additional taxi companies. The cost of car rentals also escalated because of a city-imposed surcharge. Rental cars originally were to have been located conveniently on the lower level of the parking garages but were subsequently moved parallel to Peña Boulevard, requiring a commute via bus or van. While an efficient design for hub-connecting passengers, DIA leaves something to be desired from the perspective of the O&D passenger.

Denver is unique among hub airports in the extremely large percentage of O&D traffic (58 percent) vis-à-vis connecting traffic (42 percent). In contrast, a good hub, from the dominant airline's perspective, produces at least one-third or one-quarter O&D (on whom higher ticket prices may be imposed). Denver's extraordinarily high proportion of O&D traffic stems from its geographic isolation from the rest of the nation, with Denver in the center of a vast underpopulated region. Other major cities are hundreds of miles away, making air transport the preferred means of travel for Denverites or others coming to Denver. In contrast, many eastern cities are only a few hours away from other major cities by automobile or train. Hence, those deficiencies resulting from trade-offs in airport design inconvenience more passengers at DIA than they might elsewhere.

Airside concourses and their spacing

DIA's master plan calls for four parallel east-to-west 40-gate airside terminal buildings (concourses) linked to the landside terminal via underground rail. The concourses have a maximum length of 4000 feet (approximately twice the length of the longest concourse at Atlanta's Hartsfield International Airport) and ultimately can accommodate up to 60 gates (depending on the size of the aircraft), as well as a number of commuter gates. Wide separation was built into the airport's design to allow aircraft to maneuver in and out of gates, while also allowing two lanes of wide-bodied aircraft taxiing between the concourses. As a consequence, Concourse C is nearly a mile north of the main terminal building. Airport spokesman Chuck Cannon emphasized the importance of adequate space:

> *One mistake at Atlanta we corrected: The concourses at Atlanta were too close to each other. Two 747 jumbo jets can't push back at the same time [from neighboring concourses] at Atlanta. Here, we left enough room so not only can two 747s push back at the same time, but also taxi both ways at the same time (Searles 1992).*

In 1989, the city signed a contract with the joint venture of Seracuse Lawler and Partners/TRA for the design of "three identical concourses" to be separated by dual taxiways and dedicated aircraft push-back zones. Originally, international arrivals were to be at gates directly connected to the main terminal building (McCagg 1993). As we shall see, the design of two of the concourses was significantly changed after construction began.

The design process began with the development of a project book that defined all quantitative and qualitative design requirements consistent with the DIA master plan, design standards, budgetary limits, and scheduling requirements. Because the airlines were not fully participating in the process, the task was to anticipate their requirements. Edward McCagg of TRA identified these as the goals that guided the design process:

- *Function*
  - ~ *Design facilities to be efficient in the movement of passengers and baggage, airline operations, concession operations, and management/maintenance of building facilities.*
  - ~ *Provide an optimum environment for the activities, systems, and equipment essential to each concourse.*
- *Flexibility*
  - ~ *Accommodate growth and change in terms of passengers served and quantity, size and type of aircraft used by airlines.*
  - ~ *Accommodate the evolution of key systems technologies and operational needs.*

- *Convenience*
  - ~ *Design components and systems with clarity of expression to facilitate their use. Access to facilities, services, and systems must be clearly and easily understood by passengers and visitors.*
  - ~ *Design circulation so wayfinding decisions are limited to two choices, up or down, left or right.*
- *Integration*
  - ~ *Respond to the overall airport complex in terms of function and aesthetics.*[5]

Each concourse includes a multistory central core atrium allowing rail access on the lowest level with escalators and elevators leading up to a food court and retail level. It is four stories above grade and approximately a city block in diameter. The fourth story has ample room for airline offices and club rooms. The concourses were designed to be sufficiently wide to accommodate an internal automated People Distribution System (movable sidewalks) to assist passengers in making connections at full build-out (McCagg 1993).

As we discuss in Chapter 9, because the airlines largely boycotted the original design and planning process, two concourses subsequently had to be tailored to the needs of their tenants, causing significant cost increases. Concourse A, at 1550 feet and 26 gates (22 domestic and four international), was designed and built for Continental Airlines and international carriers; Concourse B, at 3199 feet and 44 gates (plus 20 commuter gates) was designed and built for United Airlines; Concourse C, at 1500 feet and 20 gates, was designed and built for all the other airlines (Green 1992). Of course, with Continental's elimination of its Denver hub, the airport opened with a large number of empty gates.

But the idea was to create a design that could easily accommodate expansion as air traffic growth warranted it. Denver had learned from Stapleton Airport how difficult it was to extend concourses to add gates, building long concourses out from the terminal that resemble tapeworms from the air. At DIA, sufficient space was reserved so that each concourse can be extended to 4200 feet, and the main terminal building can also be significantly expanded. Poten-

5. Architectural goals of the concourses are discussed in Chapter 8.

tially, there is room for two additional concourses (D and E) to the north, although United Airlines has built a massive aircraft hangar precisely where the east wing of the E (fifth) concourse would have been built. Although United is contractually obliged to move the hangar if the airport needs the space (O'Driscoll 1994a), the airline could always threaten to locate the jobs to another city. But the day when a fifth concourse is needed at DIA is long over the horizon.

DIA's first year of operation

As DIA approached the end of its first year of operation, numerous technological problems were revealed:

- DIA's automated baggage system failed to keep pace with the surge in its first Christmas/New Year's holiday traffic, delivering bags to the wrong locations or damaging them, causing United Airlines to shut down the system on December 22. More than 500 bags were damaged in a two- or three-week period.
- The underground train system shut down several times, delaying flights and stranding thousands of passengers. One train stalled between stations, requiring passengers to deboard the train and walk the tunnel catwalk to the next station. No public address system had been installed in the train terminals, causing passengers to suffer hopeless confusion. The city went out and bought a half dozen bullhorns as a temporary solution. Because DIA installed no parallel underground walkways, passengers had to be bused between concourses.
- Surges and spikes in electricity were believed to have damaged the motherboard of a computer on Concourse C, which fed the main computer inaccurate information, causing it to shut down the system. A power surge caused escalators to stop abruptly, injuring one woman.
- It was revealed that the FAA suffered a 75 percent failure rate on equipment during a six month period—temporary breakdowns of various systems (most often the ground radar and weather detection systems) occurred during 135 of 181 days from May through October 1995. Of course, some of the FAA equipment was entirely new, not yet installed or operated at other airports. Sometimes air radar shut down,

causing the FAA to halt departures. The FAA criticized these reports as suggesting that some relatively minor equipment lapses had caused a safety problem at DIA.

Time will tell whether these complicated, high-tech systems will work satisfactorily. Most of the bugs likely will be worked out over time. But one thing is clear—by stretching the technological envelope over such a wide array of essential systems, in an arrogant attempt to have the world's most modern and sophisticated airport, DIA's planners increased the risk that essential systems would malfunction. The baggage system, the people-mover system, the radar systems, and the communications systems—all essential to the successful operation of an airport—at various times have collapsed at DIA. The failure rate has created tremendous inconvenience for thousands of passengers, for whom DIA ostensibly was built.

DIA: Conclusions

It is the runway and airfield configuration that makes DIA potentially the world's most efficient hub airport (Fohn 1994). It is also an airport clearly designed for the future. The 21st century may see DIA realize its full potential. If air traffic continues to grow, DIA can be expanded more easily, and probably at less cost, than any other major airport in the world, while other airports will be hemmed in by land and noise constraints.

The major advantage is the enormous land mass that DIA can grow to fill as it matures. At 53 square miles, DIA is about half the size of Denver prior to DIA's annexation and larger than the city limits of Boston, Miami, or San Francisco.

DIA has been widely lauded by both national planners and commercial users. According to FAA Administrator David Hinson, "Compared to other airport development projects around the world, DIA is a bargain" because of the following features:

- Five new runways with a sixth planned
- All-weather operation capable of triple simultaneous landings and takeoffs
- A state-of-the-art facility designed for maximum safety and efficiency, with no crossing runways and the latest technology

- Increased capacity and efficiency of approximately 5 percent annually throughout the entire national aviation system (Hinson 1995).

Similarly, although United was an early and vigorous opponent of the new airport, United's CEO, Gerald Greenwald, praised the new airport's advantages:

> *The bottom line is that Denver is destined to become one of the world's great airline hubs. DIA will be a critically important anchor for United . . . . Our airline operates in 150 airports in 30 countries. None of them compares to the efficiencies and amenities we expect to experience at Denver International Airport. From DIA's superior air traffic control capabilities to the many comforts and conveniences along the concourse, DIA will do more to get travelers from Point A to Point B with less hassle along the way—and in five-star comfort (Greenwald 1995).*

As in other chapters we have examined many of DIA's weaknesses, but here we have found enormous strengths. The huge mass of land reserved and the airfield design and layout, coupled with the terminal location and design (albeit plagued by costly midstream changes) and its technologies (despite the gremlins that plague it) make DIA uniquely capable of expanding to meet the traffic needs of the 21st century. Of course, unexpected economic, demographic, or technological changes—such as short-takeoff-and-landing, or STOL, aircraft—or the demise of hubbing as too costly a distribution system, may one day make DIA as obsolete as many railroad union stations are today. Nevertheless, Denver built the largest and most modern airfield in the world (Broderick 1995a).

Other recent international airports

Let's examine how these problems are being handled at other new major international airports being constructed in the 1990s around the planet.

Munich's Franz Josef Strauss Airport

The terminal building at Munich's Franz Josef Strauss Airport, opened in 1992, stretches 1 kilometer (3170 feet) perpendicularly between the runways (Fig. 7-20), nearly the same distance as DIA's Concourse B, which is 3300 feet long (O'Driscoll 1994c). The terminal has 20 jetway gates, and an adjacent ramp can accommodate 28 aircraft, from which passengers will deplane and board buses to the terminal (Fig. 7-21). All check-in, security, and waiting-room areas are in close proximity (50 meters, or 164 feet) from the curb to the nearest gate for departures (Munn 1994). The terminal is designed to allow passengers to make their connections within 35 minutes. It was designed to accommodate about 15 million passengers a year (Mecham 1992). In 1994 Strauss handled 13.5 million passengers, 26% more than its predecessor in its final year (O'Driscoll 1994c).

The basic design uses a modular approach. Four terminal modules have been built—one for domestic flights, one for Lufthansa's international flights, and two for charter and nondomestic flights (Mecham 1992). However, as in most European and Canadian airports, specific gates and ticket counters are not tied to a designated carrier, allowing the airport the flexibility of transferring use of these facilities to another carrier during those periods of the day when the primary carrier's operations are dormant, much enhancing efficiency in utilization of fixed infrastructure. The airport also provides all ground and baggage handling. Enhanced use of fixed infrastructure requires less infrastructure to be built. Adding further

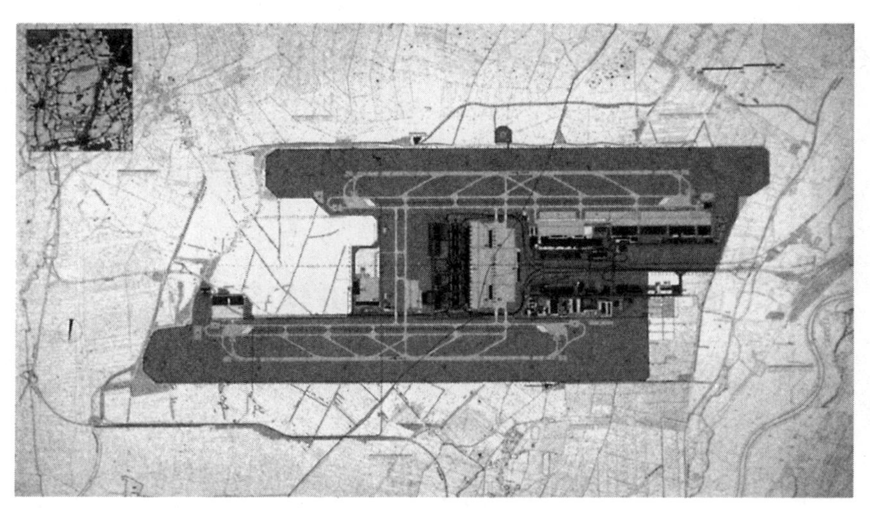

Fig. 7-20. *Runway configuration at the new Munich Franz Josef Strauss Airport, with two parallel runways sandwiching the terminal area.*

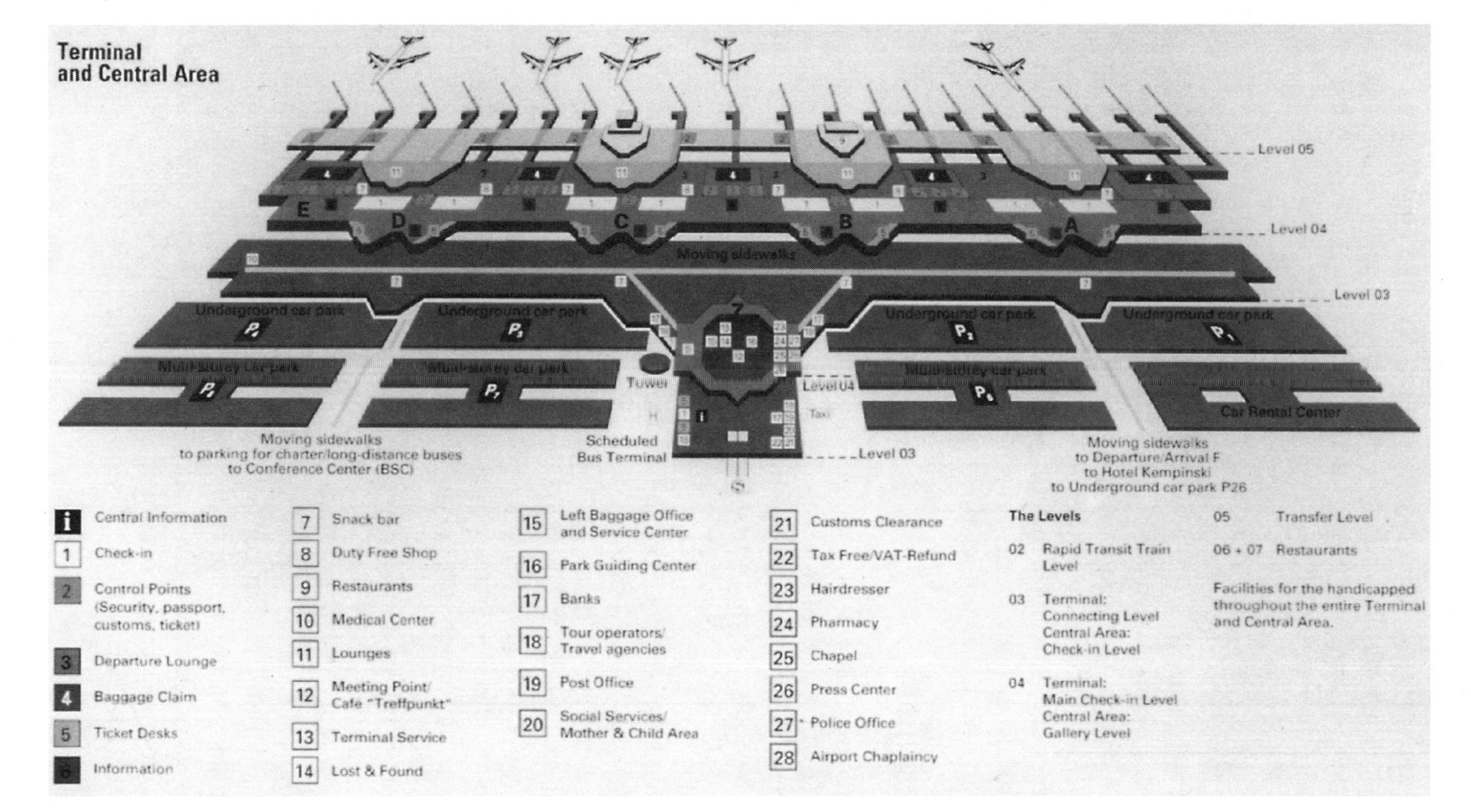

Fig. 7-21. *Munich's new terminal has several levels snugly fit beside a multistory car park.* Flughafen München

to the airport's flexibility is the provision for future expansion. An open area south of the general-aviation facility can accommodate two additional terminal modules, which could provide capacity for an additional 6 million passengers per year (Mordoff 1988). It is anticipated that a second terminal will be built no later than 2005 (*Aviation Europe* 1995).

Munich's new airport has two staggered east-west runways (compared with DIA's five), each 4000 meters in length (13,120 feet), which its planners consider to the minimum for efficient operation. Alfons Wittl of the Munich Airport Authority noted, "Munich 2 will be open at all times in winter. While one runway is being cleared, the other can remain open." As we shall see, a number of the new airports are designed with only one runway. A third, short, general-aviation and corporate-jet runway was planned for the north end of the airport, as well as a separate general aviation terminal (Mordoff 1988).

Careful attention was also paid to environmental issues. The spacing and length of the runways is designed to disperse the noise and eliminate the need for aircraft to reverse engine thrust on landing. They have also been located so that aircraft do not fly over nearby towns (Mecham 1991).

The maximum capacity of the two parallel Category IIIC runways is 275,000 aircraft movements a year; about 170,000 were anticipated for its first year of operation (Hill 1991). Each runway can accommodate a takeoff or landing every two minutes under instrument flight rule conditions (*Aviation Week* 1991). A year after it opened, Strauss handled 63 aircraft per hour during peak periods, up from 32 at Riem, the former airport (Hill 1993a).

The new airport incorporates an automated computer-controlled system developed by Siemens for aircraft parking, which will eliminate the need for "follow me" vehicles or human aircraft directors and thereby reduce delays caused by aircraft waiting for gates (Mecham 1991). Apron controllers in the 197-foot tower illuminate green lights in the center of the taxiway to guide pilots to their assigned gates. Amber lights mark the point at which pilots turn into their gates, and a sign confirms the assignment by relaying the flight number. Loops laid in the apron detect the nose wheel location (*Aviation Week* 1992).

Safety is enhanced during winter by a computerized early-warning system with 39 data-gathering points identifying crucial ice and snow surface conditions. The de-icing stations are positioned to de-ice a running aircraft in four to five minutes just before it enters the runway (Trautmann 1994). Half of the de-icing solution used on aircraft is recaptured and recycled (Hill 1993a). In fact, to assuage environmental interests, nearly 40 percent of all airport refuse is recycled.

Although the airport incorporates many modern technologies, Munich's airport planners were sensitive to the inherent risks involved. They learned from the experience at Frankfurt, which in the early 1970s was the first airport in the world to install the individual cart-and-track baggage system that DIA chose. Like Denver two decades later, Frankfurt had enormous problems with the baggage system. Munich chose a baggage system with more conventional and reliable tried-and-true technology and made it a priority that the baggage system would be working on opening day. The Dutch firm Vanderlande Industries installed the same system it had installed at Amsterdam's Schiphol Airport, a system of conveyor belts with saucer-shaped wood containers attached to the belts holding bags. Furthermore, Munich insisted the system be completed a full year before opening, so that there would be sufficient time to eliminate any bugs. As a consequence, Munich's baggage system worked perfectly from day one. In addition, unlike DIA (which gave the maintenance contract to a firm other than the company that built it so as to further expand minority participation in the project), Munich insisted that Vanderlande maintain the system for its first five years of operation (O'Driscoll 1994d).

Munich airport operations director Peter Trautmann summarized his orientation to airport technologies as follows: "We only needed three things to work perfectly. The first was the baggage system, the second was signage and monitors, and the third was the telephone system." He therefore insisted that all critical systems be tested for months prior to opening. As noted, the Dutch-built conveyor-belt baggage system was tested a full year before the new airport opened. Munich also had seven months to tweak its flight and baggage information display systems ("FIDS" and "BIDS," respectively) using television display monitors and giant railroad-style arrival and departure boards (O'Driscoll 1994b).

Osaka Kansai Airport

Given the intensive land use in the area, finding enough space for a new airport at Osaka was no simple matter. The problem was solved by building a new island in Osaka Bay. Kansai International Airport's (KIA) 511 hectares make it 1/27th as large as Denver International Airport. But because it is on landfill in the ocean, KIA cost several times as much as DIA (Ogawa 1993).

Before construction began, the developers sought advice from six airport operators on design, construction, and operation. The six were BAA (formerly British Airports Authority), Aeroport de Paris, Amsterdam Airport, Frankfurt Airport, the Port Authority of New York and New Jersey, and Dallas/Fort Worth International Airport. After they studied the design concept, their input was incorporated into the detailed design plan (Brown 1987). Paul Andreu of Aeroport de Paris established the airport's basic layout, including a 5600-meter concourse with a passenger terminal building at the center.

Although the new airport has but a single 11,400-foot (3500-meter) runway, it is ultimately expected to be able to handle up to 454 arrivals and departures a day, accommodating 68,000 passengers and 3000 tons of cargo (*Kyodo News* 1994). Ultimate capacity is 160,000 takeoffs and landings, transporting 30.7 million passengers and 1.4 million tons of cargo per year (*Aviation Week* 1994c). Because aircraft approaches are over the ocean, Kansai is Japan's first 24-hour-a-day airport (Lassiter 1994) which will facilitate cargo utilization, for cargo landings can be cleared at night and transported through Japan when highway traffic is light, thereby saving warehousing expenses (Black 1994).

The terminal uses trams to serve a mile-long line of gates (*Aviation Week* 1994d). In a peculiar setup that no doubt confuses connecting passengers, international passengers arrive on the first floor and depart on the fourth floor. The second floor is for domestic arrivals and departures, while the third floor contains the concession/shopping mall area—an idea contributed by Aeroport de Paris (Rush 1995)—which includes 27 restaurants, 48 shops, and the automated guideway transit system (Moorman 1994). KIA is Japan's only airport handling both international and domestic flights (Ogawa 1993). The terminal has a capacity of 30.7 million passengers a year.

Studies are under way for stage two, to double its capacity (*Aviation Week* 1994b).

How to do so, however, is a matter of debate. The conventional approach would be to expand the island by 1730 acres to add two new runways (one parallel and one crosswind) and a second terminal building. The cost would be a staggering $21.4 billion. A less-conventional alternative would be to connect 200 steel boxes, each 1000 feet by 200 feet, into a floating runway, an alternative that would cost an estimated $17 billion (Fiorino 1994).

The new airport utilizes modern technology to enhance security. Five hundred sensors installed on the island, and another 200 installed along roads and rail links, are able to detect hazardous objects. This sensor array allows management to field a security staff of only 600 officers, compared to Narita's 3600 (*Kyodo News* 1994). Like Munich, however, Kansai opted against using the bar-coded computerized baggage system that plagued DIA. And, again like Munich, and unlike DIA, the baggage system at Kansai was tested repeatedly and was working smoothly at opening (Lassiter 1994).

Macau International Airport

Once again, a shortage of land led to the construction of an artificial island at Macau, using hydraulic sandfilling (*Aviation Week* 1994a). Four hundred hectares were reclaimed between the islands of Taipa and Coloane, which will be the venue for the airport and airport-related businesses (Donoghue 1995). This expanded the tiny Portuguese colony's usable real estate by more than 20 percent (*Asian Wall Street Journal* 1994).

The master plan for the new airport was developed by Flughafen Frankfurt Main AG, Airconsult, in conjunction with Aeroport de Paris (Donoghue 1995). The airport has a 3360-meter runway with CAT II ILS-equipped access (*World Airport Week* 1995e). Its terminal is a three-level, 54,000-square-foot facility (*Aviation Week* 1994a) with only four loading bridges, with additional bus-loading sites on the tarmac (Donoghue 1995). The terminal apron can handle 22 aircraft. Arrivals enter on the ground floor, and departures enter on the first floor. The mezzanine above the ground floor is dedicated to airline lounges and concessions (*Travel Weekly* 1995; *World Airport Week* 1995e).

The new airport is anticipated to handle 2.7 million passengers during its first year of operation, 4 million by the year 2000, and 6 million by 2010 (*Travel Weekly* 1995). But delays in constructing Hong Kong's new Chek Lap Kok Airport suggest that these numbers should be revised upward (*Phillips Business* 1994). CAM (Macau Airport Company) anticipates that the MIA will gain up to 15 percent of Hong Kong's passenger traffic (*World Airport Week* 1995a). Initial annual capacity of MIA is 4.5 million passengers and 120,000 tons of cargo. International flights are anticipated to account for 72 percent of operations (Donoghue 1995). By virtue of the fact that the airport is built on landfill in the bay, noise impacts on residents are anticipated to be modest, raising the possibility of 24-hour departures (*Travel Weekly* 1995).

The runway has been lengthened and the ramp area increased since the initial plan, both to accommodate anticipated cargo business (Donoghue 1995). The second phase of the airport is scheduled to open in 1996 and is focused on developing Macau's potential as a cargo hub and as an aircraft maintenance facility (*Aviation Week* 1994a). Site preparation on a 10,000-square-meter cargo terminal began in 1995, while doubling the apron area for aircraft parking, which should allow capacity of 160,000 tons of cargo per year (*Air Cargo* 1995). The anticipated tripling of cargo volume by 2011 will require more land reclamation, to be completed by 2001 (Donoghue 1995). A second terminal is scheduled to open in 2010 (*Aviation Week* 1994a).

Hong Kong's Chek Lap Kok Airport

When completed, the 4.82-square-mile island airport site at Hong Kong's Chek Lap Kok will be four times the size of Kai Tak and twice the size of Osaka's new Kansai Airport (Mufson 1994). The island will be large enough for two 12,464-foot (3800-meter) runways with a 1525-meter separation (*Aviation Week* 1994b). The project will also include the world's largest railway suspension bridge (Becker 1995). Of the airport's 1248 hectares, 938 were reclaimed from the sea and 310 from leveling Chek Lap Kok and Lam Chau Islands. Site preparation cost $1.26 billion and involved the largest fleet of dredgers the world has ever seen (*Airports* 1995b).

Unfortunately, soft mud, in some places 27 meters thick, lies above the bedrock. The consortium that won the contract—Japan's Nishimatsu, Britain's Costain, America's Morrison Kundsen, Holland's Ballast Nedam, Belgium's Jan De Nul, and China's Harbour Engineering—opted to engineer the problem out, rather than engineer around it. This meant sucking the mud and clay out and dumping it in deeper water. A layer of sand was laid on top of the bedrock. Then granite was blasted from the islands and laid on top, forming the foundations for the runways and taxiways. At completion, the island will be 6 to 7 meters above sea level, and 350 million cubic meters of material will have been moved. Sea walls 12 kilometers long are also being built to protect the island against typhoons (*The Economist* 1995).

CLKA will open with only one runway, although its location suggests potential 24-hour utilization (Fig. 7-22). Kai Tak's maximum slot-controlled utilization rate was 28 flights an hour, although it sometimes reached 36. First-year volume at CLKA is anticipated to be 28 million passengers and 1.4 million tons of cargo (Mok 1993). Chek Lap Kok is anticipated to take as many as 43 movements an hour. The new airport is expected to handle 35 million passengers a year, demand that is expected to materialize by 2002 (*Aviation Week* 1994b). The airport's second runway is to be commissioned two years after opening, in 1999 (Mok 1993). CLKA's ultimate capacity is anticipated to be 87 million passengers and 9 million tons of cargo annually (Darmody 1993).

Between the two parallel runways will be two passenger terminals and aircraft maintenance facilities. The initial terminal is a four-level half-mile-long, 1-million-square-foot, Y-shaped building, with a capacity of 30 million passengers a year (Mok 1993; see Fig. 7-23). About 8 percent of the terminal will be devoted to 150 shops, restaurants, and services, exceeding the retail space at Singapore's Changi and Tokyo's Narita Airports. The terminal has an energy-efficient roof. The building can be expanded in 120-foot modular sections to expand capacity to 45 million passengers. A second X-shaped terminal is planned to be built by 2010, to provide needed capacity until the year 2040, when the airport is anticipated to peak at 85 million passengers (*Aviation Week* 1994b). Figure 7-24 shows the airport at full build-out.

What has been described as the "world's most advanced communications galaxy," which will integrate various airport computer systems while letting each stand alone as a separate computer system, is being built at Chek Lap Kok. Flight display information, security,

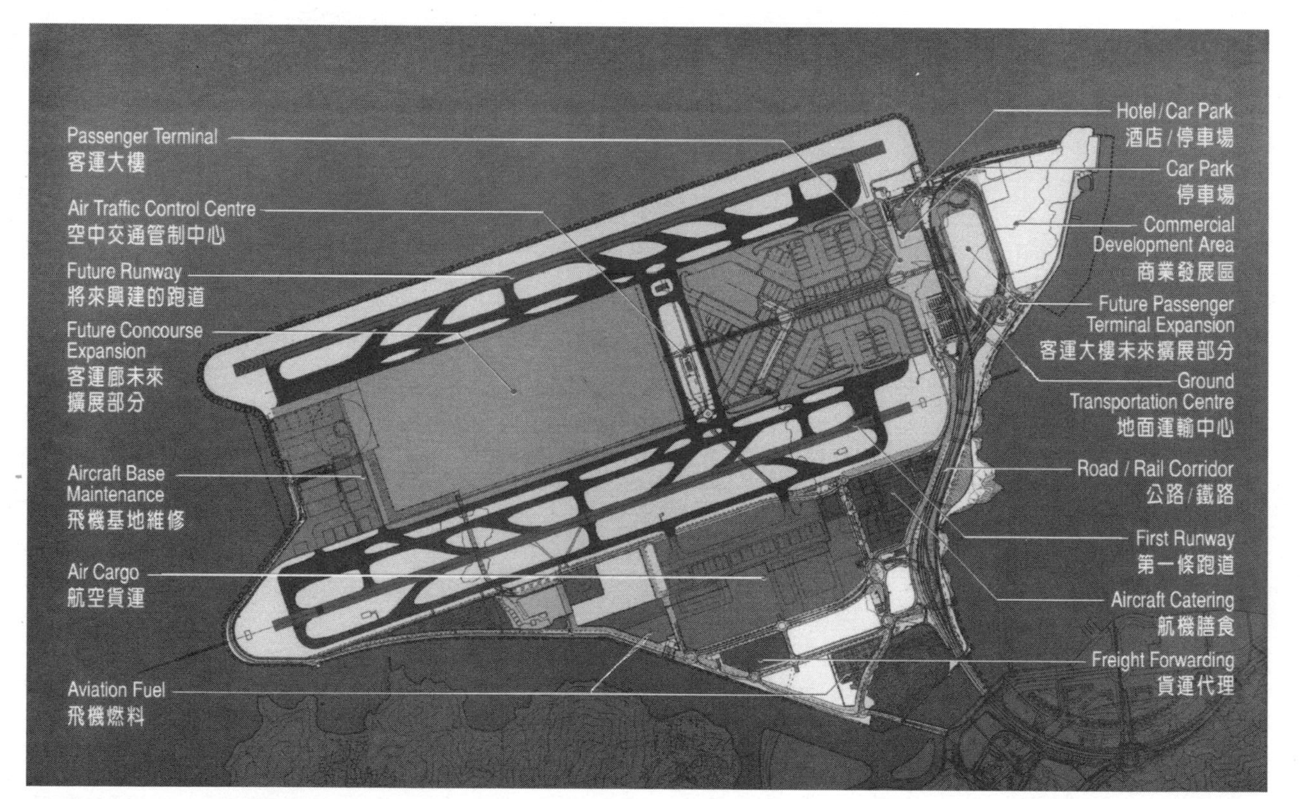

Fig. 7-22. *Site plan of Hong Kong's New Chek Lap Kok Airport at opening.* Airport Authority of Hong Kong

Fig. 7-23. *Artist's rendering of main terminal building at Hong Kong's Chek Lap Kok Airport.*

Airport Authority of Hong Kong

Fig. 7-24. *Hong Kong's new airport at full build-out, in the year 2040.* Airport Authority of Hong Kong

telephone, baggage handling, public address, building management, fire alarm, mobile radios, time generation, and display will all be integrated (Fiorino 1995).

Kuala Lumpur Sepang International Airport

Phase 1 of Kuala Lumpur's Sepang International Airport includes two parallel runways capable of handling simultaneous wide-bodied aircraft takeoffs and landings, and main and satellite terminals. With 25,000 acres, the airport has sufficient room for expansion to five runways (Hill 1993b).

Unsuitable soils led to the decision to open the airport with only one runway. Forty percent of the second runway lies on peat swamp and marine clay. The design and construction of the airport's main terminal, contract pier, and baggage-handling system will be built by a joint venture of four Japanese companies, Perspec, at a cost of $670 million (Jayasankaran 1995).

A contract has been awarded for installation of a total airport management system, with airport-wide communications and information technology, allowing airport tenants and operators to communicate and use information from shared databases (*Aviation Daily* 1995).

Contracts for the design and construction, as well as operation and maintenance of the terminal, were given to a six-member Malaysian-Japanese consortium, led by United Engineers Malaysia Bhd (*Wall Street Journal* 1995). The airport will have 45 loading bridges.

New Seoul International Airport

The site for the New Seoul International Airport (NSIA) is predominantly landfill in an area of Inchon Harbor near Seoul between the islands of Young-jong and Yong-jong, creating nearly 50 square miles of space (Fentress 1995), almost as much as DIA (Fig. 7-25). The new airport's original master plan included a single 3750-meter runway equipped with CAT-IIIA navaids equipment. It will be capable of handling 170,000 aircraft movements, 1.7 million tons of cargo, and 27 million passengers annually. Two additional runways have since been added to the plan to handle the anticipated 10 percent annual growth in traffic (*World Airport Week* 1995b). Four runways will be built at ultimate completion, scheduled in 2020 (*World Airport Week* 1995d). With expansion, the airport may be able to handle 100 million passengers by 2020 (*World Airport Week* 1995c). Airlines have urged that the new airport be opened with two runways, instead of one, as is designated in the airport master plan (*World Airport Week* 1995d).

The terminal will be 357,000 square meters, with an expansion to 1.3 million square meters by 2020 (Figs. 7-26 and 7-27). It will sit in the middle of the four parallel runways ultimately planned. The cargo terminal will be 175,000 square meters, with available expansion to 805,000 square meters, and will be capable of handling 7 million tons of freight (*World Airport Week* 1995c). When ultimately built out (in about 2020), the airport will have four runways with a capacity of 700,000 aircraft movements, 100 million passengers, and 7 million tons of cargo (Shin 1993).

Bangkok's Nong Ngu Hao International Airport

In January 1995, the U.S. consortium Murphay Jahn/TAMS was chosen to design the terminal for Bangkok's Nong Ngu Hao International Airport's first phase. Thai politicians backing the bid of rival Aeroports de Paris began pressuring Airports Authority of Thailand (AAT) to reverse its decision, leading to a diplomatic rift with the United States. The controversy over the contract, coupled with the refusal

Fig. 7-25. *The new airport at Seoul, Korea, is the largest construction project ever attempted in Korea, with land filling the bay from two surrounding islands.* Korea Airports Authority

of some squatters on the airport land to move, placed the airport nearly a year behind schedule (*Airports* 1995a).

The terminal will be 5.4 million square feet (500,000 square meters) on four or five levels. It will have 46 gates and 25 hardstands (*World Airport Week* 1995d). It is billed as larger and more modern than Singapore's Changi International Airport, one of the world's best (*Interavia Air Letter* 1993).

The new airport will have two 12,210-foot (3700-meter) runways in phase one, with a second pair to be built later, along with a second terminal. The first phase of the new airport will accommodate 30 million passengers. The second stage of the airport will probably include a concourse parallel to the main terminal, connected by an underground people-moving system. The airport is projected to be able to handle 38 million passengers in 2010 and 100 million at full build out, in 2023 (*World Airport Week* 1995d).

Don Muang, which is also being expanded, has a 12-million passenger capacity. Its international terminal will be doubled in size and its domestic terminal will be tripled, raising capacity to 25 million by 1997, three years before SBIA opens (*Travel Trade* 1993). The five-story international terminal will cost $100 million, and the domestic terminal

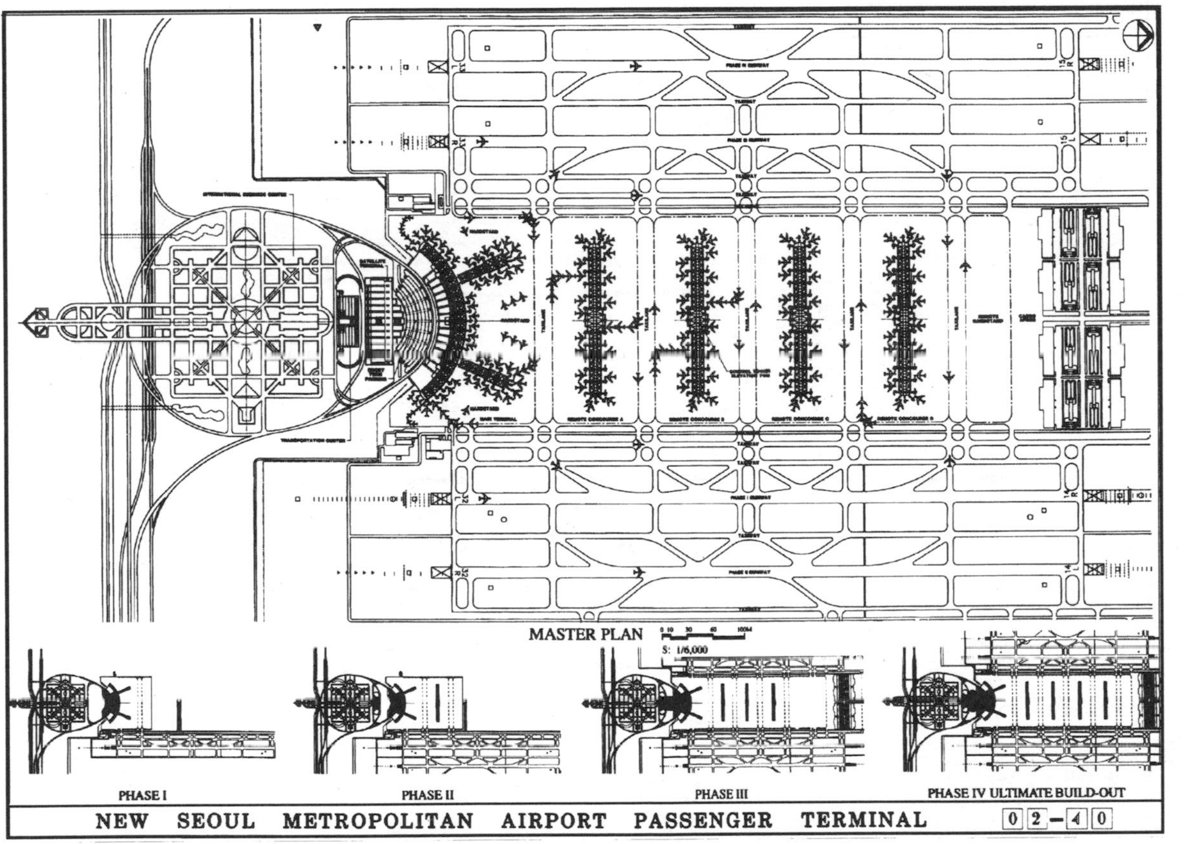

Fig. 7-26. *Schematic revealing terminal, concourse, and runway configuration at the new airport near Seoul, Korea. The bottom shows the phasing of construction to full build-out.* C.W. Fentress & J.H. Bradburn.

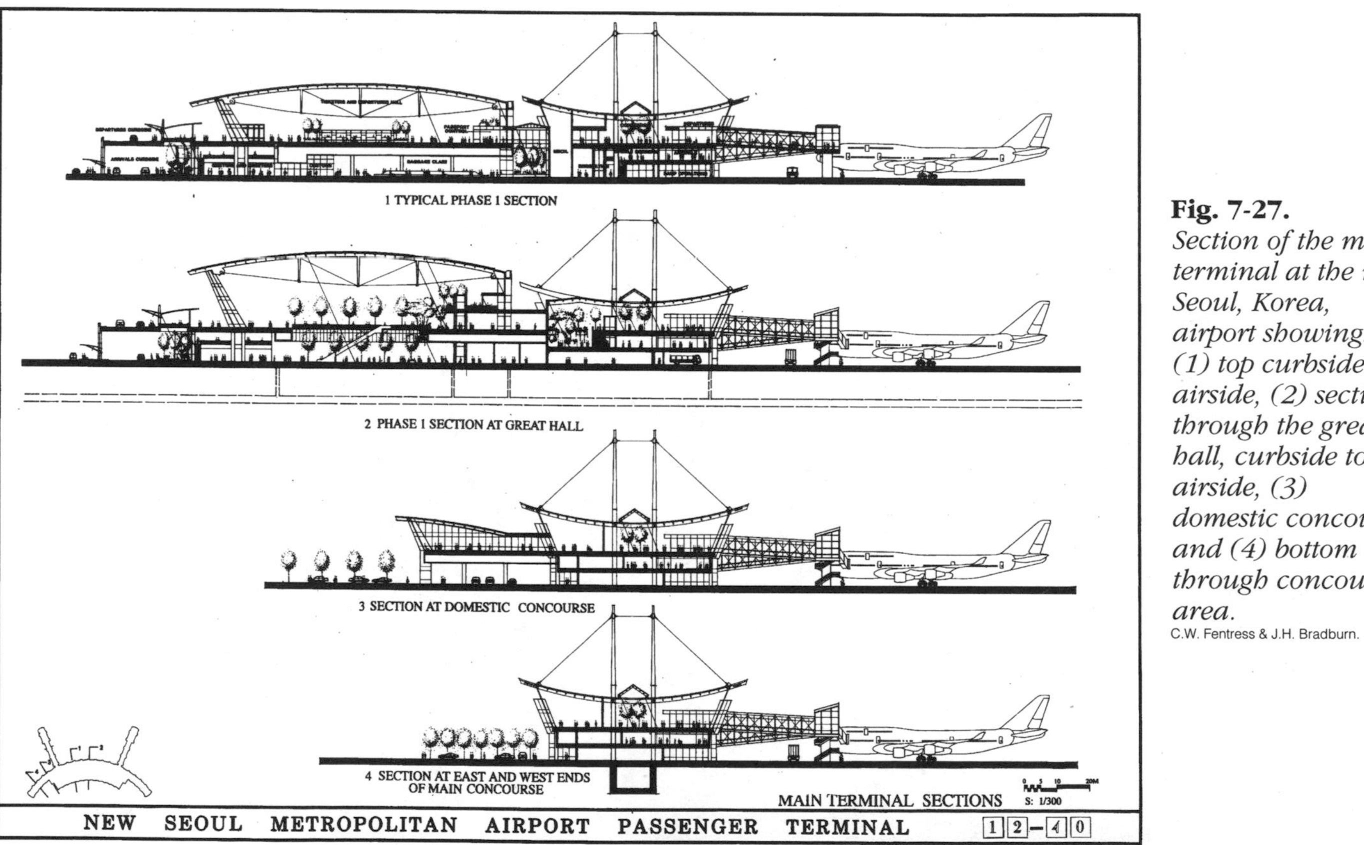

Fig. 7-27. *Section of the main terminal at the new Seoul, Korea, airport showing: (1) top curbside to airside, (2) section through the great hall, curbside to airside, (3) domestic concourse, and (4) bottom through concourse area.* C.W. Fentress & J.H. Bradburn.

will cost $478 million (*Aerospace* 1993). Such additional capacity at Don Muang led AAT to call for limiting SBIA to only its first phase.

Conclusions

This succinct survey permits us to draw several conclusions. First, the elements of airport planning and design are the same throughout the world. Planners everywhere have to deal with the noise impacts, use modern technologies, use modular approaches, and create as large a space as possible. Second, each has to deal with unexpected developments, whether physical ones such as soil conditions, or analytical ones such as future demand. Third, each site presents specific problems that are unique and that are dealt with according to particular local and national considerations in the most appropriate way. In the case of DIA, one major problem involved the way in which the design should accommodate the requirements of O&D and hubbing passengers.

Lessons learned

- *Acquire as much land as possible at the outset.* Over time, surrounding real estate development tends to hem in an airport perimeter, making future expansion expensive and politically difficult. Denver had the foresight to acquire a massive piece of property (53 square miles) adequate to accommodate capacity needs well into the 21st century. The airfield itself can accommodate 12 runways, more than any other airport in the world, which are separated by enough distance to effectively make DIA four separate airfields. DIA's designers also planned the terminals for future expansion, with adequate land left south of the landside terminal to double it, adequate land east and west of the three airside concourses to expand them to 60 gates, and adequate land north of Concourse C to build two additional concourses. Among the most astute and far-sighted airport decisions was that of the government of Thailand to set aside ample land 25 years ago within proximity of Bangkok for future airport development. Munich attempted to set aside 17 square miles, which unfortunately was whittled down to 5.4 square miles by German courts.

- *Design the airfield first.* DIA's planners recognized that aircraft safety and airline efficiency could be enhanced significantly with an airfield design that allowed multidirectional takeoffs and landings and expedited taxiing to and from gates. Residential noise might also be diminished with proper runway alignment, although at DIA, this was a secondary objective.
- *Mid-field terminals and adequately spaced concourses enhance hub efficiency.* From the perspective of the hub carriers at DIA, its runway, taxiway, and apron layout, with a center-field terminal location and with concourses spaced to allow two full lanes of access, plus ample room for parking large aircraft and backing them out, is the most efficient design. For connecting passengers, mid-field concourses with gates on both sides, connected by moveable sidewalks, is the most convenient layout for aircraft connections.
- *Decisions made to enhance hubbing efficiency may compromise O&D convenience.* While DIA works splendidly from the perspective of the connecting passenger, its distance from downtown and numerous decisions made with respect to ground access make DIA a much less convenient airport from the perspective of the O&D passenger than the airport it replaced.
- *Keep passenger flow in mind when building terminals and concourses.* At DIA, a multistory parking deck surrounds the landside terminal building, allowing a relatively short walk to the ticketing or baggage-claim area. Four lanes of moveable sidewalks were placed in the longest concourse. One shortcoming of DIA was the planning that went into the underground train system. Not enough cars or access gates were incorporated into the design. And when the train breaks down, as it occasionally does, there is no parallel underground tunnel for passenger access (e.g., moving walkways) between the concourses and the main terminal building.
- *Design terminals and concourses in modular form, with sufficient reserved space to allow future expansion.* DIA's landside and airside terminals are designed in modular form to allow additional expansion in the number of gates, ticket counters, and baggage claim areas as capacity demands. Sufficient land was reserved to such future expansion as

well. However, according to United's demand that a large maintenance hangar be built where a future additional concourse should be located was a short-sighted decision.

- *Get the national government involved early in the design process.* DIA was wise to invite the FAA to engage the project in a hands-on relationship from the outset, and the FAA was wise to participate. This made planning much more efficient, reducing the likelihood that any significant construction would have to be redesigned and rebuilt because of federal objections.
- *Get the principal airline tenants involved early in the design process.* At DIA, the major hub airlines opposed the new airport, and although they participated constructively in the airfield layout, they obstructed the terminal and concourse design process. At the cost of persuading them to sign leases, they were allowed to dictate major scope changes after construction had already begun and to insist on exclusive use of gates, ticket counters, and baggage facilities. Their scope changes and additions caused massive cost escalation and an airport that was much larger and less efficient than necessary.
- *Develop a streamlined pecking order so that responsibility is clearly defined.* The chain of command on the DIA project changed midstream because of several changes in essential personnel—Mayor Peña to Webb; aviation director Doughty to DeLong; chief engineer Smith to Evans. Moreover, criticism has been levied that the ability of the outside management team (Greiner/MKE) to manage DIA was undercut by city officials. Efficiency would have been much enhanced had key city personnel not rotated midstream, had they deferred to the management team on day-to-day construction issues, and had the city not attempted to merge internal and external management in the PMT.
- *Incorporate security concerns into airport planning and design early.* Doing so results in systems that are more efficient, cost-effective, and less intrusive than would otherwise be the case.
- *Locate the airport to minimize adverse noise impacts.* Several Asian airports are being built on the ocean, not only because of the dearth of suitable level land, but because of the noise

impact on dense population clusters. Landfill on sea bed is among the world's most expensive and complicated engineering feats. But once constructed, 24-hour takeoffs and landings may be possible. DIA, of course, was blessed with wide-open prairie within 24 miles of downtown Denver, something few other cities of its size have available.

- *Incorporate mass transit into the terminal design.* Most of the world's major new airports incorporate rail directly into the airport terminal (DIA is the obvious exception), allowing efficient, high-speed, environmentally sound ground access. At DIA, sufficient land was left between the parallel lanes of the access boulevard for a future rail line from downtown Denver. Highway congestion and automobile pollution (already a significant air quality problem for Denver) will compel its construction sometime in the next century. Unfortunately, no rail line or rail station was incorporated into the landside terminal building. Airport construction likely could have been accelerated and economized had a rail spur been built from the nearby Union Pacific Railroad line. The airport terminal could have been built around the rail head, enabling it to be used as a passenger line for Denver CBD and Amtrak intermodal connections.
- *Develop a comprehensive intermodal and multimodal transportation hub.* At Seoul, Hong Kong, Macau, and Osaka, we see efforts to link all modes of transport together and a key desire to accommodate the rapidly growing cargo sector. DIA was designed to be an efficient passenger hub. But originally locating the cargo facilities away from essential ground interstate highway access was foolish, no matter what the political dynamics. Air freight is inherently intermodal in nature, requiring efficient airport access by pickup and delivery trucks.
- *Engineer out the problems if possible.* Both Osaka Kansai and Hong Kong Chek Lap Kok were built on ocean landfill. Kansai opted to install hydraulic jacks to prop up sinking portions of the island, while Chek Lap Kok chose the more expensive route of properly preparing the seabed. Only time will tell, but the short-term cost may result in long-term benefit.
- *Adopt proven technology for critical systems.* In other words, keep it simple, stupid (KISS). Munich learned from

Frankfurt's dismal experience with a high-tech sophisticated baggage system and adopted a tried-and-true, off-the-shelf technology for critical systems. While it might not have all the bells and whistles of cutting-edge technology, it works fine.

- *Set aside ample time to test-run the technology.* The most critical failure of DIA was the decision to install a highly sophisticated baggage system throughout the airport on a fast track without time to test it. Munich and Hong Kong both set aside a sufficient time period to test the airport's systems and fine-tune them in order to get the bugs out. Had DIA done this, it might have been spared the agony of a dysfunctional baggage system.

References

Aerospace (Singapore). 1993. Fewer arrivals so airport faces scale-down. September:26.

Air Cargo Report. 1995. April is start of preparation for Macau International cargo facility. March 30.

Airports. 1995a. Politicians, squatters slow new Bangkok international airport. March 7.

———. 1995b. Hong Kong Completes Site Preparation for Chek Lap Kok Airport. June 20:245.

Air Transportation of America (ATA). 1977. Planning and Design Considerations for Airport Terminal Buildings. Report AD/SC #4. July.

Allison, G.L. 1995. Frequent flier dumps a load of complaints About DIA. *Denver Post.* August 6.

Amole, Gene. 1995. A chiller, a thriller, and it's here now. *Rocky Mountain News.* October 26.

Asian Wall Street Journal. 1994. Airport to be Macao's gate to region. October 31.

Auguston, Karen. 1994. The Denver airport: A lesson in dealing with complexity. *Modern Materials Handling.* October.

Aviation Daily. 1995. Harris wins contract for Kuala Lumpur airport. June 28:514.

Aviation Europe. 1995. Munich airport expansion plan includes second terminal by 2005. July 27.

Aviation Week & Space Technology. 1991. Denver airport. March 11:50.

———. 1992. Europeans upgrade ATC system. May 4:63.

———. 1994a. U.S. funds feasibility study for second stage of Macau airport. June 6:36.

———. 1994b. Growth outpaces Asian airports. August 29:47.

———. 1994c. Japanese carriers launch new service. September 5:33.

———. 1994d. Kansai International. September 12:19.

Becker, Stuart. 1995. Airports: Chinese opera. *Far Eastern Economic Review.* April 6:54.

Bidwell, Carol. 1995. Delayed Denver airport to open February 19, 1995. *Los Angeles Daily News.* February 19.

Black, Alexandra. 1994. Trade and tourist boom from new airport. *Inter Press Service Global Information Network.* September 9.

Booth, Michael. 1995. The paper chase. *Denver Post Magazine.* March 5:10.

Bozman, Jean. 1994. United to Simplify Denver's troubled baggage project. *Computer World.* October 10.

Brewer, George. 1995. Interview. October 26.

Broderick, Christopher. 1995a. Top-flight runway design DIA's best feature. *Rocky Mountain News.* February 26.

———. 1995b. Airport boasts latest in technological tools. *Rocky Mountain News.* February 26.

Brown, David. 1987. Japanese building international offshore airport to serve Osaka. *Aviation Week & Space Technology.* July 13:38.

Darmody, Thomas. 1993. The design and development of world class airports. Paper presented to the IBC International Conference on Airport Development & Expansion, Hong Kong, October 28.

de Neufville, Richard. 1976. *Airport Systems Planning.* Cambridge, MA: MIT Press.

———. 1995. Testimony before the House Transportation Subcommittee on Aviation. May 11.

Del Rosso, Laura. 1994. New Denver International Airport combines several state-of-the-art systems. *Travel Weekly.* February 21:20.

DeLong, Jim. 1995. Interview. October 26.

Denver Post. 1995a. DIA needs a sixth runway to perform as promised. March 1.

———. 1995b. DIA 'blind spots' being corrected. October 25.

Di Fonso, Gene. 1993. Advanced baggage technology for the new Denver International Airport. Paper presented at the International Airport Conference on Aviation and Airport Infrastructure, December 7.

Donoghue, J. A. 1995. The Pearl-Y gateways. *Air Transport World.* February 1:75.

Doughty, George. 1995. Interview. October 23.

The Economist. 1995. How to avoid that sinking feeling. February 4:73.

Eddy, Mark. 1995. Runway layout makes for all-weather landings. *Denver Post.* March 2.

Fentress, Curtis. 1995. Revitalizing the excitement of travel. *Passenger Terminal '95.*

Fiorino, Francis. 1995. Airport galaxy. *Aviation Week & Space Technology.* June 12:33.

————. 1994. Airline outlook yen for growth. *Aviation Week & Space Technology.* September 5:31.

Flynn, Kevin. 1993. Countdown to takeoff. *Rocky Mountain News.* November 28.

————. 1995a. Who botched the airport baggage system? *Rocky Mountain News.* January 29.

————. 1995b. City awards $14.1 million runway contract. *Rocky Mountain News.* October 4.

————. 1995c. DIA control tower roof intact, but the walls are leaking. *Rocky Mountain News.* October 26.

Fohn, Joe. 1994. Denver airport provides look at future. *Phoenix Gazette.* February 2.

General Accounting Office (GAO). 1991. New Denver Airport 10.

Gesell, Laurence. 1981. *The Administration of Public Airports.* Phoenix: Coast-Aire.

Goldstein, Alan. 1993. Airport for the 21st century. *St. Petersburg Times.* October 17.

Green, Peter. 1992. Big ain't hardly the word for it. *Engineering News-Record.* September 7:28.

Greenwald, Gerald. 1995. Future bright for DIA and Denver. *Denver Post.* February 28.

Gutierrez, Hector. 1995. Wind hits like a whip across Colorado. *Rocky Mountain News.* October 27.

High Country News. 1995. Megamess: "We've got it all." January 23.

Hill, Leonard. 1991. Countdown on Munich 11. *Air Transport World* June 1:70.

————. 1993a. Beyond expectations. *Air Transport World.* June 1:182.

————. 1993b. Asia's newest "dragon." *Air Transport World.* September 1:66.

Hinson, David. 1995. Letter to the editor. *Washington Times.* March 15.

Horonjeff, Robert. 1975. *Planning and Design of Airports.* New York: McGraw-Hill.

Hoversten, Paul. 1995. Denver's runway to the future. *USA Today.* February 22.

Howard, Robert B. and Donald M. Rochon. 1993. Controlling airport access: Airport security systems and transportation security. *Security Management.* November 1993:53–59.

Imse, Ann. 1995. Airport designed for safety, efficiency. *Rocky Mountain News.* February 26.

Interavia Air Letter. 1993. Thailand plans new Bangkok airport. March 17.

Jayasankaran, S. 1995. Thanks anyway. *Far Eastern Economic Review*. February 8:61.

Kerwin, Katie. 1995. DIA traffic spurs $90,000 study. *Rocky Mountain News*. April 12.

Kissler, Greg. 1994. The future of airport communications. *Telephony*. January 31, 1994

Knill, Bernie. 1994a. Flying blind at Denver International Airport. *Material Handling Engineering*. July 1994.

————. 1994b. New baggage handling solution. *Material Handling Engineering*. September 1994.

Kowalski, Robert. 1994. Air traffic control a hi-tech production. *Denver Post*. March 6.

Kyodo News International. 1994. Kansai International Airport inaugurated. September 5.

Laird, Burgess. 1994. Planning issues in civil aviation security. Unpublished paper. May 31.

Lassiter, Eric. 1994. Japan to open much-delayed Kansai airport. *Travel Weekly*. August 29:4.

Lawrence, Ronald. n.d. Moving passengers under Denver's translucent roof.

Leib, Jeffrey. 1995. Airline predicts bag bottlenecks. *Denver Post*. February 8.

Lopez, Greg. 1995. As opening day nears, past fades into future. *Rocky Mountain News*. February 26.

McCagg, Edward. 1993. New Denver International Airport: Evolution of a concourse. Paper delivered before International Conference on Aviation & Airport Infrastructure, Denver, CO, December.

Meadow, James. 1995. DIA has close call in snow. *Rocky Mountain News*. October 24.

Mecham, Michael. 1991. Germany struggles to meet airport needs. *Aviation Week & Space Technology*. August 26:38.

————. 1992. Munich turns out lights on Riem. *Aviation Week & Space Technology*. May 25:20.

Mok, John. 1993. The development of Hong Kong's new international airport. Paper delivered at International Conference on Aviation & Airport Infrastructure, Denver, CO, December 8.

Monroe, John. 1994. Denver starts off right with top ATC system. *Federal Computer Week*. July 11.

Moorman, Robert. 1994. Osaka to me. *Air Transport World*. October 1:62.

Mordoff, Keith. 1988. Air Transport Munich's new international airport expected to begin operations in 1991. *Aviation Week & Space Technology*. February 22:92.

Mufson, Steven. 1994. Accord boosts Hong Kong airport. *Washington Post*. November 5.

Munn, Felicity. 1994. Airports of the future are here and they're user friendly. *Montreal Gazette.* May 8.

Nordwall, Bruce. 1993. Delay likely in Denver opening. *Aviation Week & Space Technology.* September 6:40.

O'Connor, Leo. 1995. Keeping things moving at Denver International Airport. *Mechanical Engineering-CIME.* July.

O'Driscoll, Patrick. 1994a. DIA still just half there. *Denver Post.* March 6.

————. 1994b. Munich's airport: A case study for DIA. *Denver Post.* July 31.

————. 1994c. Munich's facility reflects future of DIA. *Denver Post.* July 31.

————. 1994d. Bag system no. 1 priority for Germans. *Denver Post.* July 31.

————. 1995. DIA Passes Storm's Icy Test. *Denver Post*, April 11.

Ogawa, Zenjiro. 1993. Kansai International Airport projects. Paper delivered at International Conference on Aviation & Airport Infrastructure, Denver, Colorado, December 8.

Park, Robert. 1994. The new high for Denver. *Business and Commercial Aviation.* February 1994.

Phillips Business Information. 1994. New Macau airport set to open July 1995. January 28.

Phillips, Edward. 1995. FAA program aims to reduce runway accidents. *Aviation Week & Space Technology.* April 24.

Rifkin, Glen. 1994. What really happened at Denver's airport. *Forbes.* August 29.

Rink, Fred. 1993. Deicing at Denver. Paper presented before the International Conference on Aviation & Airport Infrastructure, Denver, CO, December.

Rocky Mountain News. 1995. Rate of DIA flight delays lower than at Stapleton. April 1.

Rue, Dean. 1993. Airfield layout and design of the new Denver International Airport. Paper presented before the International Conference on Aviation & Airport Infrastructure, Denver, CO, December 1993.

Rush, Richard. 1995. Buy now, fly later. *Progressive Architecture.* April 1:70.

Russell, James. 1994. Is this any way to build an airport? *Architectural Record.* November:30.

Schwartzkopf, Emerson. 1986. Making it fly. *Rocky Mountain Business Journal.* September 1.

Scott, William. 1995. Storm challenges DIA opening. *Aviation Week & Space Technology.* March 6:28.

Searles, Dennis. 1992. Debate flies over Denver's big, new airport. *Los Angeles Times.* July 19.

Sellars, Don. 1993. Airport site and runway configuration: Safety, efficiency and capacity. Paper presented before the International Conference on Aviation & Airport Infrastructure, Denver, CO, December 6.

Shin, Jong-Heui. 1993. Airport developments in Korea. Paper delivered at International Conference on Aviation & Airport Infrastructure, Denver, CO, December 8.

Spensley, James. 1991. Airport planning. *Law & Public Policy* 63:73–75.

Thomas, Jeff. 1994. New Denver airport a high-tech showcase. *Colorado Springs Gazette Telegraph*. January 10.

Thurston, Scott. 1993. Denver's new airport set for takeoff amid turbulence. *Atlanta Constitution*. October 1.

Trautmann, Peter. 1994. "High tech" in total management: A case study of a world class airport—The new Munich airport. Unpublished address before the IBC Conference on Asia-Pacific Airports '94, Singapore, July 25–27.

Travel Trade Gazette Europa. 1993. Bangkok to add gateway. April 8.

Travel Weekly. 1995. Far East facility slates December opening. July 3:29.

U.S. Department of Transportation (DOT). 1976. The Apron and Terminal Building Planning Manual. FAA Advisory Circular 150/5360-7A. October.

————. 1985a. Airport Master Plans. FAA Advisory Circular 150/5070-6A. June.

————. 1985b. Airport Environmental Handbook. FAA Advisory Circular 150/5050-4A. November.

USWest Communications. n.d. Denver International Airport Communications System.

Veazey, Dick. 1995. Interview. October 18.

Wall Street Journal. 1995. Malaysia gives pacts to group for design, building of airport. March 27.

Wells, Alexander. 1992. *Airport Planning and Management*, 2d ed. NY: McGraw-Hill.

World Airport Week. 1995a. Macau air officials see bulk of first year traffic coming from Kai Tak. May 23.

————. 1995b. Air traffic growth forces expansion of Seoul airport construction. May 30.

————. 1995c. NSIA privatization opportunities. June 20.

————. 1995d. NSIA considers opening with extra runway. July 1.

————. 1995e. Macau's presence in Pearl River Delta will be as reliever airport. July 18.

8

Architecture and aesthetics

"Architects, designing toward the 21st century, have moved beyond the synoptic discourse of postmodernism and are now developing a transcendent, other than modern, other than postmodern synthesis of art and engineering. In the Denver International Airport this process is modified by the knowledge that for decades visionaries and futurists have seen the airport as a primary center, even a definer of culture."—ROGER A. CHANDLER, ARCHITECTURAL HISTORIAN[1]

Airports often provide the first impression that any traveler gets of a city or a country. They are among the most visible symbols of a community's sense of self. Increasingly, airport planners and designers have sought to build airports that make a major aesthetic statement. Historically this has not been the case. Airports have tended to stress function rather than beauty. Rare are those that combine both. The most distinctive and aesthetically pleasing of the world's airports are a short list—Eero Saarinen's landing-bird TWA terminal at New York's John F. Kennedy Airport or his magnificent swept-arch terminal at Washington's Dulles International Airport (both completed in 1962), or Helmut Jahn's futuristic, high-tech, glass-and-steel, art deco "train shed" United Airlines' terminal at Chicago's O'Hare Airport (Sinisi 1995). The wave of new airports being built around the world promises to significantly enlarge that pitifully small number. After examining the architecture and aesthetics of DIA, we review several examples of the way in which aesthetic considerations have become integral elements of the design process in new airports built around the world in the 1990s.

1. Fentress (1995).

DIA's translucent roof

From the beginning, DIA's planners dreamt of an airport that would make a major aesthetic statement, one that would be both beautiful and functional. Mayor Peña said that the design "must be memorable"—it should give the city an unforgettable visual image, like the Sydney Opera House (Fentress 1995). One also thinks of Paris' Eiffel Tower, San Francisco's Golden Gate Bridge, St. Louis' Arch, or New York's Statue of Liberty. A protégé of I. M. Pei, Curt Fentress, thinking of "Colorado's dramatic and powerful landscape," produced a striking white mountain-peaked terminal (Sinisi 1995). See Fig. 8-1.

Fentress was not the first architect to put a Teflon-coated roof on an airport terminal. Such a roof adorns the Haj Terminal at Jedda, Saudi Arabia (Wright 1994). DIA's, however, is much larger. The white, multipeaked roof is so large that, visually, it dwarfs the 33-story air traffic control tower, the world's highest. Each of the 34 white fabric peaks is approximately 120 feet high (Rothrock 1994). The 400-ton, 376,332-square-foot roof is supported by 26 masts that are 104 feet high and eight that are 126 feet high (Wright 1994).

Fig. 8-1. *Artist's drawing of main terminal building showing the proximity of the terminal to the airfield, access roads, and parking garages.*

Originally the terminal building was to have an attractive, tiered, pyramid-like glass-and-steel-roof designed by New Orleans firm Perez Associates. But the Denver architectural firm of Fentress & Bradburn and Associates was tapped in 1991 to give the airport, and its city, a distinct signature. It produced a building related to place and time. The snow-capped Rocky Mountains, or Indian tepees, or massive 747s—one can read all three images in the roof lines.

> *The new DIA sits in the middle of 53 square miles of erstwhile winter wheat, its white Teflon peaks mimicking those of the nearby Rocky Mountains (Fig. 8-2). The forms are apparitional, a congeries of 21st-century technology floating in a 19th-century frontier landscape. Not only are the fabric peaks remarkably beautiful—with their masts almost literally airships—they create a mood of excitement and adventure that is inseparable from travel but virtually absent from modern airports (Dillon 1995).*

Perhaps not surprisingly, such accolades are not universal. One source wrote, "The terminal's critics say it looks more like an enormous spiny caterpillar sprawled across the plains or, as the *Wall Street Journal* suggested, 'a circus tent draped over a big hedgehog'" (Howe 1995).

Fig. 8-2. *A view of main terminal building showing the large number of peaks rising above parking garage envelope.*

The original architectural firm, the Perez Group, was hired because the city was pleased with the attractive design work it had done at Stapleton on concourse E. Perez was commissioned to create standards that would unify the entire airport, taking the design of the main terminal through schematics and design development and developing a signature image for the airport that would capture the uniqueness of Denver and Colorado. Perez designed a stepped-pyramid, exposed truss system, with generous use of glass over a huge central atrium. The design evokes images of a railroad terminal, merging Denver's colorful history as a rail hub with contemporary jet transport (Figs. 8-3 and 8-4). The concourses were designed to mirror the main terminal design (Russell 1994).

Mayor Peña and the city council were concerned that the design was not sufficiently powerful to be a uniquely Colorado image and formed a blue-ribbon panel to evaluate the design. Eventually, the city council embraced the modified Perez design. But Fentress & Bradburn, which had been retained to evaluate the design for technical compliance, concluded that it would take the project $48 million over budget and off schedule. Fentress & Bradburn was then given three weeks to come up with an alternative and came up with

Fig. 8-3. *Exterior view of the Perez Associates step-pyramid main terminal, showing multilevel ground transport access designed to maximize curb space.*

Fig. 8-4. *Interior view of the Perez main terminal design, with a strong emphasis on glass and steel and a railroad terminal atmosphere, reflecting Denver's historical role as a western railroad hub.*

the fabric roof (Fig. 8-5). According to Fentress, "The fabric roof was the only alternative that was buildable and met the schedule. We were able to take out something like 300,000 tons of steel and 1000 linear feet of sheer walls underneath" (Russell 1994). Unfortunately, the concourses and other surrounding buildings were never redesigned to match the new terminal design. Thus, DIA, which could have realized an integrated vision, has the same eclectic architecture that afflicts virtually every other major U.S. airport.

The fabric is translucent, which, while shielding passengers from ultraviolet light, allows natural illumination of the terminal, without glare and without the need for artificial illumination during the day (Tulacz 1995). See Fig. 8-6. The roof admits 11 percent of the sun's light, providing 200 foot-candles of light at the winter solstice (Wright 1994). The interior hue is sort of a dull almond. At selected positions at the apex of the tallest pillars (Fig. 8-7), as well as around the edge of the roof, slits of glass admit direct, crisp sunlight, which slithers across the interior of the terminal as the day progresses. From the interior, one can see Colorado's deep-blue sky. Denver is blessed with more than 300 days of sunshine a year, and the roof is designed to capitalize on that phenomenon (although not nearly as

Fig. 8-5. *The Perez step-pyramid main terminal profile (top), contrasted with the Fentress/Bradburn fabric roof profile (bottom).*

much as the Perez design it replaced). It is especially impressive at night, when the roof glows softly from within. A luminescence emanates from the interior light—12 percent of interior light is transmitted (*Architectural Record* 1993)—making an impressive visual image for people cruising up 12-mile Peña Boulevard or peering from taxiing aircraft (Fig. 8-8). One airport official said, "The roof will give the building a luminescent vibrancy and warmth that no conventional structure can offer" (*DIA Newsletter* 1991).

The cable-tensioned, Teflon-coated-fiberglass fabric roof was fabricated and installed by Birdair, a subsidiary of Tokyo's Kogyo Corporation (Tulacz 1995). The roof consists of a thick outer membrane and a thin inner uncoated fiberglass layer, designed to hide the 10 miles of supporting cable (Wright 1994). Together, at 660,000 square feet, both layers would cover 15 acres if spread on the ground. The catenary cable system that supports the fabric is similar to that used to support the Brooklyn Bridge. The roof weighs less than two pounds per square foot, and, according to the architects, "relies on design curvature and equalization of the internal stress fields in the fabric for its stability and ability to support imposed wind and snow loads" (Fentress & Bradburn 1995).

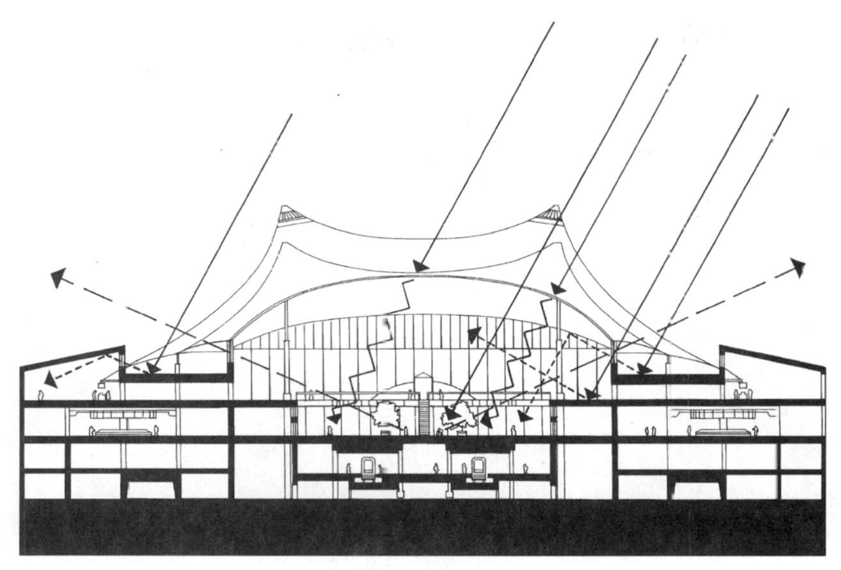

Fig. 8-6. *Artist's drawing of light admission and diffusion from the main terminal's fabric roof at DIA.* C.W. Fentress & J.H. Bradburn.

Fig. 8-7. *A closer view of large peaks on roof of main terminal showing the windows left for direct sunlight access to interior of terminal.*

Fig. 8-8. *Evening view of main terminal building, revealing interior light shining through the fabric roof.*

Before the roof was built, the installer built a scale replica of the fabric assemblies, testing alternative methods and sequencing of fabric packaging, handling, rigging, and hoisting. The model was subsequently sent to the construction site to refine procedures and educate the construction crews. After a general plan was developed, computer models were built to provide essential quantitative structural analysis of the roof's design, construction, and fabrication detailing (Brown 1993).

The fabric roof was believed to cost less than a conventional heavy steel-and-glass roof. The conventional roof would have required 300,000 tons of structural steel; the fiberglass roof required only 30,000 tons of structural steel (*Aviation Week* 1991a). The expected savings led the city to upgrade the main terminal floor to granite and double the size of the parking structure (Russell 1994). But the original estimate of $12 million for construction and insulation of the roof rose to $37 million by 1993—insulation was deleted at a savings of $2 million (D. Chandler 1993a). But everything is relative. United's terminal building at Chicago O'Hare, designed by Helmut Jahn, cost $370 per square foot. DIA's terminal was less than $100 per square foot (Sinisi 1995).

The roof is not without its critics. Fears have been expressed that the roof might collapse from Colorado's heavy snow, or be blown to shreds by its fierce winds, or disintegrate from a combination of ultraviolet light and bird droppings (Pierson 1994b). Critics pointed to the Birdair-covered Minneapolis Metrodome, which collapsed in 1990 under the weight of snow, and collapsed again in 1991 because of rips. A tent roof atop San Diego's convention center ripped apart in 1991 in 35-mile-per-hour wind, despite the fact that its manufacturer, Owens-Corning, had claimed it could withstand 90 mile-per-hour gusts (Amole 1992). Critics also decried the DIA roof's poor insulating qualities (a mere R1.85 in the winter and R2.2 in the summer—the insulation value of clear glass or plywood), which would allegedly freeze passengers in winter and roast them in summer. On the other hand, DIA chief engineer Ginger Evans believes that it is a superior design from an energy standpoint and for drainage (Amole 1992; D. Chandler 1993a).

To date, no such catastrophe has occurred; the facility has survived severe snowstorms and high winds. Water does come rolling off the fabric roof during thunderstorms, drenching passengers trying to enter the airport from automobiles, taxis, and vans—the result of thoughtless design (Fig. 8-9). Fortunately, precipitation is rare in the semi-desert environment of the plains east of Denver (Dillon 1995). Snow gathers in the valleys of the roof, but it melts. In short, it appears that the roof has performed well; the fears of the critics were unwarranted. Knock on wood. Minor repairs to DIA's roof can be made by welding on another layer of fiberglass fabric (*Architectural Record* 1993) and, although most such roofs are guaranteed for three years, DIA's has a 10-year warranty (Russell 1994).

Nor does it seem likely that another type of disaster will strike the terminal. With a sprinkler system installed only around the periphery of the Great Hall (which would likely be inadequate should the roof catch fire), it has been alleged that a fire would create "fumes from the burning Teflon, Fiberglass and resins [that] would be more lethal than mustard gas" (D. Chandler 1993a). We tried to light a sample piece of the roof with a match to no avail. Apparently, melting the roof would require intense heat. Given all the steel, concrete, and granite in the main terminal, it is not clear how a fire could grow large enough to melt the roof. Further, airport fire trucks could spray the roof with foam if a fire erupted.

Fig. 8-9. *The area over the passenger check-in level includes a smaller aerodynamically shaped roof.*

The main terminal building

Describing the main terminal building, one source wrote, "From the gleaming terrazzo floors to the cloud-like ceiling that stretches 900 feet from one end to the other, DIA is a trip unto itself" (Kelly 1995). Another compared it to the great transportation terminals of the United States: "New York's LaGuardia is about crowds and rudeness, while DFW is all bland efficiency. But Denver, like the great railway stations of the past, is about the drama of arrival and departure" (Dillon 1995).

During the airport's open houses preceding the opening, visitors were awed by the size of the facility (Fig. 8-10). DIA's terminal is enormous, covering 3.5 million square feet over seven levels—the equivalent of 35 football fields (Sinisi 1995). Almost six times the size of New York's Grand Central Station, the Great Hall (Fig. 8-11) is three football fields long and 250 feet wide with two levels—an upper floor dominated by a food court and a lower floor leading through security to the underground trains (Hight 1994). See Fig. 8-12. Two rows of 17 masts separated by 150 feet line the Great Hall (Fentress & Bradburn 1995). Despite its size, it does not make people feel small. Said one visitor "[I]t doesn't take long to get comfortable here. For such a huge airport, I think it's a very friendly place"

(Kelly 1995). Contributing to this feeling are the short distances that passengers have to travel (see Chapter 7).

The terminal is named after aviation pioneer Elrey Jeppesen, the Coloradan whose firm (Jeppesen-Sandersen) dominates the aviation navigational map and chart industry. A large bronze sculpture of Jeppesen leaning against an airplane propeller stands at the north end of the Great Hall (Fig. 8-13). At the base of the statue, a stainless steel and granite floor pattern depicts Denver's landing pattern from Jeppesen's "little black book" (Fentress & Bradburn 1995). It was that book, in which he listed, during the early days of aviation, terrain and other features to guide him, that won Jeppesen fame and fortune. Jeppesen's maps are today an essential feature of air navigation.

The floor of the main terminal is both magnificent and revolutionary, consisting of six types of colorful granite in a sweeping V design (Fig. 8-14). Originally, the floor was to have been a less-expensive two-inch thick "terrazzo" floor of polished marble chips and cement (D. Chandler 1993b). Typically, ¾-inch granite is required for the kind of heavy passenger traffic an airport gets. But at DIA, the granite is only ⅜-inch thick, because a vacuum-impregnated, epoxy-infused, polypropylene-fiber mesh reinforcement was adhered to the underside of the granite to give it strength (Wright 1994). The 9-millimeter granite slabs were

Fig. 8-10. *The entrance of main terminal building as seen from Peña Boulevard.*

Fig. 8-11. *The interior of the Great Hall of the landside main terminal, showing the south glass wall with a view of Pike's Peak.*

Fig. 8-12. *The interior of the main terminal building showing the security screening area leading down to underground trains.*

Fig. 8-13. *George Lundeen's statue of Denverite Elrey Jeppesen, aviation cartographer, after whom the main terminal building is named, standing against an airplane propeller at the north end of the terminal.*

comprised of granite cut from quarries around the world—North America, South America, Europe, Africa, and Asia—and epoxy-pregnated, cut (using laser and high-pressure water jet technology), and assembled in Italy (Fentress & Bradburn 1995). Unfortunately, these tiles have proven costly and troublesome. The city surveyed the floors after DIA opened and found cracks in 1.35 percent of the tiles examined, while a reporter found 7.3 percent had cracks; the city argued the cracks were the result of poor installation by PCL-Harbert, but PCL argued the cracks were the result of the controversial thin-cut of the granite used (Flynn 1995). The cost of the floors more than doubled, while its manufacturer (the Italian firm Tecnomaiera) went bankrupt so that, although the tiles are guaranteed, no recourse for defective tiles is possible, at least against the manufacturer. The marble on the walls of the terminal was quarried in Marble, Colorado, the same quarry that produced marble for both Washington, D.C.'s Lincoln Memorial and Tomb of the Unknown Soldier (Fentress & Bradburn 1995).

The main terminal building was originally designed to include a multistory 1000-room hotel at its north end (Fig. 8-15). No tenants were found, and other projects took higher priority, so the hotel was put on the back burner until the airport was opened. As a consequence,

Fig. 8-14. *The floor of the main terminal is a uniquely designed thin layer of polished granite from quarries around the world, assembled in Italy.*

passengers had to travel beyond the 12-mile Peña Boulevard to find a hotel or even a gas station. A Marriott Fairfield hotel and a gas station along Peña Boulevard opened in 1996.

East and west of the terminal building are huge boxy parking structures (Figs. 8-16 and 8-17). The parking decks are highly functional. Although they have little architectural merit, the parking garages were mercifully tucked below the roof line so as not to obstruct the view from the west side of the departure level. Unfortunately, the terminal itself was not designed to accentuate the magnificent view passengers might otherwise enjoy of Colorado's Front Range mountains—from

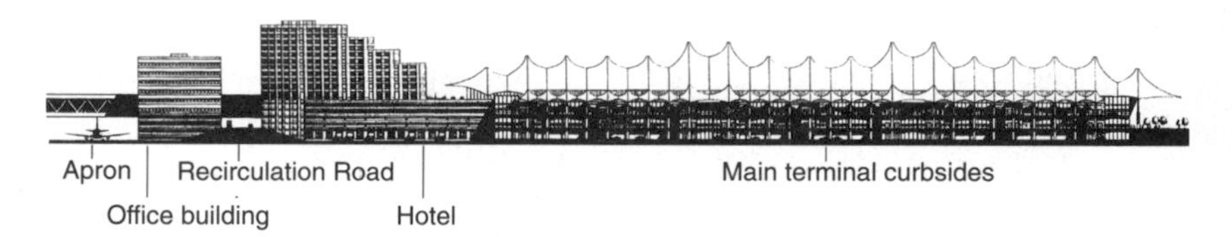

Fig. 8-15. *A horizontal view of DIA's main terminal building from the west, with an office building and proposed hotel on the left.*

C.W. Fentress & J.H. Bradburn.

Rocky Mountain National Park to Pike's Peak, as well as the downtown Denver skyline. If the terminal's aesthetics are to be faulted, it is that insufficient glass was incorporated into its design to capture that captivating view. The observation decks are at awkward places, the main one located at the far south end of the terminal building, away from most passenger traffic. From there, the view of most of the mountains (with the exception of Pike's Peak) is obstructed by a highway viaduct leading up to the top level (Fig. 8-18). Even the south patio was closed soon after the airport opened because its $232,000 granite floors buckled (O'Driscoll 1995b). The Rocky Mountains are visible outside the upper west side ticketing and baggage claim areas; but again, one must exit the terminal to take in the view. One critic commented:

> *Ironically, the best visual art of all—the Front Range of the Rocky Mountains—is blocked from view by parking garages and elevated roadways. An unparalleled vista, which says Denver and the West like nothing else, has been sacrificed to the automobile and the inertia of the modern traveler. DIA is not alone in its odd priorities, but in Omaha or Indianapolis what difference does it make? In Denver, it's unforgivable (Dillon 1995).*

Fig. 8-16. *Sketch of the main terminal building showing parallel garages on both sides and "Lorenzo's bridge," which leads over the airfield taxiway to Concourse A through a planned multistory hotel.*

Fig. 8-17. *A south view of DIA's main terminal building, with parking garages to the east and west.* C.W. Fentress & J.H. Bradburn.

Fig. 8-18. *Highway viaducts providing automobile access to the multilevel main terminal building.*

The concourses

Each of DIA's three concourses was designed by a different architect in what one source aptly described as "formulaic airport modern" (Dillon 1995). Again, generous installation of glass lets in the bright Colorado sun; the concourses were raised a full story of glass to enhance the inflow of light.

Edward McCagg of TRA, the initial designer, described the architectural objectives of the air-side concourses:

- *Represent physically the civic values of Denver and Colorado.*
- *Express the essence of the Colorado context without being replicative or literal in order to provide an exciting gateway to Colorado.*

- *Highlight the amenities of the region: sunshine, views of the mountains, colors and textures of the natural environment.*
- *Acknowledge and address the relationship of the new facility to the technology of flight.*
- *Derive esthetic strength of simple elegance from clear and sensitive expression of functions, particularly those which are essential to efficient passenger movement (McCagg 1993).*

He described the exterior appearance of the airside terminals as follows:

> *The new Denver airport unlike many, will have numerous opportunities to look back on itself. For this reason, exterior continuity is important. The concourses will share green tinted glass and warm colored concrete with the terminal. The concrete towers which visually separate the central core and subcores from the Holdrooms are cast with a special mix that includes warm colored aggregates of indigenous stone, in varied finishes and rustication. These towers will complement the buff color of the terminal parking structure. The exterior metal panels and window mullions will be a warm white metallic color with natural anodized aluminum used for air handling louvers, apron lighting poles, and fixtures. The light colors of the exterior materials lend another level of continuity with the white roof of the terminal (McCagg 1993).*

Unfortunately, as noted previously, the concept does not mesh neatly with the terminal design, and the opportunity to build an airport whose campus was a homogeneous and integrated vision was lost.

At the center of every concourse is the rail station that links it to the main terminal building. To avoid the subterranean claustrophobia of most such subway-type systems, the central concourse area was designed to be a multistory, 90-foot-tall, city-block, wide-open atrium topped by a clerestory, offering light and space to the train boarding area (Fig. 8-19). Again, much effort was taken to allow the generous Colorado sunshine to light even the most remote recesses of the cavernous airport.

Unlike Atlanta's Hartsfield International Airport (after which the basic structure of DIA is modeled), there is no parallel moving sidewalk pedestrian tunnel at DIA. Concerned about the possibility that

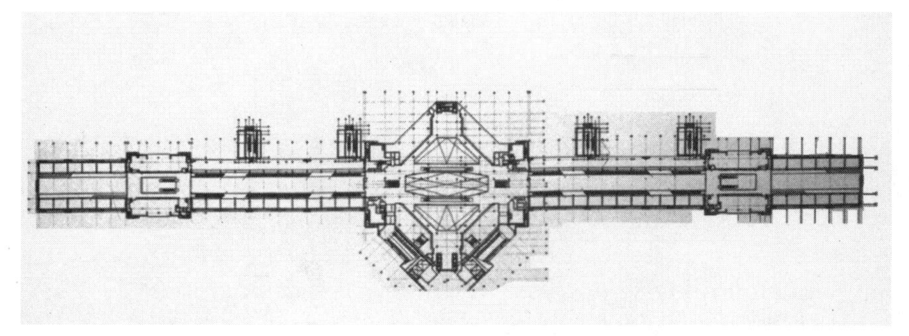

Fig. 8-19. *Schematic revealing the central atrium of Concourse C, approximately a full city block in diameter.*

the train might malfunction and seeking to gain a marketing edge over United Airlines, Continental Airlines' Frank Lorenzo insisted that a bridge be built between the main terminal and his terminal, terminal A.

The 350-foot bridge is black steel and glass, arching over the taxiway between the main terminal and Concourse A (Fig. 8-20). It is the only bridge spanning an operational aircraft taxiway in the world, and is sufficiently high to let a Boeing 737 aircraft pass underneath (Wright 1994). United objected to the fact that the bridge would not be high enough to allow its wide-bodied aircraft to pass underneath; but, of course, its planes were not destined for Concourse A. The bridge is also one of the few places in the airport that offers a panoramic view of the Rocky Mountain Front Range, a fact which United Airlines feared would give Continental an added competitive aesthetic advantage. Accordingly, United Airlines contractually obligated the city to obliterate much of the view by requiring most of the glass to be opaque. Mercifully, Ginger Evans, the head of construction breached this provision, and United Airlines (with a reputation of being among Denver's most aloof corporate citizens) chose not to risk the public relations price it would have to pay to prevent passengers from enjoying the magnificent view.

Art and glitz

An important element in DIA's aesthetics was the inclusion of a wide range of art works. As Denver's mayor, Federico Peña in 1988 issued an executive order requiring that 1 percent of hard construction dollars for projects of $1 million or more be dedicated to public art

Fig. 8-20. *"Lorenzo's Bridge" over the taxiway between the main terminal and Concourse A. It is the only bridge crossing an active airport taxiway in the world.*

(Paglia 1995). The city subsequently approved an ordinance to that effect (Rosen 1995), and, in 1990, a 13-member artist team submitted a report to the city, "Art Journey," which proposed to make art a part of the fabric of the airport's design (Pierce 1994). As the director of the mayor's Office of Art explained:

> *We wanted [visitors] to get a feeling for Denver and for Colorado. People are usually very tired and they need some sort of a respite. We tried to make the art a distraction—something, sometimes provocative, but something to change the pace of what they were going through (Moore 1994).*

DIA's public art program was assigned a budget of $7.5 million, which made it among the largest single-facility public art projects in the nation. It was co-chaired by Charles Ansbacher, head of the New World Airport Commission, which sought to supervise and oversee all activities surrounding the opening of Denver's new airport.

The commission was originally to be funded from a "combination of corporate and public contributions, admission fees, commissions, sponsorships and income from an official merchandising and licensing program," including selling DIA T-shirts, but it soon was consuming the city's money (D. Chandler 1993c). Ansbacher wanted the

President of the United States and the Pope to be aboard the first two aircraft to land at DIA on its opening day, but neither accepted his invitation (Van De Voorde 1993). Given the delays and cost overruns, Denver's new airport was becoming an embarrassment for the Clinton administration. By the time the airport opened, polls showed Federico Peña among Colorado's least popular politicians.

The New World Airport Commission put on an air show, several open houses, programs for school children, and an eclectic black-tie opening gala in the unfinished airport terminal. The gala was flush with Broadway show music, with dancing by Ben Vereen and singing by Rita Moreno, comedy by Art Buchwald, and speeches by DOT Secretary Peña, Mayor Webb, and Governor Romer. But among the speeches, there was precious little talk of aviation. No one even mentioned that it was only 90 years earlier that the Wright Brothers first went aloft (Calhoun 1994). Mostly, the gala was a thin veneer of glitz without substance. The event was widely publicized and one TV station broadcast a special program devoted to it. Shortly after the gala, Ansbacher followed his wife, Swanee Hunt, to Vienna, Austria, where she was to serve as U.S. ambassador. Before leaving, he presented the United Way with a $300,000 check as proceeds from the gala's silent auction; but, according to one report, he left the city unpaid bills for the gala and air show (Calhoun 1994).

Meanwhile, the process of art selection continued, albeit in a manner that seemed designed not to acquire the best art possible. Mayor Webb's wife, Wilma, chaired the Mayors' Commission on Art, Culture and Film, and abruptly called a moratorium on public art so that more commissions could go to minority artists. (Wilma Webb also co-chaired the New World Airport Commission with Charles Ansbacher.) However, the commission conducted a national competition for the art, not limiting it to local or regional artists (Paglia 1995). The process was handled by a cumbersome structure with more than 200 people involved in various committees. It has been reported that those with even marginal art background and knowledge were outnumbered by those without it (including politicians and activists) by 10 to one (Paglia 1995). Quality was but one of the criteria that drove the choices. One source wrote:

> *When the program went off its tracks, seeking to appease community groups more than consider artistic merit, it really went off: During the Peña administration, a promise*

was made to include American Indian artists in a project, which was administered outside of normal channels and highly restrictive (Rocky Mountain News 1995).

Had it produced the best art in America, no one would have objected. But that was not the case. One critic concluded: "Only one painting out of a half dozen canvases and many photos has something real to say" (*Rocky Mountain News* 1995). Furthermore, much of the art selected lacks a western, or Rocky Mountain pioneer, flavor (Paglia 1995). But Mayor Wellington Webb claimed the opposite:

Travelers are in for a uniquely Colorado experience when they pass through Denver International Airport. The works of art that grace the airport create a journey through our state's history and diversity. Like all successful public art, the program at DIA exemplifies an expression of ourselves and provides an opportunity to educate others. With the unveiling of this nationally acclaimed program, the cultural growth of our city and the quality of life for our citizens will be enriched (DIA Art n.d.).

The city's brochures best describe what was intended by DIA's art program. In *DIA Art: Result of a Unique Program*, Jennifer Murphy (n.d.) wrote:

Issues such as function, passenger convenience, ambient sound level, maintenance, air quality, fire and safety requirements, and vandalism played a major role in determining how art could enhance the environment of a space that is first and foremost a transportation facility. To ensure the greatest possible success, artists were selected early in the design process and were encouraged to integrate artworks into the actual structural and systems of the building. . . . It is this fusion of art and architecture that makes the Denver International Airport unique among the nation's airports and the DIA Art Program unique among the nation's public art programs.

Another DIA brochure, *Journey of the Imagination*, proclaims:

The art at Denver International Airport represents one of the largest and most impressive public art programs in the entire United States. 39 artists created original works for the project, integrating the art itself into the design and structure of the airport.

"Journey" is the unifying theme of the program, relating to the concept of travel that the airport itself embodies. Yet the idea of journey also plays on another level, as we constantly move from one place to another in our lives.

The art works

The commissioned art included all forms—sculpture, mobiles, floor tile, light shows, ceramic railings, photographic murals, and paintings (Skiba 1995). See Figs. 8-21 through 8-24. Beauty is in the eye of the beholder. Some of DIA's art would be critically praised, but most was rightfully condemned. One art critic found only "[f]ive truly first-rate pieces produced from a budget of more than $7 million" (Paglia 1995). Another wrote, "Some of the pieces are successful, a few are candidates for being swept under a rug—except there are few rugs that large" (M. Chandler 1995).

Among the most amusing pieces is "Notre Denver"—a bronze gargoyle sitting in an open suitcase on a pillar (Fig. 8-25) looking down

Fig. 8-21. *Betty Woodman's "Balustrade" pillars on the main terminal skybridge.*

Fig. 8-22. *Leo Tanguma's "The Children of the World Dream of Peace," a mural in the main terminal.*

Fig. 8-23. *Large paintings adorn the international arrival area in the main terminal.*

Fig. 8-24. *A few pieces of American Indian art were commissioned for the main terminal building.*

on the baggage claim area (M. Chandler 1995). It has been criticized as "badly modeled and unbelievably stupid" (Paglia 1995). An Indian-inspired design commissioned for the floor of the Great Hall is a tremendous disappointment compared with the magnificent polished granite floor that runs throughout the terminal (Rosen 1995). In fact, the granite polished floor is far more attractive than the floor "art" that was commissioned. A work that draws the most attention is also one of the wittiest. A pair of large colorful plywood maps of the United States, hung on opposite walls of the Great Hall of the Main Terminal and titled "America, Why I Love Her," highlight exotic tourist attractions, ranging from the world's largest Holstein cow (New Salem, North Dakota) to the boll weevil monument (Enterprise, Alabama). A large white arrow points to Denver with the caption "You are here . . . but your luggage is in Spokane." The arrow on the other map announces that the luggage is in Pittsburgh.

The train itself is a trip. It is flush with art, from carnival-type music that announces the concourses, to 5280 chrome propellers (Fig. 8-26), one for each foot the "Mile High City" is situated above sea level, and fluorescent light shows in the tunnels (Fig. 8-27), some depicting mine shafts, a cave, and deep space; to 140 purple, blue, and

Fig. 8-25.
Overlooking the baggage claim area, Terry Allen's bronze sculpture, "Notre Denver," depicts a gargoyle emerging from a piece of luggage.

yellow metal "paper airplanes" leading passengers through a tight bottleneck corridor from the train up the escalators to baggage claim (Pierce 1994). Thankfully, the tunnel art is difficult to see unless one is standing in the front train car, and there's not much to be awed by. But one must confess, it is better than looking at the walls of a dark concrete tunnel.

As noted previously, each concourse was designed with a central multistory open atrium the size of a department store, from which passengers depart from the train and move up the escalators to the gates. In each atrium there is a major piece of art. Concourse A's consists of a huge transportation sculpture of train tracks and a space ship (Fig. 8-28). It has been described as looking like a "Runaway Mine Train at Six Flags" (Dillon 1995), but a slim majority of this book's authors gave it a "thumbs up."

For efficiency's sake, United Airlines insisted that a walkway cut across its Concourse B atrium, because with a ¾-mile-long concourse (Fig. 8-29), passengers trying to connect between planes parked at the far ends would welcome the shortcut. The art piece here consists of neon-lit aluminum and colored glass that create giant arches (Fig. 8-30). One critic has passed harsh judgment, noting

Fig. 8-26. *The underground train tunnel, showing William Maxwell's and Antonette Rosato's "Kinetic Light Air Curtain," with 5280 chrome propellers, which rotate as the passing train changes the air pressure.*

Fig. 8-27. *The underground train tunnel, showing Leni Schwendinger's "Deep Time/Deep Space," a light and metal sculpture.*

Fig. 8-28. *The atrium of Concourse A, showing David Griggs' "Dual Meridian," a transportation sculpture with swirling railroad tracks.*

that it "serves as little more than 'street furniture' to direct and focus pedestrian traffic across the walkway. The powder blue and white colors make the large piece look as architecturally insubstantial as a Styrofoam water toy. . . . Further, the illumination doesn't look all that different from the neon signs at the nearby food court" (Rosen 1995). Quite right. The floor at Concourse B includes tile and metal dinosaur bones (M. Chandler 1994b). One source described the experience as an "aesthetic jolt" when the floor art runs into United Airlines' speckled industrial-grade carpeting (M. Chandler 1994a).

Although located in the smallest and most remote of the concourses, Concourse C's atrium has the most impressive piece of art (Fig. 8-31)—Michael Singer's 5000-square-foot, $912,000 mysterious Aztec or Inca-type ruin covered with a lush hanging tropical garden (Hight 1994; Miniclier 1995). Said Singer, "My work for Concourse C is meant to provide a balance to the technology-laden airport; a connection for the traveler to human history and natural world that is most often their destination" (Miniclier 1995). In one sense, the lush vegetation is a glance at eons past. When cutting through up to 100 feet of soil in building DIA, construction bulldozers turned up large rocks containing imprints of palm leaves and extinct sycamores, some 65 million years old (Johnson 1995). However, Anasazi Mesa Verde-type

Fig. 8-29. *Sketch of the interior of Concourse B, the longest concourse at DIA, occupied by United Airlines.*

Fig. 8-30. *The atrium of Concourse B, showing Alice Adams' "Central Core Structure," a glass and neon light structure over United Airlines' movable sidewalk.*

ruins might have been more appropriate for a Colorado airport (although, then the jungle growth would have made no sense). Despite its appeal, the design would be more appropriate for an airport located in the middle of the Yucatan Peninsula. However, it is the most interesting and creative art displayed in the three concourses.

Overall, the art works in the concourses do not rate highly. As one unimpressed critic commented: "If the terminal celebrates the romance of travel, these concourses apotheosize the moving sidewalk and the Naugahyde lounge chair" (Dillon 1995).

A fountain was originally designed to be the centerpiece of the Great Hall of the terminal building, but its location over the trains caused airport officials to fear that water might leak down to the electrical system. The world's largest drip pan was installed to catch any leaking water, but the fountain was put on hold. Instead, a cactus garden was added, with a Santa Fe-style stone wall around the base (M. Chandler 1994b). This complements other stone and plant elements in the terminal building. In several places red sandstone elements have been placed to mirror the Continental Divide (Pierce 1994), and the floor of the Great Hall is adorned by an assortment of shrubs, trees, flowers, and other leafy plants, as well as captive birds (Miniclier 1995).

Fig. 8-31. *The atrium of Concourse C, showing Michael Singer's interior garden growing over "prehistoric" ruins.*

Although delayed, the fountain remains a priority. It is expected to be 60 feet long by 24 feet wide and will shoot 3000 streams of water into the air, forming a liquid mountainscape (Rosen 1995).

Outside the terminal are a few questionable pieces of "art," including a 1600-foot line of rusty farm equipment and bailing wire apparently collected from around the former farmland—reminding one of an automobile junk yard. It was aptly described as a "genuine embarrassment . . . a thin idea poorly carried out . . . " (Paglia 1995). Among the pieces commissioned but not complete at the airport's opening was New Mexico artist Luis Jimenez' garish and anemic 30-foot-high blue fiberglass sculpture, "Rearing Mustang" (Fig. 8-32) with eyes that will shoot out laser beams of red light (Rosen 1995). It will stand outside the main terminal building. If the airport wishes to celebrate Denver's football team, the Broncos, a Remington-type pioneer riding a bucking bronco might be more tasteful. Other planned additions include a herd of more than 100 buffalo (O'Driscoll 1995a), and some trees, which the desolate plains environment clearly needs (Pierce 1994).

One other criticism of the art is its location—several pieces are oddly tucked in weird places. As a headline in the *Denver Post* observed,

Fig. 8-32.
Outside the main terminal stands Luis Jimenez' anemic "Rearing Mustang," a sculpture, like the roof, made of fiberglass.

"Mostly, It's Hard to Find and Hard to See" (*Denver Post* 1995). Jennifer Murphy, director of DIA's art-program office, blames the architects. She believes that, in the future, public art should be commissioned before, and not after, the project is built, forcing the architects to design the building around the art (Paglia 1995). A cynic would argue, however, that given the questionable quality of some—if not many—of the pieces selected for DIA, it is a blessing that the airport was not designed to showcase the art more prominently. One critic would agree with this pessimistic assessment. He wrote:

> *It's tempting at this point to make suggestions as to how the process could be changed to correct its many flaws. The city could hire a curator, it could work to increase participation by those with art-world know-how, it could replace bureaucratic reality with reason. None of those approaches, though, would undo the damage already done. It may sound trite, but the airport art program was a once-in-a-lifetime opportunity. It will be a very long time before Denver again has the chance to spend millions of dollars on public art. After seeing how the money was spent at DIA, I was deeply moved—in fact, I felt like crying (Paglia 1995).*

DIA: Conclusion

That a rare opportunity, in terms of both the architecture and the art, was not fully seized is indisputable. Yet there is much to praise, and it would be wrong to conclude on a wholly pessimistic note. Many of DIA's aesthetic elements are indeed praiseworthy.

One nationally syndicated columnist, for example, wrote:

> *[Denver] has built an airport of rare architectural elegance, exquisitely matched to its surroundings, and filled with quality art reflecting Western life, travel, light, and space. This could turn out to be one of the important public buildings of our era (Pierce 1994).*

And, perhaps most importantly, those who use the airport rate it a success, though with some reservations. As one observer noted:

> *After the first month of operation, experienced business travelers are giving Denver International rave reviews for its*

spacious, elegant appearance. But they rate it less highly in a few key areas [including functionality] (Lindgren 1995).

Other major new international airports

Given Denver's mixed achievement in art and aesthetics, it is appropriate to examine the aesthetics of other major new airports around the world.

Munich's Franz Josef Strauss Airport

More than 100 architectural firms competed for the design of Munich's new Franz Josef Strauss Airport, opened in 1992. Its main terminal, designed in the mid-1970s, was heavily influenced by U.S. airport terminal design and its reliance on the automobile as the primary mode of landside access (*Aviation Week* 1991b). See Figs. 8-33 and 8-34. The style is a long linear blend of white girders and glass (O'Driscoll 1994). Large amounts of glass face the runways. Even the escalators have clear sides, revealing working parts. The airport also uses a new bench system (Tubis) made of an aluminum frame with perforated steel parts covered with upholstery (Pierson 1994a). Although the design has drawn some criticism, the public likes it:

A surprisingly sleek, yet incongruous apparition in concrete, steel and glass stretches into view as motorists near Munich Airport, sprawling spaciously over the moor near rural Erding. Cynics refer to it wryly as a "designer" airport . . . .

Nearly 80 percent of Munich passengers and visitors polled . . . responded favorably. Random remarks: "Ultramodern facility," "attractive, functional layout," "stunning architecture; clear, bright design features" (Hill 1993a).

Osaka's Kansai Airport

The international design competition held for Osaka's new airport, opened in 1994, was won by the Italian firm Renzo Piano and Building Workshop, one of 15 entries from the United States, Europe, and Japan (Branch 1989). Piano, one of the architects of the Georges Pompidou Center in Paris, designed an ultramodern main terminal building—a four-story structure with a rolling, cloud-like roof (*San Diego Union-Tribune* 1994), shaped like the leading edge of a giant

Fig. 8-33. *A view of Munich's Franz Josef Strauss Airport with the main terminal (left) and parking garages (right).* Flughafen München

Fig. 8-34. *A view of Munich's new airport main terminal and air traffic control tower.* Flughafen München

airfoil (Moorman 1994). An architectural magazine described it as follows:

> *Piano's design, which . . . was chosen mainly for aesthetic reasons, employs a roof of aerodynamic curves reminiscent of an aircraft fuselage. The curves are meant to assist ventilation throughout the terminal. The other major features of the design are strips of plantings—Piano calls them "valleys"—running the length of the terminal. The theme of technology and nature coexisting is reinforced with an extensive planting scheme for the man-made island (Branch 1989).*

Kuala Lumpur Sepang International Airport

Kisho Kurokawa, a Japanese architect, came up with a "forest in a forest and forest in an airport" concept that incorporates tropical vegetation and a lush Malaysian landscape into the new airport at Kuala Lumpur, Malaysia, scheduled for completion in 1998 (Hill 1993b). The airport will be the first in the Asia/Pacific region to install glass-walled passenger loading bridges (*World Airport Week* 1995).

New Seoul International Airport

The terminal at New Seoul International Airport was designed to reflect Korean tradition, with the incorporation of pillars resembling Korean temples and the use of foliage found in Korean gardens (Fig. 8-35). The winning design, chosen by blind grading, was submitted in 1992 by C. W. Fentress, J. H. Bradburn and Associates, of Denver, Colorado, the same architectural team that designed the main terminal building at Denver International Airport (Shin 1993). At both airports, the "Great Halls" are "greened." At Seoul, this includes a hanging garden that is several stories high (Fig. 8-36). Curt Fentress described his concept as follows:

> *The Korean Airport Authority wanted an airport that was reflective of the local culture, past as well as projected. Both [Denver and Seoul] sought gateways to their cities. Both wished to provide a definable and memorable image that rekindled the lost excitement of travel as well as furnishing a portal for arrival into a new place. In both cases, this desire produced the largest public works projects in their respective countries within the last generation. . . .*

> *Combining the ideas of both temporary and permanent occupation revealed a design in Denver in which the roof top, the sky gate, became the visual gateway for arrival from both the ground and the air. The idea of using a tensile structure for Denver and the historical form of a sacred temple precinct in Seoul, replete with imagery derived from the colors, patterns and shapes of native animals, ecosystems, and even costume, to house a purely utilitarian plinth became economically plausible for some and architecturally essential to integrate these spaces. Thus, the airports assumed forms from which many things were intuitively communicated; at once contextual, regional, and instinctively derived by, of, and for human beings (Fentress 1995).*

Like Denver, Seoul attempted to embrace an architectural signature that reflected the region in an unforgettable way.

Lessons learned

- *Design the airport's architecture as an aesthetic statement of the cultural heritage or the natural features of the region.* Airports should be dynamic architectural statements, because they are the first and last impression most travelers will

Fig. 8-35. *Artist's rendering of the landside entry to main terminal at Seoul, Korea's new airport.* C.W. Fentress & J.H. Bradburn.

Fig. 8-36. *Artist's rendering of great hall in main terminal building of Seoul, Korea's new airport.* C.W. Fentress & J.H. Bradburn.

receive about a city, and sometimes a nation, they are visiting. The desire to put a signature on a city with a massive public works program is a noble one, particularly to incorporate a statement about the region. DIA's terminal building does suggest the Rocky Mountain horizon. As the first and last doorway through which many people see the city, the visual impression is unforgettable. Airports at Seoul and Kuala Lumpur, in particular, attempt to acquaint the visitor with a pleasant first and last glance at the cultural and natural beauty and heritage of the nation. This positive impression will likely enhance tourism and other business sectors, as well as giving local residents pride in their heritage.

- *Art should be incorporated into the budget.* Because an airport is a place where millions of people traverse, tasteful art should be sprinkled about to entertain, to enlighten, and to inform the traveler about the region.
- *The airport's art should reflect the unique cultural heritage and natural beauty of the region it serves.* Given that the

first, last, and sometimes, only impression a traveler might have of a region is via exposure to its airport, the opportunity should not be lost to educate the traveler about the region's unique cultural heritage and natural beauty. An airport serving a tourist mecca like Colorado should introduce the traveler to the Rocky Mountains, the Great Plains, as well as their American Indian, Hispanic, pioneer, and mining heritage.

- *Tasteful art should be selected by art experts, not politicians and bureaucrats, with public input.* Denver's art was commissioned mostly by local politicians and community activists, rather than art experts. The result was hit and miss, with several embarrassing misses. The common people have enormous common sense about the art that pleases them. As was done for Chicago's beautiful new library, models of the experts' choices should be provided, and the people allowed to vote.
- *Incorporate blind grading into the design process.* The likelihood that foreign firms with superior designs will receive design contracts is much enhanced in a process of blind grading, as occurred at Seoul. Blind grading involves judging proposals on merit alone, irrespective of the nationality of the contestant. The superior design at a lower cost may be available abroad because of foreign experience, expertise, and economies of scale.

References

Amole, Gene. 1992. A flimsy excuse for an airport roof. *Rocky Mountain News.* February 9.

Architectural Record. 1993. "Snow-capped" symbol. June:106.

Aviation Week & Space Technology. 1991a. Denver airport: International hub of the future. March 11

————. 1991b. Germany struggles to meet airport needs. August 26:41.

Branch, Mark. 1989. Piano wins Osaka competition. *Progressive Architecture.* March 1:33.

Brown, Martin. 1993. Denver International Airport tensile roof case study: The fabrication and construction process. Paper delivered before International Conference on Aviation & Airport Infrastructure, Denver, CO, December.

Calhoun, Patricia. 1994. The party's over. *Westword.* March 9.

Chandler, David. 1993a. Terminal condition. *Westword.* July 21.

————. 1993b. The bottom line. *Westword.* July 21.

————. 1993c. High society. *Westword.* November 17.

Chandler, Mary. 1994a. It's a design to match the vision. *Rocky Mountain News.* May 8.

————. 1994b. Onward, upward and delays with the arts at DIA. *Rocky Mountain News.* October 30.

————. 1995. Public art's rough flight lands with mixed results at DIA. *Rocky Mountain News.* March 26.

Denver International Airport Newsletter. 1991. New airport will light up the night. Winter 1991.

Denver International Airport. n.d. *Journey of the Imagination.*

————. n.d. *DIA Art: Result of Unique Program.* Flyer distributed by DIA.

Denver Post. 1995. Mostly, it's hard to find and hard to see. March 19.

Dillon, David. 1995. Soaring main terminal overwhelms the rest of Denver's troubled new airport. *Dallas Morning News.* February 12.

Fentress, Curtis. 1995. Revitalizing the excitement of travel. *Passenger Terminal '95.*

Fentress, C. W. & J. H. Bradburn. 1995. *Terminal Facts!*

Flynn, Kevin. 1995. City wants plan for fixing cracks in DIA floor. *Rocky Mountain News.* May 23.

Hight, Bruce. 1994. Will Austin learn lessons from Denver's problems? *Austin American-Statesman.* May 15.

Hill, Leonard. 1993a. Beyond expectations. *Air Transport World.* June 1:182.

————. 1993b. Asia's newest "dragon." *Air Transport World.* September 1:66.

Howe, Kenneth. 1995. Lost in space at Denver airport. *San Francisco Chronicle.* June 24.

Johnson, Kirk. 1995. New airport hides 65 million years of history. *Rocky Mountain News.* March 8.

Kelly, Guy. 1995. DIA visitors take a trip and never leave the ground. *Rocky Mountain News.* February 12.

Lindgren, Kristina. 1995. Denver airport: A beauty with a few flaws. *Los Angeles Times.* March 30.

McCagg, Edward. 1993. The new Denver International Airport: Evolution of a concourse. Paper presented before the International Conference on Aviation & Airport Infrastructure, Denver, CO, December.

Miniclier, Kit. 1995. Airport's green interior an oasis on the dusty plains. *Denver Post.* February 24.

Moore, Martha. 1994. Art landing in terminals. *USA Today.* June 14.

Moorman, Moorman. 1994. Osaka to me. *Air Transport World.* October 1:62.

Murphy, Jennifer. n.d. *DIA Art: Result of Unique Program.* Flyer distributed by DIA.

O'Driscoll, Patrick. 1994. Munich's airport: A case study for DIA. *Denver Post.* July 31.

————. 1995a. DIA to give buffalo a home on its "range." *Denver Post.* April 12.

————. 1995b. Cracks close DIA patio. *Denver Post.* June 10.

Paglia, Michael. 1995. A site for sore eyes. *Westword.* March 8.

Pierce, Neal. 1994. Denver airport may be folly, but it's impressive. *Houston Chronicle.* March 6.

Pierson, John. 1994a. The Europeans invade U.S. airport lounges. Associated Press. December 9.

————. 1994b. Form & function. *Wall Street Journal.* December 23.

Rocky Mountain News. 1995. Airport art. March 26.

Rosen, Steven. 1995. A critical look at DIA art. *Denver Post.* March 19.

Rothrock, Millicent. 1994. Fentress: Designing symbols of reality. *Greensboro News & Record.* October 16.

Russell, James. 1994. Is this any way to build an airport? *Architectural Record.* November:30.

San Diego Union-Tribune. 1994. Japan opens airport on man-made island. September 5.

Shin, Jong-Heui. 1993. Airport developments in Korea. Paper delivered at International Conference on Aviation & Airport Infrastructure, Denver, CO, December 8.

Sinisi, J. Sebastian. 1995. Soaring into the future. *Denver Post Magazine.* March 19:12.

Skiba, Katherine. 1995. Denver's new airport doubles as art gallery. *Milwaukee Journal Sentinel.* April 9.

Tulacz, Gary. 1995. Making markets for membranes. *Engineering News-Record.* August 21:68.

Van De Voorde, Andy. 1993. Ignorance is bliss. *Westword.* March 31.

World Airport Week. 1995. KLIAS to install glass-wall passenger loading bridges. July 18.

Wright, Gordon. 1994. Denver builds a "field of dreams." *Building Design & Construction.* September 1:52.

9

How did a $1.5 billion airport become a $5.3 billion airport?

Mike Keefe, *Denver Post*. Reprinted by permission.

> *"The only way DIA would cost $2.7 billion is if you gold-plate the terminal building."*—GEORGE DOUGHTY, DENVER AVIATION DIRECTOR[1]

Large projects are notorious for their cost overruns. Even though they might be subjected to detailed analyses of all kinds, these often are based on assumptions that turn out to be as valid as a noninstrument night landing in a Colorado blizzard. One unexpected event after another forces changes to the careful and detailed plans that were originally made; one assumption after another is shattered by new developments. The result is skyrocketing costs—overall,

1. This August 1988 statement was reported in High Country News (1995a).

megaprojects cost, on average, 90 to 150 percent more than their original estimates (Merrow 1988, p. 31; Murphy 1983, p. 19). The DIA case corroborates such findings, although its cost escalation was well above the average. But other examples abound, including two major new airports.

Recent international airports: Cost overruns and delays

Osaka's Kansai Airport took seven and a half years to build. Originally scheduled to open in the spring of 1993, its construction was overshadowed by terrorist attacks and technical problems, especially the sinking of parts of the manmade island because of the soft seabed. The unanticipated work required to solve the problem delayed the opening by 18 months and raised the cost of the airport by 40 percent over initial projections. Along the way, the United States insisted that more American firms be involved in the construction process and alleged some measure of bid-rigging in favor of Japanese construction firms. The dispute caused something of a diplomatic rift. But steering construction contracts to higher-cost local firms no doubt also contributed to KIA's high costs. As a consequence, Kansai will levy some of the highest airport charges in the world (Lassiter 1994; *Kyodo News* 1994).

Bangkok's Nong Ngu Hao International Airport's experience is similar. The first phase of the new airport was estimated in 1993 to cost $3.2 billion. This figure does not include land costs, because the Thai government has owned the land for 25 years, purchasing it for about $40 a hectare (*Business Week* 1995). Since then, more than 2000 families moved onto the swampy land and made a living farming and fishing there (*World Airport Week* 1995). Removing the families delayed construction for 18 months. The controversy over the contract, discussed in Chapter 7, added to the delay (*Airports* 1995a).

Problems with megaprojects

Like DIA these projects were all public ones. But it is simplistic to assume that such problems arise because of the public nature of these projects. In reality public, private, and mixed megaprojects all are characterized by cost overruns.

For example, a careful quantitative study of 52 megaprojects costing an average of $2 billion carried out by the Rand Corporation demonstrated that public projects grew by a factor of 1.9, mixed projects by 2.4, and private projects by 1.7; much of the increased costs of the public projects are caused by the willingness of public authorities to spend more to complete the project on time (Merrow 1988, pp. 37, 55). Overall, cost growth is most heavily influenced by several factors, two of which are most relevant to the DIA case—technological innovations and governmental regulations dealing with health, environment, safety, and minority contracting procurement controls (Merrow 1988, p. 34). Essentially, risk increased when the megaproject was executed expeditiously, was managed by government, or employed new technology or innovative design; the more complicated the project, the more likely that something (or many things) would go wrong (Merrow 1988).

The model in Fig. 9-1 provides a useful way of analyzing the reasons why costs escalate.

This model suggests that increases are due to four basic factors. First are cost-estimating errors. Estimates almost always tend to be optimistic because the standard methodology does not adequately anticipate

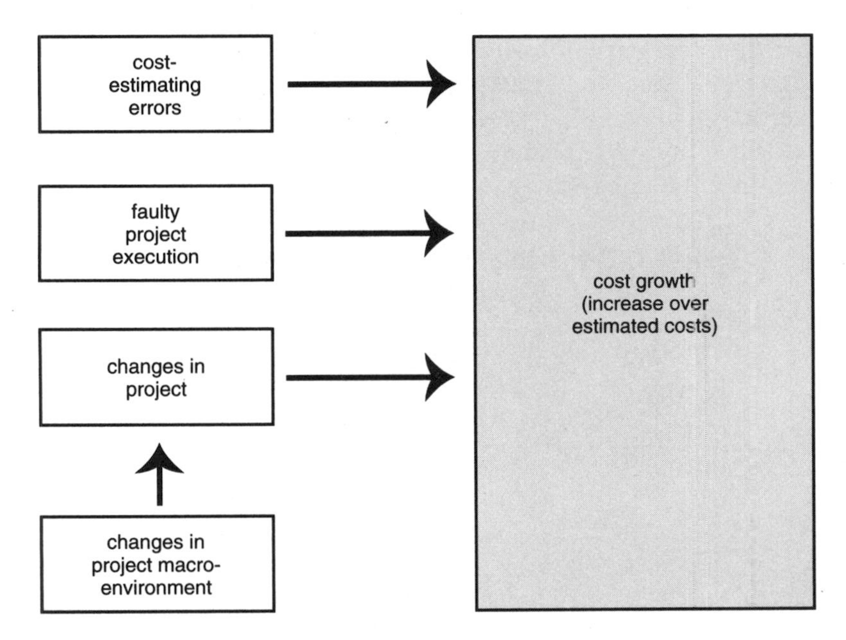

Fig. 9-1. *A conceptual model of cost growth.* Merrow 1988

unexpected contingencies. Also, other cost factors include poor project definition, erroneous economic assumptions, underestimation of inflation, and the nature of the project. The more complex a project, the greater the tendency to underestimate costs. Further, politicians who want to sell a project may intentionally understate its costs. Second comes faulty project execution. Seldom the result of major errors by the project's managers, this problem tends to stem from underlying causes such as inadequate understanding of the manpower practices and requirements that lead to strikes and other problems. Third, changes are often the single most important reason why costs escalate, since the project that was originally conceived and estimated is seldom the project that is finally completed. Changes can take many forms and be due to various factors. They may involve either the expansion or the contraction of the project, the addition of new elements, or the substitution of one element for another. Particularly troublesome are technologies that may be relatively or entirely new. Many of the changes are caused by a fourth factor—the misunderstanding of or changes in the environment (Merrow 1988, pp. 20–29).

Delays in completing a project are closely related to cost growth and are often caused by the same factors. Schedule slippage, however, is usually less than the increase in costs. Slippage can contribute to increasing the costs of the project, since decision-makers may place a high priority on timely completion and be willing to spend freely to achieve that goal or, as in DIA's case, may authorize major scope changes after construction has begun. This tendency is most pronounced in public projects, as Table 9-1 demonstrates.

Table 9-1. Ownership, cost growth, and schedule slippage

Ownership	Cost growth	Schedule slippage[1]
Public	1.9 (13)	1.05 (13)
Mixed	2.4 (9)	1.22 (7)
Private	1.7 (25)	1.22 (26)

Source: Merrow 1988

1. # of projects shown in parentheses.

As we shall see, DIA fits this pattern. In the sections that follow, the specific elements that contributed to the cost escalation are discussed and then analyzed within the model's framework.

Schedule slippage: The timetable

Cost projections on DIA were revised many times as the scope of the project changed and as essential systems were not ready on various scheduled opening dates. Clearly, attempting to accelerate construction imposed its own costs. Even with the delays, the entire project was designed and built in just five years. Considering the size and scope of the project (DIA is, as we have seen, one of the largest airports ever constructed), this was an expeditious timeline even with the missed deadlines. Kansai, for example, took half again as long.

Site preparation and construction at DIA began in September 1989, with a projected opening date of October 1993. Massive changes in the scope and design of the airport demanded by United Airlines in December 1991 made that clearly impractical, although Mayor Webb refused to reschedule the opening. Said Webb:

> *I had two options. One, we could either continue to move on the path of opening the airport on schedule and not go for more money, or I could announce to the world we're going to delay the airport by a year. And so I thought about it for no more than 10 seconds. Politically, how can you go out and tell people that you're going to delay the project by a year, when everything in the newspapers that's been publicly debated is that the project is on schedule, on budget, even though that was a smaller airport? (Flynn 1995b).*

But politically, how does a mayor continue to embrace an unrealistic deadline and risk a public-relations nightmare when the nation's largest public works project won't work? This was characteristic of Webb, who tended to view the short-term political costs as worse than the long-term political risk, even if graver. Empirically, his decision to stick with the October 1993 opening date has been defended on grounds that pushing it back would be a self-fulfilling prophecy. Unless the contractors' feet were held to the fire, they would be tempted to drag them, dragging out the project and increasing costs elsewhere, such as interest on issued bonds (Veazey 1995). Furthermore, changing the deadline would unnerve the bond houses, which apparently did not want any information released that might dampen enthusiasm for DIA's bonds. Thus, the timetable that became unrealistic (particularly after the massive scope changes and additions United Airlines insisted on in 1991, well after construction had commenced) was not amended.

Mayor Webb has also been accused of ignoring long-term benefits when he took the Park Hill community's position opposing keeping one north-south runway open at Stapleton for Continental Airlines to continue using its maintenance base. Over the longer term, it might have encouraged Continental to maintain a more significant competitive presence in Denver. To be fair, however, Webb's position on this issue can be defended at two levels. First, an agreement with the Park Hill community guaranteeing that Stapleton would be closed had been explicitly—and deliberately—granted legal standing by the Peña administration (see Chapter 3). Second, if Continental had been granted the right to use the runway, other carriers might have asserted a legal right to use the facility, a situation that would pose a serious threat to the financial viability of DIA. But on the other hand, if the passenger terminals were closed at Stapleton, it could not maintain commercial operations. A chain-link fence was all that was needed to keep passengers from boarding or departing aircraft.

In any event, DIA's first deadline (October 31, 1993) was missed because of construction schedule delays resulting from scope changes in the project, many due to airline tenant requests, and delays in completing essential integrated systems testing (ASRB 1994). Basically, construction of the airport was not complete. Then December 19, 1993, was set as the opening date, although that fell during one of the busiest air travel periods of the year and was manifestly implausible. The second delay, to March 9, 1994, was caused by delays in BAE's manufacturing and installation of the automated baggage system (power surge, mechanical, and software problems plagued the baggage system in late 1993), changes in the project, and delays in the installation of the specialty electronics system, for example, the security systems (ASRB 1994). But as the rest of the airport was completed, the baggage system remained dysfunctional, causing the opening date to be rescheduled, first to May 15, 1994, and then postponed indefinitely. DIA finally opened on February 28, 1995.

But was the baggage system the real reason for the delays? Ginger Evans, the head of construction at DIA, is one of many skeptics. In her words, "The airport could have opened early in 1994 with the automated baggage system running 40 bags per minute, but United consistently demanded that the terms of its agreement, which called for 225 bags per minute, be honored, even though it could meet its connect times with the slower rate" (Evans 1996a). Furthermore, a manual backup outbound system designed to handle a three- to four-hour jam in the BAE system had been installed by United and a manual inbound system was also in place by the fall of 1993 (Evans 1996b).

Evans's position is supported by others. "Virtually every design and construction professional [who] was involved directly or as a consultant . . . believed at that time that the project, including the BAE automated baggage system, could have been completed by October 21, 1993" (*Airports* 1995c).

If a malfunctioning baggage system was not the fundamental cause of the delay, what was? One plausible and widely accepted explanation has been provided by George Doughty, who served as Denver airport director under Federico Peña and held over in the early Webb administration, until June 1992.

> *. . . United Airlines did not want to go to DIA. United could have cooperated with the City to work out options for manual bag handling, but they did not. . . . As to exactly what United's rationale [was] one can only speculate, but a few things are clear. United had no incentive to move in 1994. They had just increased their operations at Denver in order to capture an even greater market share that would eventually force Continental to dismantle its hub. It was to their advantage not to move until that was assured. . . .*
>
> *Mayor Webb was simply outmaneuvered by United. They applied significant pressure and had previously made contributions to his political campaign and sponsored fund-raising events. He was not in a position to make a decision counter to their wishes. Therefore, the project was forced to absorb a $15 million a month delay cost until United said it was okay to move (Doughty 1995a).*

Thus, the schedule slippage was due to several factors—changes in scope, an innovative technology, and, most importantly, the decision of a powerful external actor who was able to impose its will on the key decision-maker. As we shall see, these factors contributed in significant ways to the escalating cost of the project.

Cost estimates: Sticker shock

DIA's total cost—including the construction, land acquisition, capitalized interest, and FAA and tenant improvements—is more than $5 billion. Of that, approximately $3 billion represents construction costs (Fig. 9-2), $1 billion more than the first firm estimates in May 1990 (GAO 1995). See Fig. 9-3.

Land Acquisition	**$ 241.6**
Pre-1990 Planning & Administration	**19.1**
Construction Costs	**3,003.9**
Cost of Issuance	**43.4**
Capitalized Interest	**923.8**
	$ 4,231.8

Fig. 9-2. *Breakdown of DIA costs (in millions) excluding final cost overruns and tenant expenditures.*

Denver International Airport

Construction Cost History

November 1988	$1.339 Billion (conceptual)	Four Runways, minimal apron, two concourses, 78 gates, two-module terminal
May 1990	$2.079 Billion	Five runways, expanded apron, three concourses w/basements, 94 gates. three module terminal, basic airline tenant improvements, conventional spine baggage system
October 1991	$2.663 Billion	Five runways, three concourses, 94 gates, three module terminal, automated baggage system, UAL scope increases, airline support grading and access, expanded cargo, sixth runway grading, expanded parking, commuter buildings, basement expansions
February 1993	$2.700 Billion	Five runways, expanded apron to accommodate commuter buildings and parking, sixth runway grading, three concourses (Concourse B lengthened and widened at UAL request), 94 gates, 41 commuter positions, three module terminal with additional finished space, expanded parking structure, airline tenant improvements, airport-wide baggage system
September 1994	$2.953 Billion	In addition to above, scope changes on UAL Systems, South Site additions, Backup Baggage System, additional Capital Costs.

Fig. 9-3. *Explanation of early DIA cost overruns and cost escalation.*

Early projections of the cost of DIA were indeed conservative, focusing only on construction costs (excluding the cost of land, and capital, or tenant and FAA expenditures). In 1986, Mayor Peña's new airport coordinator, Skip Spensley, indicated that financing the new airport's modest cost would be a simple matter:

> *Finding the financing for the $1.5 billion project may be one of the easier parts of building the airport. Half of the costs will come through special-revenue bonds with the balance split between federal funding and developing land and other assets of Stapleton after the old airport closes . . . . (Schwartzkopf 1986).*

In October 1988, Peña announced that DIA would have 120 gates and cost $1.5 billion, of which one-third would be paid by existing airport funds (principally, the sale of the 7-square-mile Stapleton property), one-third by the federal government, and one-third by bonded debt (Chandler 1993a). Thus, the city would need a mere half a billion dollars in revenue bonds.

In November 1988, Denver developed a "conceptual estimate" of $1.34 billion for the proposed airport (GAO 1995). By 1989, the city announced that DIA would open in mid-1993 with 94 gates, at a cost of $1.7 billion (Chandler 1993a). It was on that basis that the voters of the city and county of Denver approved the new airport at a referendum in May 1989.

During its construction, critics insisted that the cost projections were understated. By mid-1988, United and Continental were alleging that DIA might cost as much as $2.7 billion. Stapleton Airport director George Doughty rebutted that the only way DIA would cost $2.7 billion "is if you gold-plate the terminal building" (Fumento 1993). But by 1991, DIA was projected to cost $2.1 billion (Flynn 1991). By 1994, it was projected to cost $3.2 billion (*Denver Post* 1994a). DIA ultimately opened in February 1995 at a cost of more than $4.8 billion, and gold-plating was nowhere to be seen. Doughty had long since left Denver, complaining that the new mayor (Wellington Webb) could not "just say no" to belated airline scope changes that were driving up the cost of DIA by about $900 million directly (not including delays associated with the dysfunctional baggage system). During the 16 months of delay (from the original opening date of October 1993), construction costs went up about $300 million, to $3 billion; financing costs increased an additional $250 million, to $919 million. The cost of the land, coupled with facilities built by tenants, such as the airlines and rental car companies, brings the total project cost to $5.3 billion (Flynn 1995p).

Table 9-2 reveals how the price of DIA grew while the scope of the project was scaled back.[2]

Table 9-2. Changes in the cost versus the size of DIA

		Projected number		
Date	**Projected cost**	**Runways**	**Gates**	**Source**
10/88	$1.5 billion	6	120	Mayor Peña
5/89	$1.7 billion	6	120	Mayor Peña
2/91	$2.1 billion	6	90	Mayor Peña
6/91	$2.7 billion	5	90	Mayor Peña
5/92	$3.7 billion	5	90	City of Denver
10/92	$3.8 billion	5	80–90	City of Denver
2/93	$4.4 billion	5	80–90	City of Denver
2/95	$4.8 billion	5	87	City of Denver
2/95	$5.3 billion	5	87	City of Denver

On a per-passenger basis, DIA's costs also escalated. Stapleton International Airport cost less than $6 per passenger. In 1991, the per-passenger cost at DIA was estimated to be $11.18 by opening (Flynn 1991). By 1994, the per-passenger cost had been revised upward to $15.10 (*Denver Post* 1994b). By 1995, the per-passenger cost was $18.10 (*Rocky Mountain News* 1995a).

As was noted in Chapter 5, these costs have important implications for the airlines. Texas Air's Frank Lorenzo, as brilliant, yet erratic, a financial man as the airline industry had ever produced (he engineered leveraged buyouts of Continental and Eastern Airlines, although he subsequently marched them through bankruptcy several times) made one of the most prophetic estimates of these costs (Dempsey and Goetz 1992). In a November 1988 speech to Denver civic leaders, Lorenzo warned:

2. Much of the data were derived from a similar chart published in Chandler (1993a).

[Continental Airlines'] per-passenger cost in Denver . . . would, we estimate, rise from over $4 a passenger today to $12 when the new airport is scheduled to open. . . .

If we try to pass these costs on to our customers, then a significant portion of those who connect here will choose another Western hub. . . . Some have suggested that we exempt the connecting passengers from the additional cost and place it entirely on the one-third of Stapleton's traffic that is either bound for Denver or initially boards here. That would mean an extra $36 per ticket—and you can imagine what that would do to Denver's attraction as a convention city and to your traffic generally.

To which Mayor Federico Peña responded, "Their documentation of this claim is weak at best." Actually, Lorenzo was too conservative; he predicted $12 per passenger whereas more than $18 is the real number. Lorenzo predicted an extra $36 a ticket on local passengers; the real number is $40.

By 1995, DOT Secretary Peña was saying, "Frankly, nobody thought the cost would be $18 per passenger. Had we known that, we would have restructured the size of the airport" (Kerwin 1994). Nobody, it appears, but the CEO of one of Denver's two hub airlines. Actually, the size of the airport fluctuated dramatically over time. As we have seen, the airport—the number of concourses, gates, and runways—was designed in a modular manner that permitted expansion or downsizing as market conditions warranted. As Stapleton Airport director George Doughty said in 1990:

The airport on opening day [in 1993] . . . will be sized to meet 1995 traffic levels only and to respond to airline facility requests. No expansion will be done unless it is justified by future demand.

In short, if traffic doesn't grow, neither will the airport. On the other hand, if demand does increase as projected, we can respond to that (Doughty 1990).

And they did; the airport was downsized from 120 gates originally to 88 gates, 20 fewer than at Stapleton Airport (Fumento 1993).

Aside from Peña, Colorado Governor Roy Romer is most responsible for getting DIA off the ground. He too suffered from sticker shock as DIA opened. A month before it opened, Romer said, "Had somebody told me that the per-passenger cost was going to be $18, I would not have supported the airport. . . . They went Cadillac."

Although neither Peña nor Romer heard them, airport critics were insisting that costs would be significantly higher than the city's estimates. Frank Lorenzo, for all his innumerable shortcomings, hit the nail on the head way back in 1988. But, as noted in Chapter 3, the political process was not conducive to healthy discussion of the issues. DIA's opponents were treated as Luddites. In May 1988, Governor Romer had pledged to "roll over and crush" opponents of DIA (Fumento 1993). He said, "They're like gnats. You need to brush them away."

By the time the airport opened, a majority of Denver residents (63 percent four days before DIA opened, and 51 percent two weeks after it opened) would have voted against DIA if the city gave them a chance to reconsider their 1989 referendum endorsement (Brown 1995). In 1989, 63 percent of Denverites casting ballots voted in favor of DIA; but of course, that was for a 120-gate, $1.7 billion airport located 17 miles away from downtown (Reid 1989). This change of heart highlights the shortcomings of the decision process that were discussed in Chapter 3.

One of the problems throughout the process was the unwillingness of the politicians to "come clean" with DIA's costs, probably because of fears that the project might be derailed. On many occasions, a low-ball figure was quoted, comprising only construction costs, which rose from $1.7 billion to around $3 billion with the airlines' scope changes and additions, plus the cost overruns. But these figures were net of some very real costs—land acquisition costs and capitalized interest, for example. Capitalized interest totaled $920 million, with $250 million of that attributable to delays (*Denver Post* 1995b). And even these construction cost figures did not include tenant construction costs. Hence, when the real numbers were given to the people, they suffered serious sticker shock.

Project execution: I

Land deals

In Chapter 3, we reviewed the role of landowners, developers, and businessmen in the decision to build the airport. Here, the focus is on two issues: first, whether land was purchased for more than fair market value, and, second, whether the airport site was chosen so as to benefit particular landowners.

When one thinks of the land deals surrounding DIA, one is reminded of the movie *Chinatown*, starring Jack Nicholson—they are at least as convoluted. The land transactions surrounding the site of DIA involved Neil Bush's Silverado Savings and Loan and a subsidiary of Charles Keating's Lincoln Savings and Loan, both caught up in the national savings-and-loan deregulatory frenzy that ultimately cost taxpayers half a trillion dollars (*Denver Post* 1994c; Fumento 1993; Kowalski 1994; Wilmsen 1991). Closely connected was MDC, which was, as noted in Chapter 3, the subject of an extensive investigation by the Colorado attorney general, Gale Norton. She concluded that "Probable cause exists to believe that MDC committed numerous criminal violations" in raising funds for the mayoral candidates and that "two of MDC's subcontractors were extorted" to raise contributions, but the company could not be prosecuted because Colorado's statute of limitations had expired (Colorado 1991; Fumento 1993; *Westword* 1993).

The upward escalation in land values, inflated by a desire to support secured debt, was in some instances a by-product of the nationwide savings-and-loan "land flip" scandal of the mid-1980s. S&L speculation on DIA real estate cost the taxpayers $24 million, with speculators selling land back and forth to each other so that, in some instances, the land price was as high as $107,000 an acre (*High Country News* 1995). Denver began buying land on top of inflated values, some escalated by speculation on the future location of the E-470 beltway even before DIA was planned. Dry, parched farmland that should have sold for $500 to $600 an acre was sold to Denver for an average of $6700 per acre, or a total of $227 million (Imse and Hubbard 1995). Some transactions drove land prices even above $40,000 an acre (Hubbard and Flynn 1985). More than $15 million in land was purchased by speculators in the first 10 months of 1984. Purchasers included Silverado, and MDC Holdings (Imse and

Hubbard 1995). A 39-acre tract, the Little Buckaroo Ranch, changed hands three times in two years, with a threefold price increase, from $1.5 million ($37,000 an acre) to $4.3 million ($113,000 an acre). MDC and Silverado were involved (*U.S. News* 1995a). George Doughty, the former director of aviation, conceded that the city was forced to pay inflated prices for land to expedite construction. Even before DIA was under way, the Peña administration paid $19.4 million for 644 acres, according to state Attorney General Duane Woodward, "ten times what the property was worth" (*Westword* 1993).

One reason the city paid so much for land and water rights was that three years elapsed between the time it decided to build a new airport at its present location and Denver's offer to buy the land it sought. The city was paralyzed until Adams County ceded the site. It then moved promptly to begin the acquisition process but was chastised by the FAA because the environmental impact assessment had not been finally approved. As we shall see in Chapter 10, this is a typical shortcoming of the FAA's bureaucratic approach. This additional delay gave landowners even more time to draw up plans for land development and to secure water rights, driving up condemnation values (Imse and Hubbard 1995). The law that obliged the city to pay for the land based on a valuation tied to its "highest and best use" rather than its fair market value, as determined by a board comprising real estate professionals further limited the city's ability to acquire the land at lower prices (Flynn 1996a).

The reason why this particular site was eventually chosen for the new airport is also murky. Once a decision had been made to give up on expansion onto the Rocky Mountain Arsenal, the first and second sites for the new airport in 1985 and 1986 were immediately east of the Arsenal (see Chapter 6; Carnahan and Hubbard 1995a). As runway configuration was changed to a pinwheel design, the airport site was moved north and east to lessen adverse noise impacts. But some observers suspect that the airport site was moved again, in January 1988, three miles north and four miles east to consume more land belonging to the Fulenwider family, which, in partnership with the Van Schaack family, owned thousands of acres at the airport site (Imse and Hubbard 1995), which had been purchased in the 1930s. An additional five parcels of Fulenwider land were on airport land after the shift. This shift also left the Fulenwiders prime property to develop on the traffic corridors leading into DIA. Cal Fulenwider sold 14,300 acres of land (42 percent of DIA's 34,000 acres) to the city for $52 million and retained prime tracts on Peña Boulevard for hotel development. Errol Stevens, an airport official involved in land acquisition, argued that the shift in venue was no coincidence (Hubbard 1995a), although no hard evidence of impropriety exists.

Skip Spensley, Mayor Peña's chief airport consultant and head of Denver's New Airport Development Office, took part in decisions that moved the venue for DIA. Peña became uncomfortable with Spensley's close relationship with Cal Fulenwider. Both Spensley and Fulenwider deny that Spensley gave or Fulenwider received insider information about the airport. But Peña was uncomfortable with the appearance of impropriety, dropping Spensley from the airport team (Imse and Hubbard 1995). Spensley left the airport project in 1989 and became a consultant to Fulenwider (*U.S. News* 1995a).

The routing of 12-mile Peña Boulevard is yet another mystery. It runs north several miles from Interstate 70, then turns sharply east for several additional miles. If the shortest distance between two points is a straight line, why did the city add unnecessary mileage from downtown by routing Peña Boulevard as an inverted upside down "L" instead of a diagonal straight line?

Several reasons have been advanced to explain the decision. First, the access road arguably needed to come directly from the west to be compatible with a future east-west runway south of the terminal. True, a diagonal straight-line Peña Boulevard running from I-70 to DIA's main terminal might require a tunnel under a future east-west runway. But the aggregate community costs of gasoline and automobile emissions would have made this a sensible long-term investment. DIA's passengers drive an estimated 800,000 additional miles a year vis-à-vis Stapleton (*High Country News* 1995c). Second, a straight-line access road would have complicated the land-acquisition process by requiring slices of many individual parcels, or vertical sections thereof. But that is nothing new to interstate highway construction. Some 30,000 vehicles per day drive on Peña Boulevard. If a modest two miles had been shaved off the driving distance by straightening out the highway, automobile emissions would have been reduced by more than a million kilograms of carbon monoxide, and nearly two million gallons of gasoline would have been saved annually. Had a rail line been built, the savings would likely be even larger.

Cynics have offered their own explanations for the circuitous routing of Peña Boulevard. For one thing, Fulenwider owned land in the corner of the L, land that would be extremely valuable if bordered on the north and the west by the airport boulevard. But then again, Fulenwider could have developed both sides of Peña Boulevard had it not been located immediately adjacent to the Rocky Mountain Arsenal. Another explanation lies in the city's earlier $47 million purchase of the Eastwood Estates mobile home park for expansion onto Rocky Mountain Arsenal, which uprooted 420 families (Imse and Hubbard 1995). Denver paid an astounding $30,000 an acre for an undeveloped 132-acre section of mobile home lots (Imse and Hubbard 1995). When Arsenal expansion was abandoned, the land became unnecessary. Perhaps by routing Peña Boulevard there, the city sought to eliminate criticism for the bloated purchase.

In any event, all these decisions—moving DIA north and east of the original site and giving it a circuitous highway access—put the airport miles farther from downtown. That, of course, will continue to add to the surface transportation costs, energy consumption, and environmental pollution on a daily basis.

In July 1988, the U.S. Department of Transportation received allegations that "elected officials of Adams County together with state officials, Denver city officials and various developers may have used inside predisclosed information to purchase land that was being planned for acquisition in the construction of a new airport." Specific allegations were that Denver purchased a square mile of real estate from the Alpert Corporation in 1984 to quiet its opposition to a proposed runway (land that was unnecessary for the project) and that L.C. Fulenwider Co. received inside information from a Denver airport staff member (Hubbard and Flynn 1995).

A DOT memorandum of May 1989 concluded, "Despite compelling evidence and allegations indicating criminal violations and conflict of interest by affected landowners, businessmen and public officials, no prosecutors would accept the case." And in March 1990, the DOT concluded, "The Colorado state attorney general's office [then headed by Duane Woodard] refused to consider the case for 'political reasons' and 'anticipated opposition' from the governor's office." In response, Governor Romer subsequently insisted, "If I had any knowledge that there was some impropriety about land transfers at the airport, I would have been outraged" (Hubbard and Flynn 1985). But one must remember that Romer had received generous contributions for his 1986 campaign from MDC, Silverado, and their associates

(Wilmsen 1991)—although neither of these parties was mentioned in the DOT memorandum—and that in 1988 Romer had threatened to "roll over and crush" opponents of DIA (Fumento 1993).

The DOT let the issue drop, sending supporting documentation to Dallas for storage, where the files were subsequently destroyed (Hubbard and Flynn 1995). Apparently, under Federico Peña's tenure as DOT Secretary, no efforts have been made to reconstruct those records. The state attorney general's office under Gale Norton subsequently did perform an exhaustive review of the land deals surrounding DIA and found inadequate evidence of illegality worthy of pursuit.

Aviation consultant Michael Boyd blamed the land speculators as the driving force behind the new airport:

> *There are only two reasons Denver International is being built, and those are graft and political expediency. . . . Denver shifted to the idea of a brand-new airport because of enormous pressure from land speculators where DIA is now being built. It was the speculators who pulled together the other interests—the contractors and politicians and the like. There was and is an enormous amount to be made (Westword 1993).*

There is no doubt that vast profits were earned through land deals and that additional millions will be made in the future. Still, given the available information, it is no easy matter to identify precisely the role that different actors played in key decisions. In Chapter 3 we presented the concept of a growth coalition consisting of the political and economic elites, whose members shared a common objective. Certainly, land speculators were active members of the coalition.

The fast-track approach

Once the decision to build a new airport was made, Denver opted for a fast-track design and construction approach with design, construction, installation, activation, and tenant activities being performed simultaneously (West & Assoc. 1992). Specifically, Mayor Peña, in the second quarter of 1991, signed hundreds of millions of dollars in contracts, many of which lacked complete engineering and design work and were an invitation to future cost increases, which were plentiful (*Wall Street Journal* 1995a). At its peak, the airport was consuming more than $25 million in capital a week and employing more than 9000 people (West & Assoc. 1992).

Federico Peña had the opportunity to delay DIA's construction but chose not to do so even though many persons (including one of the 1991 mayoral candidates, Don Bain) had proposed mothballing DIA for two years until passenger traffic at Denver, which fell sharply after 1986, recovered. But Peña would have none of that, and in the eleventh hour of his administration, he forcibly shoved DIA beyond the point of no return. Six days before leaving office, Mayor Peña signed a contract with United Airlines that included $600 million in airline-inspired design changes at DIA and guaranteed the airport would not open unless the infamous automated baggage system met the rigorous specifications contained in the UAL agreement (Flynn 1995a; Evans 1996a).

Several considerations drove the decision to fast-track the project. First, politically, it would have been difficult to derail DIA once significant resources had been allocated to it and it had passed the "point of no return." Political support for the airport was strong, but opposition was tenacious. A mayoral election or an electorate shifting toward green over growth values might well have mothballed the site (as several community leaders proposed) so that DIA might not be completed. Fast-tracking narrowed the window through which such airport opposition might coalesce.

Second, financing would be made via revenue bonds whose interest began to accrue upon issuance. The longer it took to build DIA, the higher the financing cost would be. As noted in Chapter 3, detailed analyses were made of the costs and risks, and these played an important role in the decision.

Third, a compressed timetable might result in a coherent design with fewer changes since there would be less turnover of managers (each one eager to put his or her fingerprints on it) and the preservation of more institutional history (Veazey 1995). In fact, however, at DIA the airlines came in late with their input on concourse design and the identity of key players in the process (the city engineer, the airport director, and the mayor) changed anyway once construction had started.

Of course, hindsight is perfect, but Peña's decision to adopt an expedited construction timetable coupled with Webb's belated scope changes came back to haunt DIA in the form of higher costs. Fast-tracking also led to hastily approved and paid change orders and belated scope changes and additions. Had it been realized that the automated baggage system could not meet the specifications in the UAL agreement by October 1993, the rest of the airport might have been constructed at a less hectic and expensive pace. Airport construction chief Ginger Evans estimated the fast-track timetable resulted in acceleration costs of between $30 and $40 million, as well as an erosion of quality (Russell 1994). However, each additional year of delay would cause approximately $200 million in capitalized interest to be incurred.

The management system

DIA's design and construction process was performed under a program manager system, discussed in Chapter 7. The city teamed with Greiner/Morrison Knudsen, a joint venture, to coordinate and ensure the quality of 61 design contracts, 134 construction contractors, and more than 2000 subcontractors (GAO 1995). This system encountered difficulties, and Ginger Evans noted that she would never choose a joint venture to manage a project because "what you've got is two kinds of animals trying to be a third kind of animal" (*ENR* 1994).

The situation was aggravated by attempts by the mayor and city council to cut up contracts into small pieces so a maximum number of minority- and women-owned firms could get a piece of the action. The result was that the number of possible disputes over who did what to whom multiplied like fleas, costing uncounted millions to resolve them. Architects have said that the key to success in a large program is to minimize the number of conflicts over issues of responsibility and cost. The airport project featured the deliberate creation of a large number of such conflicts (Knight 1994a). Tension reduction occupied an important part of Evans' time. "I spent half my time babysitting," she told a reporter (*ENR* 1994).

Still, it appears that under Mayor Peña the system worked relatively well. As Bob Crider, the city's auditor, noted in his comprehensive report on the changes at DIA: "The project was ahead of schedule and under the $2.7 billion budget when Webb took office in July 1991. A private consultant hired by the administration reported in May 1992 that DIA was 'on time and at budget'" (Crider 1994). The

GAO also noted that the "project was close to being on schedule and was under budget." The GAO went on to warn, however, that "construction was in the early stages, and problems could surface as construction progressed" (GAO 1992). That proved to be a prophetic caution; Mayor Webb's election brought drastic changes in both the management style and the design of the airport. Some of the ensuing difficulties were the legacy of Mayor Peña's decisions, many of which placed Mayor Webb in a no-win situation, but many have argued that Webb should have anticipated and been prepared to deal effectively with at least some of them.

Surprisingly, there was no central control mechanism to monitor and evaluate the requests for changes. The city council was supposed to evaluate all changes that would increase costs by 10 percent or more, but this was seldom done. Decisions were made on site so that as late as April 1994, the exact number of outstanding change orders remained unknown. As auditor Crider noted in his report, "[T]he large percentage increase in Change Orders over the past 15 months validates the views of many contractors . . . that DIA is a project out of control and without leadership" (Crider 1994). He went on to identify the principal reasons for the project's cost escalation as follows:

> *The Webb administration allowed without question thousands of changes to the project and their subsequent Change Orders. More importantly, they avoided checks and balances, required no accountability, and never established a policy on Change Orders that would have helped keep the project at a reasonable cost. This neglect shows a disturbing void in project leadership and is what perpetuated the upward spiral in cost to the project.*

Former Denver aviation director George Doughty expressed a similar view:

> *I'm very concerned about some management errors under Mayor Webb, who doesn't have the management ability or guts to manage a project like this. . . . He let the airlines and contractors roll over him. If he hadn't done that, it would have been completed on time (Sahagun 1994).*

Doughty contrasted the management system under mayors Peña and Webb:

> *[Peña] left the technical, financial, and business management of the project largely to the professional staff. Even though he was briefed frequently on the status of the project, those of us charged with carrying it out made the management decisions. That meant that each element of the program was accomplished in a timely manner, problems were identified early and solutions found before they affected the overall program. The very efficient design of the DIA airfield is a direct result of this management approach.*
>
> *Unfortunately, when Federico Peña chose not to run for a third term and Wellington Webb was subsequently elected, that approach ended. Mayor Webb was unable to make tough decisions in a timely manner or delegate those decisions to knowledgeable subordinates. As a result, the Program Management Team (PMT) could no longer stay ahead of the problems. Mundane issues became crises (Doughty 1995a).*

Doughty believed that the single most important reason that DIA opened both late and grossly overbudget was the failure of Federico Peña to run for a third term as mayor (Doughty 1995b). But is that judgment justified? *Denver Post* editorialist Al Knight certainly does not think so. He commented:

> *Inevitably . . . the press offers up retrospective articles on what went wrong with the multi-billion dollar [DIA] project.*
>
> *Tucked away in the folds of many of these stories is a recurring fiction: namely, that up until 1991 when then-Mayor Federico Peña left office, things were going rather splendidly. The project, it was said, was on time and under budget.*
>
> *To the degree that things have gone sour since, the story goes, the current problems relate to project management failures that couldn't have been foreseen when Peña was in office. Peña, who is now U.S. transportation secretary, has recently said that had he remained in office he would have down-scaled the airport further to keep costs in line with estimates. . . .*

> *This picture of Peña as a visionary whose work was later soiled by poor local decision-making is intellectual trash which is nowhere supported in the historical record. . . .*
>
> *A more apt image is of a former mayor unwittingly planting a series of time bombs before his departure which then, one by one, exploded, driving up the project's cost, creating future inefficiencies and virtually ensuring that a single airline would dominate Denver aviation to the clear disadvantage of the traveling public (Knight 1995).*

Which analysis is more correct? Certainly Federico Peña is the man who conceived, marketed, sold, and began construction of DIA. As such he deserves credit—and blame—for its location, design, and the implementation that occurred during his tenure. In earlier chapters we discussed the financial and political aspects. We have seen that Mayor Peña placed the project on a fast track and, prior to the election, signed major agreements with Continental and United Airlines, granting them numerous concessions. Furthermore, the Peña administration broke DIA into about 150 contracts, whereas three or four might have been much more efficient and cost-effective. This fracturing greatly complicated on-site logistics and coordination, which in turn created more complications, leading to millions of dollars in additional costs. The Peña administration also passed a minority-contracting ordinance that added additional expense to each contract to demonstrate compliance and caused some work to be steered to less-experienced firms (Knight 1995).

Regardless of where the blame lies, the project's management was in serious trouble, and in 1992, an airport consultant, S.G. West & Associates, recommended that authority be delegated to airport field staff to approve change orders and award small contracts for last-minute construction changes. Specifically, the consultant said the city should "institute a more aggressive, focused Design Management Program to identify and intensively manage all designs and changes that are currently critical to the project" (West & Assoc. 1992). To attempt to keep the project on its October 1993 schedule, the city endorsed the recommendation (P. Green 1992).

As we shall see, a fundamental problem was that the mayor's office and city council were micromanaging critical aspects of the contracting process. Even so, some of the construction work turned out not to have met specifications. Federal investigators found the concrete to be of poor quality, the steel tie bars to have been improperly installed, and the steel dowel bars to have been improperly aligned and attached. Forty-five percent of the panels did not conform to the contracts (GAO 1995).

The climate of uninhibited greed that DIA seemed to unleash also corrupted federal employees. Under FAA rules, employees are entitled to moving-expense reimbursement and closing costs if they are relocated to a new airport and if it is more than 10 miles further away from their homes (DIA is 17 miles from Stapleton), irrespective of whether commuting time has increased (Brinkley 1995a). Nearly 40 federal employees at Stapleton Airport collectively pocketed half a million dollars in relocation reimbursements as a result of being transferred to DIA—an average of about $14,000 each, with an additional $2 million to come for closing costs and moving expenses (Brinkley 1995b). Four were reimbursed $85,000 for moving farther from DIA than their original venues. One worker was reimbursed $64,000 even though he still lived in the house he sold (Brinkley and Imse 1995). An embarrassed FAA sent its inspector general to Denver to put a halt to all the pigs slopping at the trough (Brinkley 1995c). Colorado's U.S. Senator Hank Brown added a rider to an appropriations bill requiring the FAA to examine the "misuse of public funds" (*Rocky Mountain News* 1995e).

All the unfavorable publicity led Congress to withhold federal money for the construction of a sixth runway at DIA. The House transportation subcommittee concluded that it "remains unconvinced at this time that the runway is a high priority, and that such a project could be managed efficiently given the past management history of the overall project" (*Rocky Mountain News* 1995c).

Project changes

The airlines

If the airline industry is characterized by anything, it is uncertainty and a relentless focus on the short term. The behavior of the airlines regarding the new airport certainly conformed to this generalization.

By 1987, the major airlines that hubbed at Denver (United and Continental) had already announced their opposition to the construction of a new airport. United had spent considerable money completely rebuilding its Concourses A and B at Stapleton International Airport, and the city had added a new Concourse E (*Denver Post* 1994c). The airlines argued that Stapleton capacity was adequate and that the airlines simply could not afford to pay for a new airport (indeed, airline profitability, fair to poor under regulation, collapsed to levels of miserable inadequacy under deregulation). The implicit reason for resistance to DIA was that the dominant hub carriers had no enthusiasm for an airport that would have ample capacity for a third hub airline, as the city had promoted it. From a competitive standpoint, a duopoly was better than a triopoly, as United and Continental had learned when Frontier/People Express was competing in the market (*Denver Post* 1994c). By July 1987, both Continental and United began withholding surcharge payments, leading Mayor Federico Peña to halt Stapleton construction projects and cancel a planned new runway (*Denver Post* 1994c). From 1987 to 1989, United began moving aircraft out of Denver to establish a hub at Washington Dulles International Airport, while Continental did the same to build a hub at Cleveland.

But, as we have noted, their efforts to dissuade Denver's politicians, or the Wall Street bond houses, from going forward were to no avail. The city decided to proceed without the airlines on board, and in 1989 began to solicit bids for construction. As the major hub airlines slowly came aboard, they demanded—and received—significant and expensive design changes, many in the sophisticated mechanical, electrical, and telecommunications systems. Mayor Webb readily acceded to their demands, and Continental's changes added $200 million to DIA's costs, while United's added $600 million (Carnahan and Hubbard 1995b).

Changes were also caused by the decision to move the location of rental car facilities, the shift of the cargo terminal, and, of course, the infamous baggage system.

Continental Airlines Frank Lorenzo's Continental Airlines was the first hub carrier to sign up, taking the coveted Concourse A, the closest concourse to the main terminal building. The agreement, in April 1990, was for 30 gates, consisting of 25 narrow-body gates, 5 wide-body gates, and 15 commuter aircraft positions, for a period of 20 years. At the time, Continental operated 160 flights a day out of Denver's Stapleton International Airport (Kowalski 1994). In negotiating those leases, Continental insisted the city move the international gates from the main terminal to its Concourse A.

Perceiving both a marketing advantage and potential traffic jams on the underground railway linking the concourses to the main terminal, Lorenzo also insisted that the city build a 600-foot pedestrian bridge over the taxiways linking the main terminal to Concourse A. The two-level bridge (one for domestic traffic and one for international) spanning dual taxiways is the only such bridge of its kind in the world. The city also agreed to hold the cost of the new airport to $1.9 billion (ASRB 1990). Continental's changes pushed all three concourses further to the south. The air traffic control tower, which was to have been located on the north side of Concourse B, is now on the south side of Concourse C as a result of the move. The city also leased Continental 75 acres of land on which it built Continental a 152,000-square-foot maintenance hangar, a 63,000-square-foot air freight facility, and a 65,000-square-foot flight kitchen, all financed with $73 million in DIA revenue bonds (ASRB 1994).

In December 1990, Continental Airlines fell into Chapter 11 bankruptcy. Because it was Continental's second trip to bankruptcy court (the first occurring in 1983), some jokingly refer to it as Chapter 22 bankruptcy. Since Continental would be free, with the bankruptcy judge's approval, to walk away from its Denver leases, the city attempted to reach an agreement on the number of gates Continental would occupy, and eventually the airline promised to identify the number of gates it needed by April 1, 1992, and if no number had been designated by December 31, 1992, its agreement for gates at DIA would terminate (ASRB 1992). In the meantime, the city scaled back DIA by $150 million, from 94 gates to 85, and from six runways to five (*Rocky Mountain News* 1991).

By 1992, Continental had more than 250 flights a day in Denver. In August 1992, Continental signed leases for 20 gates on Concourse A, 15 for five years, and 5 gates for 10 years (ASRB 1994). DIA critic Michael Boyd criticized the agreement, saying "The lease deal with Continental is little more than a sham by the political leadership to mislead the public into believing that the new airport has two firm anchor tenants" (Chandler 1993a). [United had signed on in 1991.] The delay caused by Contintental's bankruptcy forced the city to incur significant acceleration costs to complete Concourse A on time.

Beginning in 1993, as DIA's costs accelerated and the market pressure of dominant United Airlines took its toll, Continental began peeling most of its aircraft out of Denver and eliminated the Denver hub from its system. United was engaging in a plethora of predatory practices to deny Continental profitability at Denver to provide incentives for the competing airline's withdrawal. Continental dropped service from 148 departures in late 1993, to 107 in March 1994, to 86 in the summer of 1994, and then to 23 in the fall of 1994, as it transferred planes east of the Mississippi to build up CALite, a short-lived Southwest Airlines clone flying short-haul linear routes (ASRB 1994; Mahoney 1994).

As it exited Denver, Continental took the position that it was free to walk away from its $59 million annual lease obligations by paying the city $5 million in liquidated damages (ASRB 1994; Sahagun 1995). Continental would renegotiate again, this time down from 20 gates to a mere 10, of which two were subleased to America West and four to the new Frontier Airlines, Continental's two code-sharing partners at Denver. After two years, Continental may opt to lease only three gates with no penalty (*Airports* 1995b).

The reduction in Continental's operations left the city with a surplus of 215,000 square feet on Concourse A. Under the then-existing airline agreement, the net concession revenues would be used to cover any shortfall in vacant space. Under the Continental agreement, all other airlines had to absorb the cost of excess airport capacity (*Aviation Daily* 1995).

Recognizing Continental's poor financial condition, the time, cost, and uncertainty of litigation on the lease terms, and anxious to get on with building a sixth runway (estimated cost: $100 million), the city came to terms with the reality that Continental would occupy only three or four gates of a 20-gate concourse that had been built specifically for it.

United Airlines UAL chairman Stephen Wolf has revealed he fought DIA as vigorously as he fought anything in his life. United did not believe the airport was needed, nor did they think they could afford it. But when United jumped aboard the train about to leave the station, everything significant about DIA changed—its size, its scope, and its cost. Testifying before a Congressional Committee investigation, "What Went Wrong?" at DIA, Denver aviation director Jim DeLong pointed the finger to the massive scope changes insisted upon by United as the primary cause of cost escalation (DeLong 1995a).

United Airlines signed up for 30-year leases for 31 narrow-body gates, 11 wide-body gates, and 24 commuter aircraft positions in June 1991 and agreed to bring some 1500 additional reservations jobs to Denver (ASRB 1992), a promise it never fulfilled. United was subsequently relieved of the jobs commitment because the city secured United's assistance in fronting construction-delay money. United wanted an automated destination-coded vehicle (DCV) baggage system for its concourse (which the city rashly decided to install throughout all three concourses) and a guarantee that DIA would not open until the baggage system conformed to its technical specifications (a condition that would allow United to unilaterally veto the opening of the new airport). This required the construction of a new mezzanine level inside the landside main terminal building, a redesign of the electrical system, the addition of structural steel to support curbside baggage check-in, and structural penetration to allow the movement of baggage between levels. Electrical, plumbing, and other fully or partially complete systems had to be torn out and replaced. We discuss the automated baggage system in greater detail later.

United also negotiated for the city to build it a 500,000-square-foot maintenance facility capable of housing 16 narrow-body aircraft or a mixture of two 747s, two 757s, and three 737s (ASRB 1994). Unfortunately, the United hangars were constructed directly north of Concourse C, blocking potential future construction of Concourse E. The city also financed various traditionally tenant-financed facilities on Concourse B. Concourse B is 8 feet wider than its sister concourses due to the addition of two parallel moveable sidewalks in each direction. Its roof was redesigned to provide a clerestory for natural light (Russell 1994). The basement of Concourse B was also widened to allow room for the automated baggage system (McCagg 1993). Among United's almost-unnoticed innovations was its insistence that a pneumatic-tube aircraft-parts distribution system be installed between its maintenance hangar and each gate, allowing aircraft replacement parts to move expeditiously to where they are needed, further enhancing the efficiency of aircraft turn-around times. As United Airlines placed the launch order for Boeing 777s, spacing of the concourses was widened (Brewer 1995).

United's contract also provided that most of the glass used on the walkway to Continental's Concourse A be opaque to obstruct the magnificent view of Continental golden-tailed aircraft against the Rocky Mountain snow-peaked panorama. Ginger Evans, the construction chief at DIA, chose to "forget" this part of the agreement and did not specify opaque glass. In her words: "If they (at United) want to put a mother of three in jail . . ." (*ENR* 1994).[3]

In early 1991, Peña responded to Continental's second bankruptcy by scaling back the airport from $2.1 billion to $1.93 billion. The scope changes demanded by United in December 1991 (more than two years after construction began) increased DIA's projected cost to $2.7 billion, the most expensive ingredient of which was the $193 million baggage system (Flynn 1995b). United's facilities were financed by special facility bonds, which carried a lower interest rate than that available to the junk bond-rated major airlines (ASRB 1994). Further, the city forgave United a $44 million debt (Chandler 1993a).

The June 1991 agreement also provided that the city would hold the cost of the new airport to $2.7 billion, and that the cost to United would never exceed $20 per passenger in 1990 dollars. If the city failed to do so, United would have the right to terminate its lease (ASRB 1992, 1994).

Again, these major changes produced significant additional costs. The terminal itself ended up costing $319 million, almost 90 percent more than originally planned (Calhoun 1993). According to architect Jim Bradburn, "When the airlines said we needed a DCV baggage system, we had to rearrange all the guts of the building to accommodate it" (Russell 1994). At most airports, baggage systems are tenant-financed and -installed. At DIA, the financial distress of the airlines (whose debt had been downgraded by Wall Street to junk) and the eagerness of the landlord to sign up anchor tenants made them city-financed systems (DIA bonds could be floated at lower interest rates than airline debt, although by the time the project was finished, DIA bonds were also listed as junk). The baggage system initially added $193 million to the project's cost. Dick Haury of Greiner estimates that related changes to the design and construction of the buildings added another $100 million to the project,

3. According to Meyers (1995), United wanted the windows painted black.

much of it in the coordination of mechanical, electrical, and telecommunications systems—ripping out and changing things that had already been completed (Russell 1994).

In a better world, the city would not have proceeded so deeply into the design and construction process without the airlines' commitment and input; or alternatively, the city would have insisted on a precise point of no return, beyond which no major scope changes would have been implemented. As it is, the midstream changes gave the airlines the airport design they wanted at a price they didn't. Further, rather than having dedicated gates and concourses, Denver could have built a much smaller, less expensive, and more efficient airport by adopting the European model of multiple-carrier gates, ticketing, and baggage facilities, allowing each carrier to consume gates, ticket counters, and baggage systems only during those periods when it has aircraft arriving or departing. If a new airport cannot be privately financed without such exclusive-use carrier leases, we should reexamine the funding alternatives, relying more heavily on governmental sources of capital.

Each change caused a major ripple effect and change order followed change order, frequently without authorization. The annual increases in change orders were as follows:

1990	.45 percent
1991	1.53 percent
1992	19.85 percent
1993	63.15 percent
1994	15.02 percent

Table 9-3 shows the specific areas involved.

The impact on costs is indicated by Fig. 9-4, which shows how the cost of the change orders climbed to $525 million by March 1994.

The major areas where costs had escalated (as of March 1994) are shown in Table 9-4.

Table 9-3. DIA change orders areas of change

1990 (.45%)	1991 (1.53%)	1992 (19.85%)	1993 (63.15%)	1994 (15.02%)
Grading/drainage	Grading/drainage	Fencing	Concourse B construction	Concourse B construction
Runways	Runways	Terminal bridge	Concourse C construction	Communications
Sitework	Sitework	Roads	Central core	Taxiways
	AGTS tunnel	Garage/parking	Electrical	Parking garage
	AGTS platform	Grading/drainage	Parking garage	Terminal
	Trailers	AGTS system	Communications	Utilities
	Concourse C construction	Concourse C construction	Apron/taxiways	Runways
	Medical support	Terminal	Concourse A construction	Aircraft systems
	Paving	Central plant	Baggage	Noise monitoring
	Roads	AGTS platform	Fueling	Electrical
	AGTS system	Fueling system	Terminal bridge	Landscaping
		Administration building	Concourse A tenant finish	Signage
		Lighting vaults	Paging	AGTS system

1990 (.45%)	1991 (1.53%)	1992 (19.85%)	1993 (63.15%)	1994 (15.02%)
		Concourse B construction		Alarms
		Baggage		Security
				Baggage
				Fueling system
				Concourse A tenant finish
				Concourse B tenant finish
				Concourse C tenant finish

Source: Crider 1994.

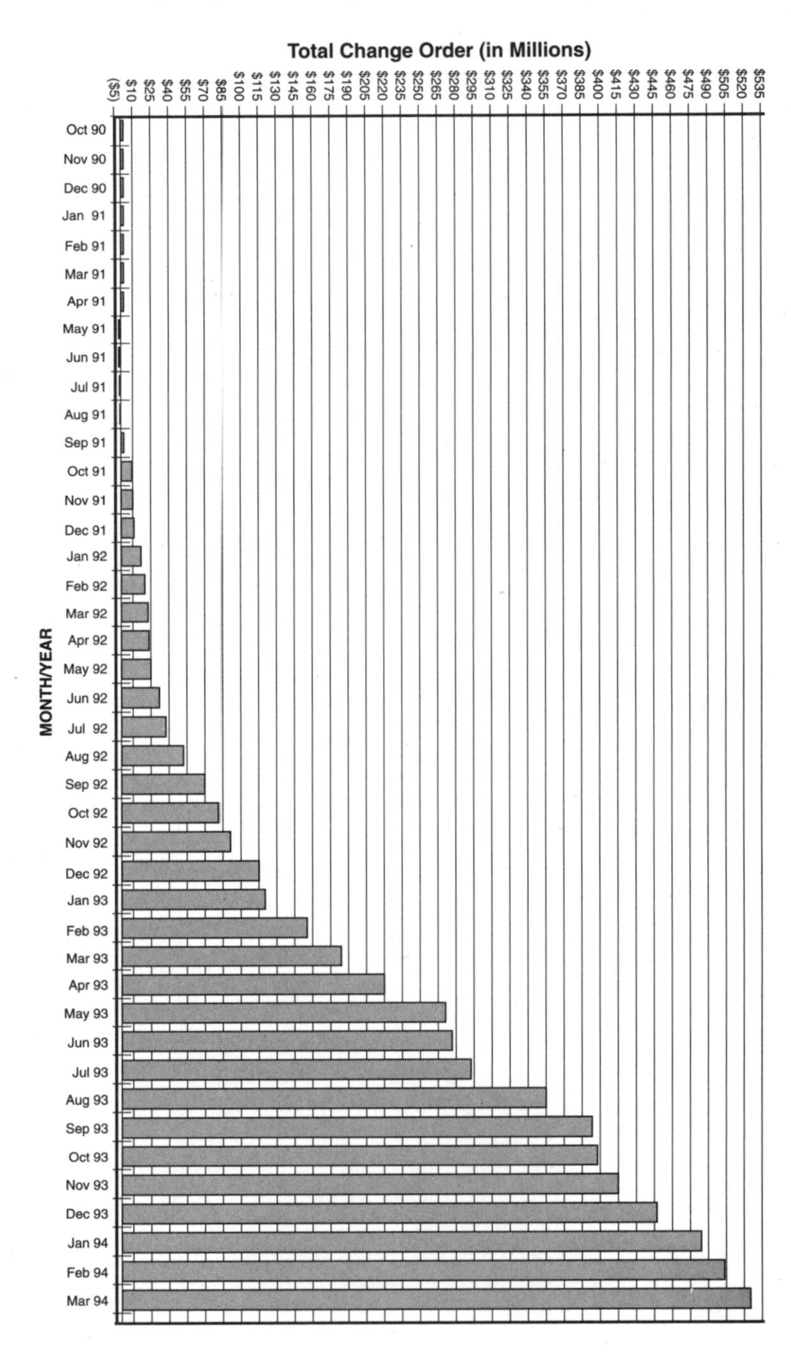

Fig. 9-4. *DIA change order incremental increases.* Source: Crider 1994

Table 9-4. Areas of cost escalation.

Project	Original contract	Total change order	Total contract amount
Concourses	$ 344,792,301	$142,331,992	$ 487,124,223
Terminal	$ 164,558,000	$154,442,000	$ 319,000,000
Runways	$ 353,253,167	$ 49,360,430	$ 402,613,597
Paving	$ 165,534,120	$ 33,672,579	$ 199,206,699
Others	$ 398,610,225	$ 64,450,304	$ 463,067,534
Total	$1,426,747,813	$444,257,305	$1,871,012,053

Source: Crider 1994

One source described the situation as follows:

> *The airport authority logged more than 19,000 change orders, ranging from new door trim to a $70 million addition for a cargo terminal. United . . . insisted on adding the $193 million computerized baggage system that so far has lost, eaten, or otherwise mauled everything that has been fed into it (Dillon 1995).*

These major scope changes or additions posed many problems for the construction process. However, the airport's designers were able to use state-of-the-art computer software to input the changes and monitor their effects on other systems so that they would retain their holistic relationships, allowing the construction to stay on a fast track (Brewer 1995).

Air cargo

The focus of DIA's initial development was on its use as a passenger facility. What appears to have eluded the city's planners is that air cargo enjoys by far the faster and more consistent growth vis-à-vis passengers. DIA's planners originally located the air cargo facilities on the north side of the airport, away from the only interstate highway in the area, I-70, presumably to placate Adams County's interest in sharing in the wealth of jobs DIA would produce (Chandler 1993b). But Adams County began promoting Front Range Airport's CenterPort project, an all-cargo airport.

In 1990, the Denver Air Cargo Association polled its members and found that 90 percent were dissatisfied with plans for the new airport and asked the city to redesign the cargo facility. Denver's aviation director, George Doughty, denied the request, citing Denver's intergovernmental agreement with Adams County, giving it certain economic benefits and spin-off development (Knight 1991).

Southeast of DIA, near I-70, and near 11,000 feet of Union Pacific Railroad track, is a small airport called Front Range (Lane 1992). Several entrepreneurs and real estate developers saw enormous opportunity for just-in-time aviation/motor carrier product assembly and distribution and began to develop it as a dedicated cargo hub. A project dubbed CenterPort would have been built on 1600 acres adjacent to Front Range. Its consultants predicted that as an air cargo hub, it would generate nearly 22,000 jobs directly and 35,000 jobs indirectly by 2014 (Lane 1991). CenterPort officials asked the FAA to pay 90 percent of the $70 million cost of expanding Front Range's two runways to allow landing by heavy cargo jets (Bronlkowski 1991).

Several cargo carriers, including Federal Express, Airborne, Emery, and UPS, signed letters of intent with Front Range, where costs would be lower and interstate highway access would be superior to DIA (Mahoney 1992). UPS insisted that high operating costs prevented it from creating a cargo hub at DIA (Leib 1994). Despite lobbying by Governor Roy Romer, in 1994 the FAA decided that UPS and Adams County could safely develop an air cargo hub at Front Range.

But Jefferson County (on the west side of the Denver metropolitan area) and United Airlines filed environmental challenges. Fearing that passenger traffic at Front Range could jeopardize the financial well-being of DIA, as well as the diversion of significant cargo revenue (DIA would be left with the "belly freight" from the combination carriers), Denver also attempted to block the project. Only after Adams County promised that no commercial jet service would serve Front Range did Denver acquiesce (Sleeth 1992). In the agreement between the two counties, Denver got the following:

- The right to continue competing with Adams County for air cargo and general-aviation business.
- Front Range Airport would not offer commercial airline service, and Denver International Airport would be the region's hub for passenger operations.

- Up to $500,000 a year in fuel tax revenue from Front Range Airport from 1993 through the year 2000.
- A seat on the Front Range Airport Authority Board.
- Two additional Adams County roads designated for construction traffic to DIA (*Denver Post* 1992).

The agreement gave Adams County the following:

- Denver would withdraw its environmental concerns over Front Range Airport, clearing the way for a major air cargo center in Adams County.
- Denver would not file an objection with the FAA to expanding Front Range's two runways to 12,000 feet, with an option to extend one to 16,000 feet after January 1, 1997.
- Denver would pay for maintenance and restoration of county roads used by construction crews building DIA (*Denver Post* 1992).

Of course, the simpler solution to Front Range Airport as a competitor to DIA and the airspace congestion problem would have been for Denver to have annexed Front Range when the city annexed the 53 square miles of Adams County. Then again, such an enormous land grab might have scuttled the deal politically. In 1993, after grading the north side of the airport for the cargo campus, the city decided to move the cargo facilities from the north to the south side of DIA, closer to I-70, as a preemptive strike against Front Range (ASRB 1994). The eventual collapse of the CenterPort project brought the air cargo carriers back to DIA. Federal Express and Airborne signed long-term leases at DIA, but UPS became the last cargo carrier to sign a lease at DIA, in April 1993, and then for only five years (*Denver Post* 1994c). The change added more than $50 million to DIA's cost (Flynn 1995b).

Rental cars

Rental cars were originally to have been located on the lowest level of the parking garages, adjacent to the main terminal. But rental car companies balked at the fees and maintained that servicing vehicles inside parking structures was inconvenient. They insisted on their own campus and were moved out to a 150-acre tract immediately north of Peña Boulevard along the way to the terminal (Russell 1994). Although this meant that passengers visiting Denver would have to board shuttle buses and vans from the baggage claim level to reach their rental cars, the decision was a blessing for local passengers, because the sorting area of the backup baggage system had consumed a good portion of a parking deck level, and parking spaces sometimes were hard to find. Moreover, the move more than doubled the space dedicated to rental car companies to 180 acres, compared with the 70 acres on the lowest level of the parking garage (ASRB 1994; DeLong 1995b).

This shift had the same kind of domino effect as other changes and clearly illustrates the cascade of change orders that such a decision initiates. It was not merely a question of erecting some new buildings. Many elements that had already been constructed, such as extensive signage and electrical wiring, now had to be changed. Redoing the signs alone cost $700,000 (*ENR* 1994).

The baggage system from hell

DIA opened on February 28, 1995. With the exception of the baggage system, the airport had been ready to open for nearly a year.

As noted, United Airlines had originally insisted on an automated baggage system for its Concourse B, but the city decided to install it in all three concourses. The scope and magnitude of the project were radically expanded by this strategic decision.

United wanted a system that could provide high-speed movement of inbound and outbound luggage over the relatively long distances between gates on Concourse B and the main terminal. The system was designed to meet United's objectives of turning around a wide-body jet in 30 minutes and a narrow-body aircraft in 20 minutes (ASRB 1994). If successful, it would make Denver capable of the most expeditious hub rotations in the world. BAE Automated Systems, Inc., was given the contract to build the baggage system.

The automated baggage system required several serious change orders to buildings already designed and under construction:

- Structural penetrations were required to allow for movement of baggage between levels.
- The system doubled the connected electrical load, requiring redesign of the electrical distribution system.
- The requirement for curbside baggage check-in required significant amounts of additional steel and extensive

coordination with the landside terminal contractor during construction of the terminal.

- A new level of the landside terminal, Level 4A was needed to accommodate the steel tracks of the system after the terminal had been constructed (ASRB 1994).

The city had been cautioned by several consultants that the automated baggage handling system was a high-risk proposition, particularly within the time frames specified. The consultants warned, "The single-bag DCV solution should not be furthered, we believe, because of throughput limitations, developmental risk, competitive availability, etc." When Wellington Webb took office in July 1, 1991, city aviation director George Doughty warned him, "This is the only major design issue remaining that could have any effect on schedule" (Flynn 1995a). Under Doughty's recommendation, the city had agreed to the DCV for all three concourses, not just United's (Chandler 1994). Concourse C is about a mile from the main terminal. Concourse B is 3400 feet from the main terminal and is 3300 feet long. The tracks for the system had to be more than 20 miles long. Nothing on such a scale had ever been attempted.

As Webb took office, the city began soliciting bids for the automated baggage system. There were two conditions: (1) the system had to be finished in 26 months; and (2) it had to move 1000 bags a minute. Neither BAE, nor its European competitor, AEG, wanted the job under those conditions. BAE warned that such a sophisticated system would take four years to build. Two bids were submitted. Sandvick Process Systems' bid was rejected because its bulky equipment would consume too much of the baggage tunnels. Harnsichfeger Engineers' bid was rejected because the firm could not complete the project until November 1994 The city turned to BAE, which agreed to complete the system in two years, provided it had unlimited access to the facility (Flynn 1995a). The city's failure to fulfill its obligation to guarantee BAE primary and expeditious access to the baggage system venues contributed significantly to delays in its completion.

As scheduled opening dates for DIA came and went, and the baggage system remained dysfunctional (with the nation's media capturing videotape of the BAE system eating, devouring, and choking on bags), the city was swallowing $1 million a day in delay costs and became desperate for a solution. It hired a German firm, Logplan, to evaluate the BAE automated baggage system to determine whether it would ever work.

Logplan concluded the BAE system would work, eventually, but recommended, for reasons that remain vague, that the city construct a "backup" baggage system even though a "completely manual inbound baggage system was designed, purchased, installed, tested, and accepted by both the airlines and the city in the fall of 1993" (Evans 1996b). In July 1994, the city decided to install a stand-alone backup baggage system, using a conventional conveyor belt system augmented with tugs and carts carrying bags in the tunnels between the concourses (Eddy 1994b). This system would consume a sizable portion (500 parking spaces) of the third level of the parking garage for human baggage sorting (ASRB 1994). In litigation between United and BAE filed in March 1996, United alleged the BAE system missorted 500 bags a day (2.5 percent of those handled) and that the $218 million system was in danger of being relegated to "scrap metal" (Imse and Flynn 1996).

Desperate to get the airport open, the city cut a deal with BAE to work with United to get the automated baggage system up and running on Concourse B and raised the BAE contract from the original $193 million to $232 million. The city would sign away its rights to sue BAE for producing a dysfunctional baggage system by the scheduled opening date. Rapistan Demag was commissioned to build a backup baggage system, which would rely on conveyor belts and natural gas-fueled tugs and carts (the original plan before United Airlines asked for a high-tech bag system), for $63 million. Thus, the city was spending nearly $300 million to handle bags for an airport with fewer than 90 gates.

United Airlines, DIA's principal tenant, wanted the BAE system finished in its Concourse B and got the city's permission to use parts of the primary baggage system in Concourse C as well. Depriving Concourse C carriers of an automated baggage system might enable United to get its bags to its customers long before most of its rivals. In return, United agreed to pay a good portion of the $1 million-per-day delay costs.

Thus, by the summer of 1995, only United Airlines was using the automated baggage system that it had originally sought for its concourse, but which the city had installed in all three concourses. But even United's baggage system, although three times faster than conventional systems, was only partially functional, working only on outbound and transfer luggage (Imse 1995c). It was not fully functioning until the fall of 1995. It collapsed again during the peak holiday season in late 1995. The BAE automated baggage system in Concourse A is wholly inoperable. Nevertheless, the city, as specified by its agreement with UAL, has billed Concourse A airlines for 35 percent of the spine portion of that system. This not only places a burden on the airlines operating there, but practically ensures that no major tenant, such as American Airlines, will ever locate there (Evans 1996b).

Webb defended his decisions to go forward with BAE and forego litigation with a "miserable camel" analogy: "You don't shoot a camel who took you half way across the desert merely because you got angry as hell at the miserable camel for taking so long to get you there; at least, not if you want to get the rest of the way across" (DeLong 1995). Unless, of course, the camel is incapable of getting you to the other side. . . .

Ultimately, the baggage fiasco cost DIA $100 million in cost overruns and $361 million in additional financing costs. Who was to blame? One investigative reporter drew the following conclusions (Flynn 1995a):

- Peña, Webb, and other city officials are not the only culprits.
- Denver gambled on an untested technology.
- Denver, United, BAE, and various experts decided that the system could be operational in half the time BAE had originally required.
- Denver and the airlines demanded major changes in the system during construction.
- Mayor Webb took a year to accept the city's engineers' recommendation that a backup system be built.
- BAE experienced schedule slippage and failed to invest adequate resources to catch up.

After DIA opened, a group of foreign airport officials marveled at its impressive cutting-edge technology. DIA Director Jim DeLong cautioned, "I'd strongly suggest that you avoid new technology and rely on proven technology" (Imse 1995b). The entire saga of the baggage system fiasco is captured in the following chronology (GAO 1994).

Chronology

April 1990	The city's consulting firm, TRA Architecture, Engineering, Planning, Interiors, concluded that there was insufficient time to design, install, and test a fully automated airport-wide system by the October 1993 opening date.
August 1990	The city hired a second consulting firm, Neidle Patrone Associates (BNP), which, after analyzing five systems, recommended against one automated system and suggested that Concourse A use a high-speed conveyor belt; Concourse B, an automated system; and Concourse C, the conventional tug and cart.
	The city decided to install a tug and cart system for the entire airport.
October 1990	The city, in a letter to the Airlines Airport Baggage Subcommittee, stated that requests by the airlines to build their own systems would be considered if they did not jeopardize a future integrated system.
November 1990	The hubbing airlines began designing systems. Continental hired BNP to implement its recommendation for Concourse A. United Airlines hired BAE Automated Systems (BAE) to design an automated system. Concourse C retained the tug and cart system.
July 1991	The aviation director decided to pursue the original plan to build a fully automated system, rather than have the three separate systems, after being assured by experts working for UAL and AMR that the October 1993 deadline could be met.
August 1991	The city issued a formal request for proposals (RFP).
September 1991	BAE submitted a discussion paper stating that an automated system could be built by the deadline, although some additional work would be required.
October 1991	Having been asked by the city for additional information, BAE provided a schedule and a cost estimate of $185 million.

November 1991	The Baggage Subcommittee wrote the city expressing concern about the project, especially since the master project schedule called for the baggage system to be completed in mid-1994.
December 1991	The proposals of the two firms that had formally responded to the RFP were rejected by the city.
	BAE, which never submitted a formal proposal because of time pressures, signed a $20 million contract and began developing the system.
February 1992	Several airlines formally expressed doubts about the operational and financial implications of the BAE project and urged the city to evaluate their concerns before proceeding further.
May 1992	The city awarded BAE a final contract of $195.6 million for the automated system.
August 1992	The city made major alterations in the system, reducing the price by $23 million.
September 1992	The city issued another change order for $5.5 million for revisions in the design and for a subsystem for Concourse A.
January 1993	The city executed a $1.6 million change order.
	The city notified BAE that its installation schedule was inadequate and had created major difficulties for planning and coordination.
	BAE responded, blaming the city for not providing it with access to certain facilities on the specified dates and for its changes, which led to the redesign of subsystems.
March 1993	The city issued a $7.2 million change order.
	The opening of DIA is postponed to late December 1993, partly due to the baggage system.
May 1993	The city wrote BAE that the projected completion of the baggage system by December 1993 should be justified and that the company should meet its contractual obligation and complete the system by October 1993.
	A city engineer who evaluated BAE's progress found that work on a critical software program had not even begun.

June 1993	BAE wrote the city that installation of the system would be completed by October 1993 and that, with the exception of certain components, it would be tested by the new opening date of December 1993.
	The city executed a $1.1 million change order.
August 1993	A city engineer monitoring BAE's progress expressed doubts about the completion of the mechanical and electrical work by December 1993 and of the testing of the system before the end of January 1994.
September 1993	Another engineer's report doubted that BAE would meet its testing schedule.
	BAE assured the city that testing of the entire system would be completed by December 1993.
October 1993	DIA's opening was postponed to March 1994 because of the baggage and fire-safety systems.
November 1993	The city acceded to UAL's request and implemented a change order for $3.1 million for equipment and spares for the baggage system and another for $2 million for operations and maintenance.
February 1994	The first major test of the system was disastrous. BNP, now UAL's consultant, reported that major "debugging" was required and that the system would not be ready for the March 1994 opening date.
	The city postponed the opening to May 1994 because of the baggage system.
March 1994	The city implemented a $350,000 change order for operation and maintenance.
May 1994	DIA's opening was postponed indefinitely.
	The city retained Logplan to study the automated system and to identify alternatives that would permit the airport to open quickly. Logplan recommended that the automated system be brought online in stages and that the city build and use a conventional tug and cart system in the meantime.

July 1994	Tests of the baggage system continued to go badly. The city decided to build the backup system.
September 1994	The city executed a $1.4 million order for systems services.
	BAE, UAL, and the city agreed to modify the system, at a cost of $35 million, to serve Concourse B at a speed of 30 bags per minute per line, less than half the projected speed of 65.
	The city hired Rapistan, Demag, Inc., to build an alternative baggage system for $63 million that would move baggage from a newly built staging area in the parking structure through the tunnels.
November 1994	The city signed three change orders, the first for $1.1 million, relieving BAE of its original obligations and providing for changes to serve Concourse B; the second for $2.1 million for system modifications; the third to formally reduce the specified speed of the system to 30 bags per minute per line.
December 1994	The city approved a change order ($605,000) for work to interface the two baggage systems.
February 1995	DIA opened with an automated system that handles outbound luggage for UAL and a tug and cart system for inbound baggage and for Concourses A and C.

Project execution: II

Midstream personnel changes

Some of the midstream design and construction changes resulted from midstream changes in essential personnel; the people who planned, designed, broke dirt, and signed contracts on the new airport were not the same people who completed DIA. For various reasons, one after another of the key actors left the stage, and these changes inevitably influenced the course of the project, since different people have differing values and personality characteristics that impact a major project such as this one. The initial architectural firm for the main terminal (Perez Associates) was replaced by Fentress and Bradburn. Mayor Peña was replaced by Mayor Webb. Moreover, airport director George Doughty's good relationship with Peña (which resulted in expeditious decision-making) was replaced by a relatively poor relationship with Webb (resulting in slower decision-making), leading Doughty to leave the DIA project midstream (Doughty 1995b). Chief DIA engineer Bill Smith died of brain cancer in 1992, and Webb was slow to replace Doughty or Smith, leaving Ginger Evans to handle two very demanding responsibilities. Each of these midstream changes in key personnel created a ripple effect in the design and construction process that caused the project to be completed less efficiently and economically than it otherwise might have been. It could have been worse. New leaders have fresh ideas and are tempted to put their fingerprints on a project they did not inaugurate. Fortunately, aviation director Jim DeLong explicitly refrained from changing the project's design when he came aboard in early 1993 to replace Doughty. The changes are summarized in the following paragraphs.

Federico Peña, the Denver mayor who conceived and began DIA, decided not to run for a third term of office and moved to investment consulting. From there, he was plucked to become head of President-elect Bill Clinton's transportation transition team and immediately thereafter (on January 22, 1993) Secretary of Transportation. Peña was succeeded as mayor by Wellington Webb in June 1991.

Bill Smith, the city's chief engineer, died of brain cancer while DIA was under way. He had served the city for 25 years and had been designated to supervise DIA. After many months' delay, he was replaced by Ginger Evans.

Dick Fleming, head of the Greater Denver Chamber of Commerce, the business community's most visible and articulate advocate of DIA, was displaced in a power struggle within the chamber. Fleming's departure was ill-timed since it immediately preceded Continental Airlines' effort to keep its maintenance facilities at Stapleton International Airport; if it had succeeded, Continental might have remained wedded to its Denver hub months or years longer. Today, Fleming heads the St. Louis regional Commerce & Growth Association.

Frank Lorenzo (the maverick chairman of Texas Air, which controlled Continental Airlines), who committed Continental to Concourse A, was ousted from Continental and replaced initially by Hollis Harris (now CEO of Air Canada) and subsequently by Bob

Ferguson, who peeled aircraft out of Denver to establish the ill-fated CALite east of the Mississippi. Lorenzo had purchased People Express, which had earlier consumed Denver-based Frontier Airlines, to give Continental a stronger western hub. Lorenzo's application to start up a low-cost Philadelphia-based airline was denied by Federico Peña's Department of Transportation subsequently on grounds he was unfit to run an airline. CALite was abandoned by Continental, and Ferguson was replaced by Gordon Bethune, though Contintental did not rebuild its Denver hub.

Stephen Wolf, CEO of United Airlines, fought DIA vigorously. After Continental took the coveted Concourse A, Wolf capitulated and signed up for Concourse B. After an employee leveraged buyout of UAL, the parent company, Wolf left to become a consultant for Air France and was replaced by Gerald Greenwald, a former Chrysler executive. At this writing, Wolf heads USAir, renamed USAirways.

George Doughty, who served as the city's director of aviation, the highest airport job, took over Denver's Stapleton International Airport in 1984. He left after becoming disenchanted with Wellington Webb and the interference his office provided in the contracting process of building DIA. Doughty became director of the Lehigh-Northampton regional airport, near Allentown, Pennsylvania. After 10 months, he was replaced in Denver by Jim DeLong, who had headed major airports in Houston and Philadelphia. The departure of George Doughty could have been avoided if Webb had been more diplomatic. Doughty felt he did not fit in with Webb's team, especially his chief of staff, Venita Vinson. Doughty believed the mayor's office was slowing down DIA, often to maneuver political friends into contracts (Doughty 1995b). Webb also apologized publicly for Doughty being in his cabinet, explaining to audiences that he could not replace Doughty because of the status of DIA (Flynn 1995b). This was truly an insult to Doughty, who left for an easier life heading a smaller airport.

The only key player who remained in office was Colorado Governor Roy Romer, who, after selling the airport to Adams County with his breakfast "oatmeal circuit" leadership in the spring of 1988 (*High Country News* 1995d), effectively dropped from sight on airport issues. Romer did nothing to assist Continental Airlines in its attempt to keep its Stapleton maintenance base, nor did he make much effort to break the political logjam that paralyzed the suburban communities from completing the E-470 beltway on the east side of suburban Denver and other transportation projects that would have given DIA the ground-access interstate highway connections it dearly needed. But, as we have seen, Romer was very effective in selling DIA to the residents of Adams County, something that former mayor Federico Peña would have had difficulty doing.

Mayor Wellington Webb took much political flak in his 1995 bid for re-election for the cost overruns at DIA. During the campaign, Webb announced he would demand the resignations of all his political appointees, reappointing only those who could do the job. Some feared he might replace DIA aviation director Jim DeLong, George Doughty's successor. But Moody's, a prime Wall Street investment house, warned, "The continuation of a unified management approach to daily airport operations as well as capital planning remains an important credit factor" (Imse 1995a). When DeLong was reappointed after the airport opened, and Webb re-elected, DIA's bonds were upgraded.

Mike Musgrave, the city's director of public works, was not reappointed. Musgrave would enter the revolving door like a number of former city employees, taking a job with the engineering firm of O'Brien-Kreitzberg, a firm with which, as a city employee, he had personally negotiated and had overseen a $2.8 million contract to consult on the installation of DIA's baggage system (Lopez 1995a).

Imprudent cost-cutting

In several respects, the planners of DIA attempted to cut expenses in ways that have and will cause increased long-term costs due to a lack of foresight. The baggage system is the most obvious example, with no backup system in place until it became evident that the primary system was dysfunctional.

Another essential system for which no backup exists is the underground people-mover train that links the terminal and concourses. Atlanta's Hartsfield International Airport has a parallel walkway to its underground train, linking the main terminal and concourses. No such alternative tunnels were built at DIA, presumably because of cost considerations. Yet when the train breaks down, passengers stranded on Concourses B and C have no means of transport to the main terminal (or vice versa) unless the city can quickly scramble a fleet of buses to shuttle them across the tarmacs. Concourse A, of course, has the infamous Lorenzo Bridge as an alternative to the train. In the summer of 1995, the train slowed to a crawl because of systems malfunctions, causing delays for passengers, many of whom missed their flights (Leib 1995a; *Rocky Mountain News* 1995d). The underground trains were shut down for two hours twice in December 1995, once because sprinkler systems accidently let loose a torrent of water in the tunnels, and a second time because a passenger had a heart attack.

Fortunately, breakdowns are rare, and debilitated trains blocking access tunnels are rarer still, because a broken train clogging a tunnel can be bypassed by the X corridors connecting the parallel rail tunnels between the concourses, allowing a working train to detour around a dead train. Moreover, unlike the baggage system, the people-mover system is fifth-generation technology (DeLong 1995).

A third rather obvious oversight was the width of Peña Boulevard, the only major highway linking Denver to its airport. Peña Boulevard is only two lanes in each direction. It will need to be expanded, and soon, since it has already been designated as "congested" by city planners (Chandler 1993b). The cost of widening the bridges alone will be significant.

Denver turned down a $75 million federal grant from the Federal Highway Administration (FHA) to build the access highway and used bond money instead. The FHA had insisted that if it financed the road, the boulevard would have only two points of access before reaching the airport gates—Interstate 70 and the future Denver beltway—E-470 (Brewer 1995). Thus, the road would have been a high-speed, uncongested express corridor, with no intermediate entry or exit ramps, similar to the Dulles International Airport access highway near Washington, D.C. The primary reason to turn down federal money to have intermediate access was to facilitate local land development, which itself will generate additional vehicles and congestion, making the four-lane highway more and more inadequate. Not that the land developers minded.

Oversight and lack thereof

Like most airports, DIA was designed by committee, starting with the FAA and including half a dozen architecture and engineering firms, an equal number of city departments, 150 subcontractors, a blue-ribbon commission, and the Denver City Council. It soon became clear that no one was really in charge and that the dominant airlines would be able to squeeze whatever concessions they wanted from a suddenly vulnerable city.

The Denver City Council's airport committee had the job of providing oversight of the airport's escalating costs. *Denver Post* editor Al Knight leveled his guns at the committee for "handling the job of oversight so poorly." He wrote:

> *It has more often than not focused on the small disputes and especially on the tawdry politics of minority- and women-owned businesses. In the process the city council has, again more often than not, been a rubber stamp for policies that have run up the costs and created the need for unwanted delays, both of which now pose a serious threat to the success of the project (Knight 1994b).*

The airport committee members operated at a disadvantage with little or no central staff help to sort through the mass of data tossed up by a huge public works project (Knight 1994a).

Ethics: Greed, cronyism, and revolving doors

In 1991, Wellington Webb ran for mayor as the man who could bring DIA in on time and on budget. But when he ran for re-election in 1995, accusations of ethical lapses and cronyism—giving jobs, contracts, and other city favors to friends, family, and campaign workers, became key issues. Mayor Webb took the position that "I am the proven leader who got the new airport built and other projects done. My opponent is trying to trash my character and downplay my accomplishments" (Eddy 1995b; Weber 1995b).[4] His opponent, city councilwoman Mary DeGroot, insisted that "My opponent has given away city hall, and its lucrative contracts, to his friends. I can do better ethically, and in managing Denver" (*Denver Post* 1995d).

4. Under Webb's leadership, the airport failed to meet scheduled opening dates of October 29, 1993, December 19, 1993, March 9, 1994, and May 13, 1994. The airport did not actually open until February 28, 1995, a few months before the mayoral election, after the city agreed to pay BAE another $40 million to fix the primary baggage system (and drop all claims against the baggage handling manufacturer) and another $63 million to Rapistan Demag to build a backup baggage system. The 16 months of delay cost $100 million in overruns and $361 million in financing costs.

Examples of questionable practices were abundant. Webb gave city jobs to his brother and sister-in-law,[5] and ethical questions surfaced when it became known that his chief-of-staff, Venita Vinson, had not filed an income tax return in years. DeGroot and other city council members alleged that the city's contracting procedures were circumvented so that Webb's friends would enjoy preferential treatment (Eddy 1995c; Kerwin 1995). And the *Denver Post* observed, "The issue of cronyism became glued to Webb when friends of his were among those awarded concession contracts at DIA" (Eddy 1995c). The *Rocky Mountain News* argued, "There should be political consequences for lackadaisical management at crucial moments in the construction of Denver International Airport, and for excessive cronyism in Webb's first couple of years as mayor" (*Rocky Mountain News* 1995b).

The link between political contributions and city contracts was also widespread. In Chapter 5, we discussed the link between bond sales and political contributions. Suffice it here to note that two months after being elected mayor of Denver, Wellington Webb assigned the key role of co-lead to the airport bond underwriting to an investment banking firm that donated $30,000 to his campaign for election, the largest single contributor, accounting for roughly 6 percent of his campaign war chest. One-third was donated under a different corporate name, apparently to obfuscate its source. In the October 1991 bond issuance, the firm earned $1.46 million from the $600 million sale, compared with the $162,476 it earned during Peña's administration (Flynn 1995i). Similarly, a Wall Street bond underwriting firm that had donated $11,000 to Webb's mayoral campaign in 1992 received $5.8 million in airport work. These types of allegations prompted widespread investigations by federal and state agencies as to "whether city officials steered millions of dollars in airport work to brokerage houses, law firms and consultants because of political contributions" (Bettleheim 1995). Overall, 22 percent of Webb's campaign contributions came from airport contractors, consultants, and concessionaires (Hubbard 1995b).

Webb also allegedly used his office to pressure DIA chief engineer Ginger Evans to hire a campaign worker, Fred Riley, for an airport job in 1993, despite protestations about the improper influence violating the career service hiring process, which is supposed to be free

5. Webb's sister-in-law, Marilyn Webb, was given a $48,000-per-year job in the mayor's office of Employment and Training.

of politics (Flynn 1995h). Riley went through the process, but wasn't hired, allegedly making Webb "very upset" with Evans (Flynn 1995f). Evans told associate aviation director Ed Trommeter that she was going to leave DIA because "she was fed up with the lack of ethics from the mayor" (O'Driscoll 1995a). Evans complained directly to Webb and to the assistant city attorney about the interference with the career service hiring process (Flynn 1995h).

When he failed to win a job at DIA, the mayor's office gave Riley a "personal services contract" until another city job opened (Flynn 1995g). When accused of trying to use his influence to secure a job for his former campaign worker, Webb flatly denied meeting with Evans, saying, "It seems amazing to me that she has a perfect memory now and she couldn't remember when we were supposed to open the airport," apparently laying blame for the multiple delays in opening DIA at Evans' feet. Evans responded by saying, "I have a crystal clear memory because I got my butt chewed. You bet I remember it" (Eddy 1995d). Evans had chewed a few butts herself as chief engineer at DIA. Both Evans and Trommeter left the city's employ in May 1994. Questions were raised about Evans' decision to go to work for CH2MHill, one of DIA's largest contractors, although she complied with the city's ethics code in making the move.

George Doughty, Denver's director of aviation, believed the mayor's office was slowing completion of DIA, "often to maneuver political influences into contracts" (Flynn 1995b). Specifically, Doughty believed that Webb refused to adopt a process for selecting concessionaires on the basis of merit and quality because he could not figure out how to give such business opportunities to his supporters.

In fairness, Webb was not the first Denver mayor to ensure his friends and supporters were favored in the city contracting process. Still, such practices seem to have been far more widespread under Mayor Webb's administration. Furthermore, his concern with these issues carried indirect costs that might have been far higher than the direct ones. As the *Denver Post* astutely noted: "The mayor spent excessive time fretting over which retail and food companies would win lucrative concession contracts; on what public art projects would grace the terminal; and on a minority-contracting program. These trivialities distracted City Hall from the really important issues, such as whether all six of the promised runways were fully funded, and if the high-tech baggage system could be up to speed on opening day" (*Denver Post* 1995a).

Minority contracting

The question of affirmative-action contracting was another issue that influenced the cost of the project. Federico Peña was Denver's first Hispanic mayor; Wellington Webb was Denver's first African-American mayor. Each felt strongly about the need to bring women and minorities into the business mainstream. During their tenures as mayor, the city awarded between a fifth and a quarter of its contracts to companies owned by minorities and women, a total of more than $300 million. During the same period, only 1 to 2 percent of the companies in the Denver Metropolitan Statistical Area were minority-owned (Flynn 1995e).

Many criticized the program as a rationale to dole out contracts to politically connected contractors (Kilzer et al. 1995a). Webb's chief of staff, Venita Vinson, was determined to give an existing concessionaire $8 million in rent relief for a deal under which it would turn over some business to selected minority enterprises. DIA bond issues were also moved to a selected minority firm. Trommeter testified at a city personnel hearing that Webb's office had accused him of bigotry and failing to maximize DIA contracts for "disadvantaged business enterprises," an allegation Webb denied (O'Driscoll 1995b). Another city airport employee, Richard Boulware, head of public relations at Stapleton International Airport from 1984 to 1993, claimed that Webb's city attorney Mike Dino directed him to give a minority-owned public-relations firm a subcontract for airport publicity (Flynn 1995c). Claiming that political meddling by the mayor's office led to his demotion, Boulware subsequently filed an EEOC complaint against the city, alleging age and gender bias (O'Driscoll 1995c). Boulware also complained that city hall directed that airport funds be used for non-airport purposes, an allegation subsequently confirmed by DOT Inspector General Mary Schiavo. And the city gave the operating and maintenance contract for the error-prone baggage system to a firm other than the one that built it, largely because of doubts that BAE would hire enough minority workers (BAE pledged to hire 36 percent local and minority workers, but apparently this wasn't enough). Chief airport engineer Ginger Evans described this as the "worst management decision" possible (Eddy 1994a).

Such decisions raised many serious questions; the *Denver Post* discussed these as follows:

> *Mayor Wellington Webb is a firm supporter of the city's effort to steer work to minority and women businesses, saying it helps level the field and gives disadvantaged companies a chance at winning city work.*
>
> *A review of the program, however, shows that many minority concerns winning city work have had close ties to the administrations of Webb and Federico Peña. Further, a few of those firms won city work despite lacking in-depth experience in areas covered by their contracts (Kilzer et al. 1995b).*

Furthermore, some of the multimillion-dollar minority "disadvantaged business enterprise" firms were accused of "[b]loated payrolls, overcharging and political influence" (Flynn 1995d). Others were alleged to be minority "front" companies, with sloppiness in the oversight process resulting in 32 ineligible minority- or female-owned companies being awarded $34 million in city contracts over three years. All of this led to a federal grand jury investigation of Denver's minority-contracting program (Weber 1995a).

Webb promised to reform the minority-contracting program by imposing harsh penalties for firms that lie about their eligibility, ensuring companies no longer eligible are not given contracts, and generally tightening overall management of the program (Eddy 1995e).

The investigations

All these charges and countercharges led to numerous formal investigations. Several Congressional committees and federal agencies—the FBI, U.S. DOT, FAA, Department of Justice, and the Securities and Exchange Commission (SEC)—launched probes, and grand juries were convened to assess allegations of construction fraud, bogus surety performance bonds, minority-contracting irregularities, altered documents, and spending airport revenue on a $10,000 trade junket to Gabon, which has no, and will likely never have, direct air service to Denver (*Denver Post* 1995c, 1995e; Eddy 1995a; Flynn 1995q, 1995r). Some of these were motivated by political considerations. Since Federico Peña was now President Clinton's Secretary of Transportation, Congressional Republicans were eager to embarrass a cabinet officer and his Democratic President. And, as we shall see, most of the allegations proved to be unfounded.

The most serious threat to the financial well-being of the city lay in investigations by the SEC and the private civil lawsuits discussed in Chapter 5. The SEC conducted extensive investigations into whether bond houses that make significant political contributions enjoy a disproportionate share of municipal underwriting business. Once focused on Denver, the SEC began evaluating whether the city misled bond investors about the state of the airport and its opening date (Purdy 1995). By October 1995, SEC investigators were urging their agency to sue Denver for violating antifraud laws in the issuance of DIA bonds (Lopez and Hodges 1995). University of Denver professor J. Robert Brown saw the probe as an attempt by the agency to "clean up bond disclosures" (Leib 1995b). The SEC ultimately dropped its probe, concluding that then-existing regulations did not bar campaign contributions by bond houses.

Denver's affirmative-action program also led to a federal probe. Under federal law, at least 10 percent of the Department of Transportation's expenditures "shall be expended with" minority businesses, "except to the extent the secretary" determines otherwise (*Wall Street Journal* 1995b). As we have noted, mayors Peña and Webb did their utmost to ensure that this target be met. Indeed, it was exceeded, although in a way that led to significant abuses and increased costs for the city (Hubbard and Flynn 1995), both directly in terms of the project and indirectly in terms of having to defend its practices. Denver spent $379,000 in airport funds defending its minority-contracting program against a suit by a company that was the unsuccessful low bidder on several non-airport contracts. The FAA agreed that spending airport money on such a lawsuit was proper, but that the airport's share was too high, requiring the city to reimburse the airport's account by $23,000 (Flynn 1995j). But another DOT investigator who worked on a DIA probe of whether the city properly spent federal money on legal fees defending its affirmative-action contracting program was fired without explanation (Flynn 1995k).

Several other investigations also were resolved with no finding of criminal wrongdoing. Attorney General Janet Reno absolved Federico Peña of unduly using his influence as DOT Secretary to secure business for his investment consulting firm. An investigation by the U.S. General Accounting Office (GAO) found no improprieties in 37 prime contracts for the construction of DIA; all were found to have been issued to qualified contractors who exceeded the city's women and minority business goals (Lopez 1995c). The GAO found that allegations by several airport inspectors that tests showing weakened runway concrete were altered, that steel reinforcing bars were left out of concrete walls on Concourse C, and that thousands of welds in the parking garage were improperly done (Hodges 1995; Hubbard 1994; Paulson 1994) were unfounded, although it had never interviewed key inspectors. In other cases, the problems were dealt with locally (Flynn 1995n). William Torrez, who had served on Mayor Webb's Hispanic Advisory Council and the New Airport Bond Assistance Program (which helped small businesses obtain bonds to perform DIA contracts), accepted a plea bargain to forgery charges. The Denver district attorney's grand jury investigations came up empty (Flynn 1996b).

Yet a taint continues to hover over the project and the city. As one commentator wrote:

> *The scandals and questions that haunted Wellington Webb before his re-election remain alive. The most troubling: whether, in financing the new airport, Webb put Denver taxpayers at risk while his political buddies benefited.*
>
> *Far from disappearing with Webb's second term as Denver's mayor, the subjects carry renewed intensity. If the investigations erupt into criminal indictments, civil penalties, or more lawsuits, the spotlight of bad national publicity will burn the city with greater heat—because it will focus on an incumbent politician, not on someone voters already had turned out of office (Purdy 1995).*

"Send lawyers, guns, and money"

It is the nature of the beast that it is difficult to measure whether one gets full bang for the buck with the work of consultants. Is the value of their insights, analysis, and advice worth the price that is paid? At Denver, there was such a feeding frenzy on fistfuls of money thrown in a multitude of directions, it appears that the city often failed to get its money's worth. Here are some examples:

- Unison Consulting Group, a Chicago-based aviation management consulting firm, was paid $250,000 by the city, of which about a third was for a 20-page report on the future of the terminal building at Stapleton Airport. That works out to $14 a word. The report itself was criticized as containing little new information, with portions appearing to having been lifted, almost verbatim, from an earlier, much less costly report (Hodges 1994a).

- Six different law firms, at any one time, were hired by the city to assist it in selling DIA's bonds. These firms billed well over $4 million in legal fees. All six contributed money to Wellington Webb's election fund (Hodges 1994b). Webb's city attorney, Dan Muse, spent $15 million on outside lawyers in just three and one-half years. To deal with SEC probes that DIA bond purchasers were fraudulently misled, the city spent more than a million dollars with a New York law firm whose billing practices included $116 per hour for a paralegal to copy materials (Calhoun 1995b). By the end of 1995, Denver had spent more than $10 million in legal fees on DIA. In 1996, another $3 million was spent on legal fees in the city's unsuccessful suit against the DIA's architects.
- The city spent $240,000 for lobbying Congress to move the annex of the Smithsonian Air & Space Museum (the world's most popular museum) to Stapleton. Ron Brown, who subsequently joined President Clinton's cabinet with Peña, was a partner in the firm (Phillips 1995).
- Denver paid Washington, D.C., public-relations firm Hill & Knowlton $60,000 to improve its image after *Forbes* magazine described DIA as a "financial crash in the making." The reader can determine how good a job the PR firm did. But among its bills was $75 for taking a half hour to determine what time the sun would rise on DIA's opening date (Hodges 1994c). The more useful question would have been whether the sun *would* rise on DIA.
- The builder of the "backup" baggage system at DIA, Rapistan Demag, charged that, as public works director, Mike Musgrave saw to it that O'Brien-Kreitzberg (which had $5.8 million in DIA contracts to manage construction on the baggage system) earned more than it should have (Lopez 1995b). Five months before leaving the city government to work for O'Brien-Kreitzberg, Musgrave insisted the city give the firm a $100,000 expert-witness contract, in a "screaming phone call" to assistant city attorney John Gross at his home

> (Flynn 1995o). A front-page story in the *Denver Post* charged O'Brien-Kreitzberg with a "pattern of lavish spending on hotels and apartments, disregard for any financial pinch the airport may have felt during construction and enough sympathetic ears from Mayor Wellington Webb's staff to keep the city's wallet open." One city inspector complained, "They were going haywire with expenses. Their wives were coming with them, and they were taking trips to Vail. We finally had to put the kibosh on that" (Lopez 1995d). The firm also billed the city for five apartments at $9000 a month. Musgrave and Evans often approved the firm's expenses (Lopez 1995d; Flynn 1995m).

Was such a frenzy inevitable? Denver airport director Jim DeLong has insisted that the answer is yes. He makes two points. First, in order to build a new airport, it was necessary to build a political coalition of business, the media, labor, and minority interests, to succeed in the public referenda on the issue. The votes in Adams County and Denver were crucial in defusing the small groups of highly motivated anti-growth and pro-environmental individuals, which in so many other cities have derailed such projects. The two votes in favor of DIA made stopping it infinitely more difficult. But the quid pro quo would be that each of these constituencies—business, labor, minorities, and the press—would expect either to have a piece of the action or (in the press' case) to have the latitude to make recommendations and have them endorsed. In other words, DIA could not have been built without rewarding the political constituencies that had supported it. And the most amazing thing about DIA, in DeLong's view, is that given the inability of any other city to build a new airport in two decades, DIA was built at all.

Second, while Peña was cast as the visionary who imagined a great airport and sold it to the people, he handed the ball off to his successor, Wellington Webb, to build it. It is significantly more difficult and less glamorous to complete a megaproject than it is to begin it (DeLong 1995b).

These arguments, however, raise two important issues. The first involves the decision process required to build megaprojects. It suggests, rather cynically, that no such project can be built without a large, broad-based, supportive coalition whose members must be "paid off." We discussed the role of political coalitions and public participation in Chapter 3. Here we would simply raise the question of whether the public is doomed to pay such heavy and wasteful subsidies for its infrastructure Our answer is that it depends on the quality of the political leadership and the structure of the political system. With weak leadership and a corrupt system, the answer is obviously yes. But such an outcome is not inevitable.

The second issue focuses explicitly on leadership. Would the payoff have been as high if Peña had won a third term? In other words, do individuals make a difference? Clearly we believe that they do. We shall return to this fundamental point in our concluding chapter.

Conclusion

How much money was actually wasted? Some would point to $3 billion cost overruns (*Wall Street Journal* 1995a). Some would peg the cost overruns at $750 million, still a substantial sum; others would argue "Hundreds of millions of dollars [of DIA's $800 million to $1 billion cost overruns] represents work redone—a colossal waste" (Russell 1994).

Throughout this chapter we have discussed the reasons why DIA cost so much more than originally anticipated. As we have seen, the factors that drive megaprojects to come in over budget were all present here—fallacious cost estimates (the original ones were for a "bare-bones" facility that many considered unrealistic), poor project execution (ranging from land acquisition to project management), and innumerable changes in the project, caused by changes in the environment. The airline industry has traditionally been unable to anticipate these factors. It is said of airlines that they order aircraft in good times and take delivery in bad. It could be said of airports, that airports sometimes are planned, designed, and constructed in good times and opened in bad. Tragically, the city of Denver built an entire concourse for an airline that ultimately needed only three or four gates. DIA was planned for three major hub airlines (United, Continental, and Frontier), built for two (United and Continental), and is occupied by only one (United).

It has been argued that not all of the increased costs can be considered as overruns. The airport that now graces the plains is not the airport that was originally planned (Fig. 9-5). Accordingly, many question just how much such a facility would have cost if it had been designed and implemented in a rigorous professional manner. To answer such a question is no easy task. Some idea of where money was wasted can be obtained from Table 9-5, which compares the 1980 (high and low) estimates of key units with the original plan (1990) and the changes over time.

Table 9-5. Summary of project costs (in millions $)

	1980		1990	1991	1992	1995
Land	40	80	210	210	223	242
Airfield	200	400	403	357	357	422
Terminal	600	1200	644	728	1239	1203
Other	160	320	1052	536	881	1331
Total	1000	2000	2309	1931	2700	3298

Sources: Peat Marwick 1980; ASRB 1995 (ASRB 1990a, 1991a, 1992b-g, 1995a-g)

→ **Runways**

Five 12,000 ft.
Full Buildout - 12 Runways

→ **Structures & Facilities**

Terminal

1.2 million sq. ft.
3 modules
Automated Guideway Transit System - train to concourses
Full Buildout - 5 modules

Concourses

3 airside concourses w/88 gates
3.3 million sq. ft.
Full buildout - 5 Airside Concourses w/ 300 gates

Terminal Access Roads and Parking

Peña Blvd - 12 mile, four lane highway from I-70 to Terminal
15,000 car garage

Car Rental Facilities

150 acres
12 car rental agencies

Fig. 9-5. *A breakdown of what Denver got for its more than $3 billion in construction costs.*

These figures provide some clues, although they must be treated with caution since the 1980 figures are not adjusted for inflation. Still, it is noteworthy that the cost of the terminal and airfield closely approximate the high estimate. Only the land costs are far greater—$240 million in 1995, triple the high figure ($80 million) estimated in 1980. Some of that difference may be attributable to the enormous size of the facility and to its location.

The total cost of the airport, however, is much more than the $3.3 billion figure, because that sum does not include the $1.32 billion that the city borrowed to pay the bondholders prior to DIA's opening, various financing and other expenditures, or the expenses incurred by the tenants or paid by the federal government, some $696 million. The following breakdown indicates where the money went and who paid for what:[6]

Construction (city)			
Basic project	$	2700	million
Net additions since March 1994 (estimated)	$	191	million
Backup baggage system	$	63	million
Total	$	2954	billion
Capital costs (city)			
Land acquisition	$	242	million
Pre-1990 planning and administration	$	19	million
Stapleton capital fund reimbursement	$	72	million
Total	$	333	million
Total Denver capital costs:	$	3287	billion
Construction (tenants):			
United Airlines facilities (maintenance, cargo, kitchen)	$	261	million
United modifications to baggage system	$	56	million
Continental facilities	$	75	million
Federal Aviation Administration tower and facilities	$	224	million

6. The GAO's figures differ slightly; it places the construction costs at $3.004 billion and the capital costs at $1.219 billion, for a total cost to Denver of $4.223 billion. The total costs to tenants is $599 million, making DIA's total cost $4.822 billion (GAO/AIMD-95-230, September 20, 1995).

Rental car facility	$	66 million
Independent flight kitchen	$	6 million
General aviation facility	$	8 million

Total $696 million

Total construction costs for all facilities, including land: $3.983 billion

Financing cost (city): $1.320 billion

Grand total: $5.303 billion

Note: Financing cost includes interest owed on bonds before airport opened and revenue begins flowing to repay bondholders; bond reserve funds; discounts and commissions. Ultimate financing, paid over 30 years by airline revenue, will total billions more.

Accordingly, the total price of DIA is about $5.3 billion, of which Denver's share came, according to the city, to $4.523 billion; according to other sources, $4.607 billion (Rebchook 1995; ASRB 1995). If we consider only the land and construction costs, we return to the $3.3 billion figure as the city's cost.[7] This sum is still $1.3 billion more than the high estimate made in 1980 and probably approximates how much was wasted through mismanagement, incompetence, and corruption, although some have placed the figure at $2 billion or more. Whatever the exact figure, it is obvious that DIA cost much more than it should have. One columnist has attempted to place the issue in perspective (C. Green 1995):

> *We should understand why we're paying nearly $5 billion for a $2.5 billion airport. But when our children's offspring look back into history, 40 or 50 years from now, they will see an airport whose design, cost and purpose will seem ridiculously obvious. And that extra $2 billion won't seem like too much to pay—even though it was.*

Lessons learned

- *Make every effort to get the principal tenants (i.e., the airlines) on board early in the planning and design process to minimize scope and design changes.* At Denver, the airlines were invited to participate in the planning and design process, although as United Airlines began to resist the new airport project, its participation became increasingly

7. For a similar analysis, see Flynn (1995l).

less constructive. The scope changes insisted upon by Continental and United Airlines as the price they exacted for signing leases at DIA, added more than $1 billion to its cost and resulted in reconstruction and delays (GAO 1995). Exclusive-use agreements for gates, ticket counters, and baggage facilities can be used as bargaining chips, although they force construction of a much larger and more expensive airport, with less efficient use of infrastructure, than multiple-carrier use of such facilities.

- *Resist major scope changes after construction has begun on essential facilities.* Had the city not acceded to the airlines' massive scope changes, the project would not have suffered such severe delays and cost overruns.
- *If scope changes must be made, amend the timetable for opening the airport accordingly.* Mayor Webb knew when he took office that the 1993 opening date for DIA was unrealistic and refused to endure the short-term political pain for moving that date. As a consequence, he endured severe long-term pain, as his political opponents blamed him for the massive cost overruns. He might have been unseated as mayor had a stronger candidate opposed him.
- *KISS (keep it simple, stupid).* In their exuberance to have the world's newest, most spectacular, and technologically sophisticated airport, the city embraced the most modern technologies. Stretching the technological envelope is always risky. Doing so with a high-tech baggage system on a scale the size of DIA was simply dumb. While less-sophisticated technology may be less glamorous, a solid airport must simply work. Bells and whistles land not a single plane.
- *Install backup systems, strategies, and contingency plans for essential sophisticated systems, particularly those employing new and untested technology.* DIA's automated baggage system was one of the largest and most sophisticated in the world. Nothing of its magnitude had ever been attempted at any new or existing airport. The mechanical and software problems that emerged were the principal reasons why the airport opened more than a year late. DIA also has no backup system for the underground train that links the main terminal with the remote concourses. If it ceases to operate, DIA will be paralyzed, at least for a while.

- *Provide for a smooth transition between key personnel in the airport construction project.* At DIA, the mayor, chief engineer, airport director, and other instrumental people involved in its planning and design left the project, and in some instances, were not replaced for months. This created a vacuum of leadership on the project and inevitably led to changes midstream.
- *Provide for a vigorous quality control and assurance program.* Nearly half the concrete allegedly was installed improperly (GAO 1995). Mistakes identified early are less costly to correct.
- *Allow airport and construction professionals to take the project to completion without political micromanagement.* DIA would have had a much better chance of coming in on-time and on-budget had the politicians stayed out of the way. Denver could have hired a general contractor with pecuniary incentives to bring the project in under budget. Instead, it divided DIA into thousands of Lilliputian contracts to ensure that minority, female, and political-patronage interests were assuaged.
- *Depoliticize the contracting process.* Denver's mayors and city council exerted their influence in the contracting process to ensure their supporters were awarded contracts. In some instances, more qualified contractors with lower bids were turned away in favor of those with political clout in city hall. Some of these awards were done under the increasingly controversial program of affirmative action or minority and female contracting. If the taxpayers want their hard-earned dollars well spent, the contracting process must be depoliticized. One solution might be the creation of a metropolitan or state airport authority, which would include representatives from the city and suburban communities. Given that many suburban residents must use Denver's airport (and pay an inflated price for the privilege), equity suggests they should have had a say in the planning, design, and construction of DIA. Suburbanites can exact no political retribution against city politicians who mismanage a project that takes such a large bite out of their wallets. Given that federal trust funds are essential for any project of this magnitude, Congress could condition grants on the establishment of a regional airport authority.

- *Honesty is the best policy.* Throughout the process of selling DIA to the public, city hall low-balled DIA's costs, focusing only on the construction costs, net of land acquisition costs or capitalized interest. Including them in the full price tag, coupled with airline scope changes and additions, as well as massive cost overruns, gave the public a healthy case of "sticker shock." Low-balling the cost and failing to inform the public that major scope changes would make the scheduled opening date unrealistic cost Denver dearly in public relations and credibility and may cost the city dearly in dollars if civil securities fraud litigation is successful.
- *Set a good example.* Denver's political leaders rewarded many of their supporters with generous construction and concession contracts. So many pigs were feeding so ravenously at the public trough that the climate of greed it unleashed was infectious. Even many traditionally dedicated public servants became infected.
- *Promulgate a vigorous regime of ethical regulation and enforce it.* The mayors' friends were showered with generous contracts and concessions at DIA, some under the rationalization of "affirmative action." If affirmative action has become a euphemism for raw political cronyism, then it cannot survive as a legitimate social objective. Moreover, Denver has never promulgated meaningful regulations prohibiting the "revolving door" of city employees leaving government to go to work for firms they directly supervised. The opportunity for corruption is too great under such a system. Several key city officials have found themselves receiving salaries or business from firms involved in the land, design, or construction contracts involving DIA. If we expect frail human beings to avoid temptation, they must be prohibited for a reasonable period of time from going to work for, or accepting as clients, individuals who have done business with the city. The federal government imposes strict prohibitions against the revolving door. It should insist that local governments that accept federal dollars do the same. Further, some limitations must be placed on campaign contributions by city contractors, including bond houses.
- *The political process can be improved to facilitate prudent management if all affected citizens have a greater political voice in electing the individuals who are responsible for the*

conception, design, and construction of the infrastructure the citizens must pay for. The City and County of Denver built a $5 billion airport for a metropolitan region whose population outnumbered it three to one. As passengers, the suburban residents would disproportionately pay the price of the massive cost overruns at DIA, yet they were denied the opportunity to exact political retribution on the political leaders who mismanaged its construction and favored their supporters with construction and concessions contracts. Multibillion dollar regional infrastructure projects should be built by regional governments. Political leaders whose re-election is dependent on approval by those groups most likely to use the facilities and pay for their mistakes will be incentivized to use the best management principles and bring the project in on time and on budget.

- *The federal government should condition use of federal airport funds on (a) the creation of a Regional Airport Authority responsive to the will of the regional electorate, and (b) the promulgation of strict ethical standards for its political leaders and all its employees.* This recommendation flows from the preceding two. Cities are not likely to give up control of an airport to regional authorities without a federal mandate that they must do so. Moreover, despite the enormous criticism the City of Denver has suffered because of ethical lapses, as of this writing, it still has not promulgated meaningful ethical regulations. Of course, it would be hypocritical for politicians to impose ethical standards on city employees they were unwilling to impose upon themselves. Federal law should be tailored to investigate and prosecute local governmental officials responsible for waste, political favoritism and corruption. Half a billion dollars of federal money was spent to build DIA. If our government wants its money spent prudently, it should provide essential oversight to ensure that waste and corruption are discouraged.
- *Consider alternatives to expensive infrastructure overdevelopment, such as peak-period pricing and multiple-use gates, ticket counters, and baggage facilities.* Peak-period pricing can flatten somewhat the demand curve of aircraft arrivals and departures to enhance the efficiency in the use of airport capacity. The European model of multiple-use,

shared gates, ticket counters, and baggage facilities allows fewer facilities to be built and more efficient use of these resources.

References

Airports. 1995a. Politicians, squatters slow new Bangkok international airport. March 7:93.

————. 1995b. Denver, Continental reach agreement over airport lease terms. April 4:131.

————. 1995c. Greiner says SEC staff case against Denver consultants not legitimate. December 19.

ASRB. 1990–1995. (See City and County of Denver.)

Aviation Daily. 1995. Denver, Continental settle lease dispute. April 4:11.

Bettelheim, Adriel. 1995. Potential influence peddling, shoddy work probed. *Denver Post.* February 26.

Brewer, George. 1995. Interview. October 26.

Brinkley, John. 1995a. $2.5 million set aside for DIA workers. *Rocky Mountain News.* June 13.

————. 1995b. 4 FAA workers paid to move away from DIA. *Rocky Mountain News.* July 15.

————. 1995c. FAA suspends moving-cost reimbursements. *Rocky Mountain News.* July 21.

Brinkley, John and Ann Imse. 1995. FAA oks moving funds for a big non-move. *Rocky Mountain News.* July 25.

Bronlkowski, Lynn. 1991. Study predicts 10,000 jobs at Front Range air cargo site. *Rocky Mountain News.* October 18.

Brown, Fred. 1995. Denver International gains some fans. *Denver Post.* April 15.

Business Week. 1995. A new airport has villagers raging. July 24:5.

Calhoun, Patricia. 1993. Airport '94. *Westword.* October 27.

————. 1995a. Blank you very much. *Westword.* February 15.

————. 1995b. Blank check. *Westword.* February 22.

Carnahan, Ann and Burt Hubbard. 1995a. Flying blind into a mountain of debt. *Rocky Mountain News.* February 5.

————. 1995b. Hard sell shoved DIA down Denver's throat. *Rocky Mountain News.* February 7.

Chandler, David. 1993a. Cash landing! *Westword.* February 17.

————. 1993b. Collision course. *Westword.* March 3.

————. 1994. Crash and carry. *Westword.* January 28.

City and County of Denver (Denver). 1990. Airport System Revenue Bonds (ASRB), Series 1990A. May 10.

________. 1992. ASRB, Series 1992B. May 6.

________. 1994. ASRB, Series 1994A. September 1.

________. 1995. ASRB, Series 1995B. June 1.

Colorado Attorney General. 1991. Report.

Crider, Robert. 1994. City and County of Denver (Denver) Auditor. DIA Change Orders. April 6.

DeLong, James. 1995a. Testimony before the House Transportation Subcommittee on Aviation. May 11.

————. 1995b. Interview. October 26.

Dempsey, Paul S. and Andrew R. Goetz. 1992. *Airline Deregulation & Laissez-Faire Mythology*. Westport, CT: Quorum Books.

Denver Post. 1992. Highlights of cargo agreement. January 22.

————. 1994a. Opening day pushed back three times. March 1.

————. 1994b. Continental plans to cut 1400 Denver flight-crew jobs. March 4.

————. 1994c. History in the making. Special section on DIA. March 1994.

————. 1995a. Many leaders share blame for DIA's slow takeoff. February 27.

————. 1995b. Denver's DIA is a bargain in a global comparison. February 28.

————. 1995c. Probe zeros in on Peña. March 12.

————. 1995d. The mayoral match-up. May 28.

————. 1995e. Agent to plead guilty in DIA case. October 19.

Dillon, David. 1995. Soaring main terminal overwhelms the rest of Denver's troubled new airport. *Dallas Morning News*. February 12.

Doughty, George. 1990. Regardless of what the future brings, Denver's new airport will work. *Denver Post*. August 31.

————. 1995a. Testimony before the House Transportation Subcommittee on Aviation. May 11.

————. 1995b. Interview. October 23.

Eddy, Mark. 1994a. BAE operating exclusion blasted. *Denver Post*. May 1.

————. 1994b. DIA bag tag: $50 million. *Denver Post*. July 28.

————. 1995a. Grand jury to probe funding of delegation's trip to Africa. *Denver Post*. February 1.

————. 1995b. Webb's cronyism woes began early. *Denver Post*. May 22.

————. 1995c. Cronyism "already an issue for the public," says pollster. *Denver Post*. May 22.

————. 1995d. Webb denies job coercion. *Denver Post*. May 29.

————. 1995e. Webb faces list of campaign vows. *Denver Post*. June 11.

Engineering News-Record (ENR). 1994. Woman of the year Ginger S. Evans. February 14:34.

Evans, Ginger. 1996a. Interview. February 13.

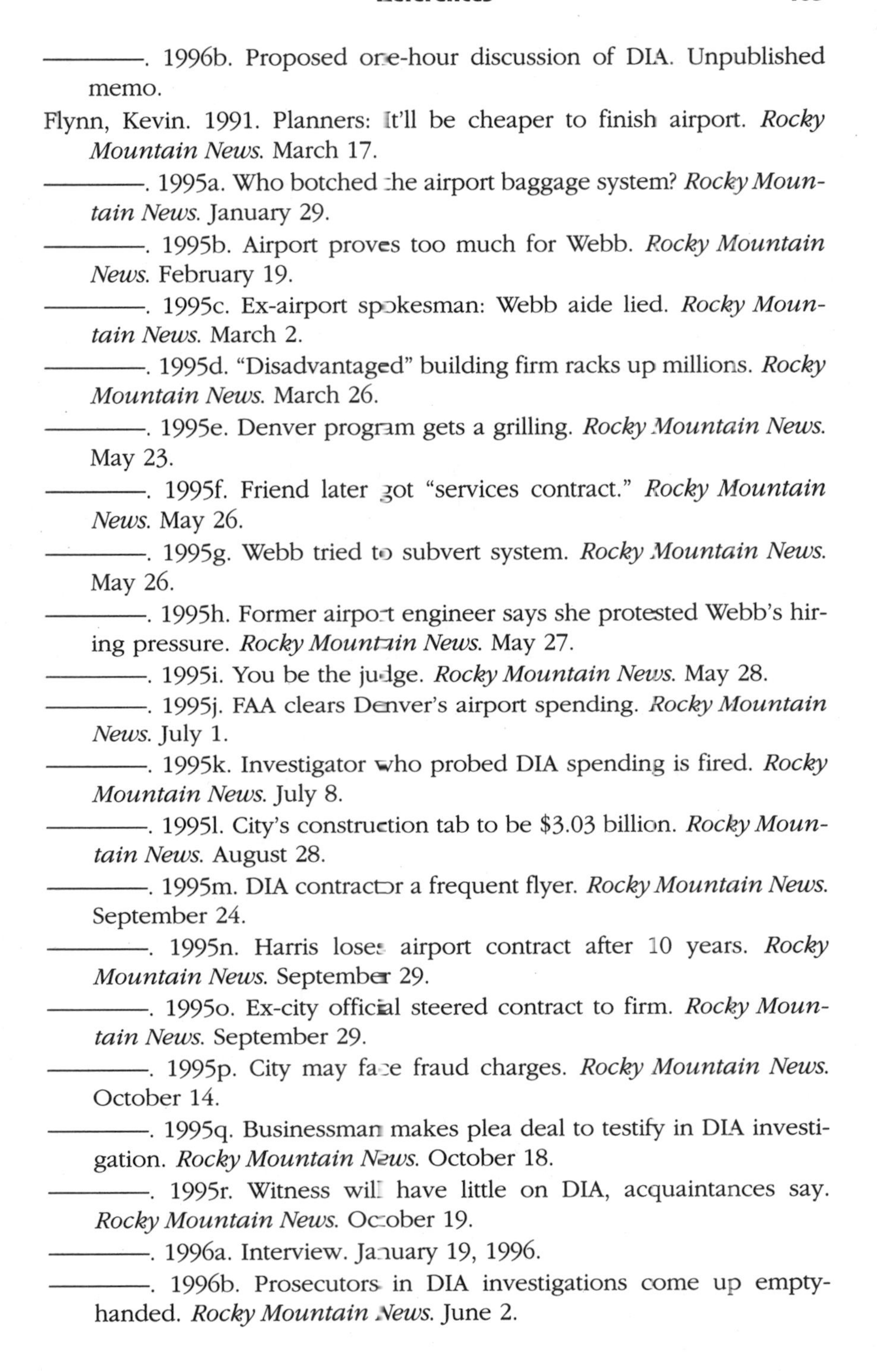

———. 1996b. Proposed one-hour discussion of DIA. Unpublished memo.

Flynn, Kevin. 1991. Planners: It'll be cheaper to finish airport. *Rocky Mountain News.* March 17.

———. 1995a. Who botched the airport baggage system? *Rocky Mountain News.* January 29.

———. 1995b. Airport proves too much for Webb. *Rocky Mountain News.* February 19.

———. 1995c. Ex-airport spokesman: Webb aide lied. *Rocky Mountain News.* March 2.

———. 1995d. "Disadvantaged" building firm racks up millions. *Rocky Mountain News.* March 26.

———. 1995e. Denver program gets a grilling. *Rocky Mountain News.* May 23.

———. 1995f. Friend later got "services contract." *Rocky Mountain News.* May 26.

———. 1995g. Webb tried to subvert system. *Rocky Mountain News.* May 26.

———. 1995h. Former airport engineer says she protested Webb's hiring pressure. *Rocky Mountain News.* May 27.

———. 1995i. You be the judge. *Rocky Mountain News.* May 28.

———. 1995j. FAA clears Denver's airport spending. *Rocky Mountain News.* July 1.

———. 1995k. Investigator who probed DIA spending is fired. *Rocky Mountain News.* July 8.

———. 1995l. City's construction tab to be $3.03 billion. *Rocky Mountain News.* August 28.

———. 1995m. DIA contractor a frequent flyer. *Rocky Mountain News.* September 24.

———. 1995n. Harris loses airport contract after 10 years. *Rocky Mountain News.* September 29.

———. 1995o. Ex-city official steered contract to firm. *Rocky Mountain News.* September 29.

———. 1995p. City may face fraud charges. *Rocky Mountain News.* October 14.

———. 1995q. Businessman makes plea deal to testify in DIA investigation. *Rocky Mountain News.* October 18.

———. 1995r. Witness will have little on DIA, acquaintances say. *Rocky Mountain News.* October 19.

———. 1996a. Interview. January 19, 1996.

———. 1996b. Prosecutors in DIA investigations come up empty-handed. *Rocky Mountain News.* June 2.

Fumento, Michael. 1993. Federico's folly. *The American Spectator*. December 1993:42.

Green, Chuck. 1995. DIA debate in the view of history. *Denver Post*. February 26. p. D-1.

————. 1996. DA's look at DIA draws ire. *Denver Post*. September 11.

Green, Peter. 1992. Big ain't hardly the word for it. *Engineering News-Record*. September 7:28.

High Country News. 1995a. Megamess: "We've got it all." January 23.

————. 1995b. Ambition becomes a megamess. January 23.

————. 1995c. Megamess: Tent, roof, wind shear. January 23.

————. 1995d. Ambition becomes a megamess. January 23.

Hodges, Arthur. 1994a. Filing a costly flight plan. *Westword*. March 9.

————. 1994b. Gentlemen prefer bonds. *Westword*. May 18.

————. 1994c. Flack attack. *Westword*. June 17.

————. 1995. Denver cleared on DIA. *Denver Post*. August 23.

Hubbard, Burt. 1994. DIA concrete test results faked, ex-workers say. *Rocky Mountain News*. November 14.

————. 1995a. Appearance of conflict spurred Spensley's exit from new airport office. *Rocky Mountain News*. February 12.

————. 1995b. 22% of Webb funds have airport link. *Rocky Mountain News*. April 10.

Hubbard, Burt and Kevin Flynn. 1985. Airport land deals filled with intrigue. *Rocky Mountain News*. February 12.

————. 1995. Grand jury joins minority-contract probe. *Rocky Mountain News*. March 19.

Imse, Ann. 1995a. DA refinancing to save city $45 million. *Rocky Mountain News*. June 9.

————. 1995b. Official calls DIA fantastic. *Rocky Mountain News*. June 15.

————. 1995c. DIA baggage system delayed again. *Rocky Mountain News*. July 28.

Imse, Anne and Kevin Flynn. 1996. United, bag firm fire suits at each other. *Rocky Mountain News*. March 19.

Imse, Anne and Burt Hubbard. 1995. Sky high. *Rocky Mountain News*. February 12.

Kerwin, Katie. 1994. Peña says he'd have made DIA smaller. *Rocky Mountain News*. March 29.

————. 1995. DeGroot blasts "cronyism." *Rocky Mountain News*. April 13.

Kilzer, Lou, Robert Kowalski, and Steven Wilmsen. 1995a. Minority firm carves out an empire. *Denver Post*. April 7.

————. 1995b. Minority contracts won despite lack of experience. *Denver Post*. April 7.

Knight, Al. 1991. Wooing air cargo, aviation's ugly sister. *Denver Post*. September 8.

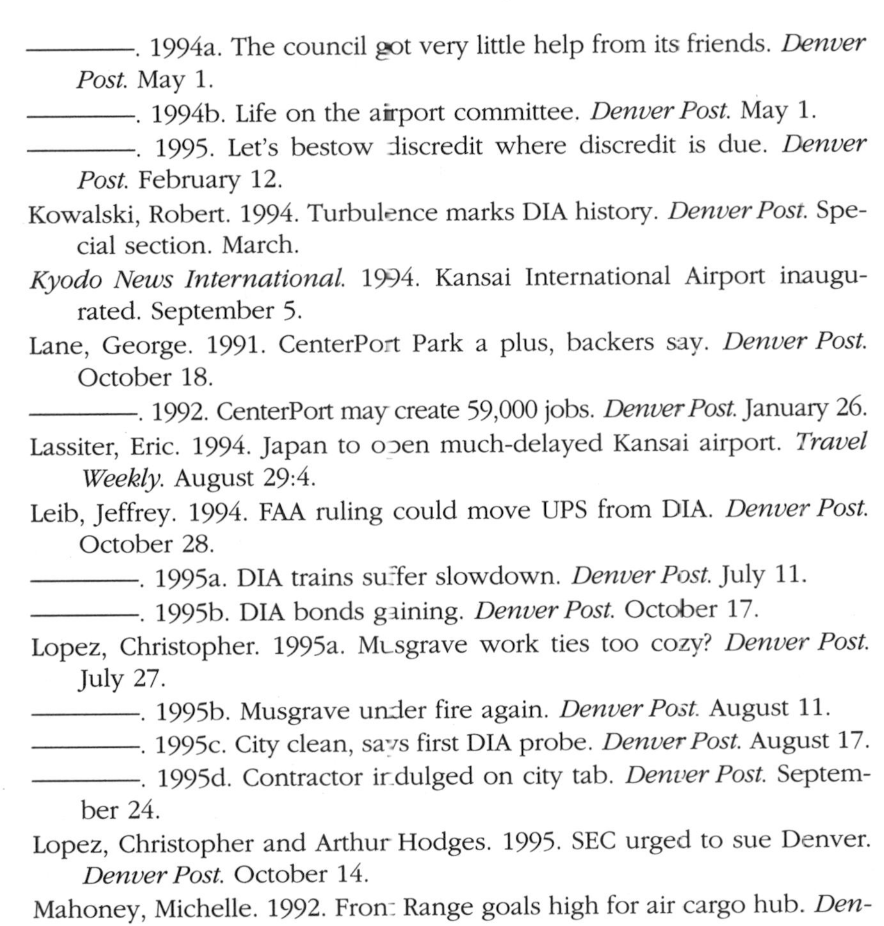

———. 1994a. The council got very little help from its friends. *Denver Post.* May 1.

———. 1994b. Life on the airport committee. *Denver Post.* May 1.

———. 1995. Let's bestow discredit where discredit is due. *Denver Post.* February 12.

Kowalski, Robert. 1994. Turbulence marks DIA history. *Denver Post.* Special section. March.

Kyodo News International. 1994. Kansai International Airport inaugurated. September 5.

Lane, George. 1991. CenterPort Park a plus, backers say. *Denver Post.* October 18.

———. 1992. CenterPort may create 59,000 jobs. *Denver Post.* January 26.

Lassiter, Eric. 1994. Japan to open much-delayed Kansai airport. *Travel Weekly.* August 29:4.

Leib, Jeffrey. 1994. FAA ruling could move UPS from DIA. *Denver Post.* October 28.

———. 1995a. DIA trains suffer slowdown. *Denver Post.* July 11.

———. 1995b. DIA bonds gaining. *Denver Post.* October 17.

Lopez, Christopher. 1995a. Musgrave work ties too cozy? *Denver Post.* July 27.

———. 1995b. Musgrave under fire again. *Denver Post.* August 11.

———. 1995c. City clean, says first DIA probe. *Denver Post.* August 17.

———. 1995d. Contractor indulged on city tab. *Denver Post.* September 24.

Lopez, Christopher and Arthur Hodges. 1995. SEC urged to sue Denver. *Denver Post.* October 14.

Mahoney, Michelle. 1992. Front Range goals high for air cargo hub. *Denver Post.* January 22.

———. 1994. Airline changes buffet Denver. *Denver Post.* February 13.

McCagg, Edward. 1993. New Denver International Airport: Evolution of a Concourse. Paper delivered before International Conference on Aviation & Airport Infrastructure, Denver, CO, December.

Merrow, Edward W. 1988. *Understanding the Outcome of Megaprojects: A Quantitative Analysis of Very Large Civilian Projects. Santa Monica, CA: The Rand Corporation.*

Meyers, Dan. 1995. Denver airport's wing and a prayer. *Philadelphia Inquirer.* February 1.

Murphy, Kathleen J. 1983. *Macroproject Development in the Third World.* Boulder, CO: Westview Press.

New York Times. 1994. Denver aide tells of laxity in airport job. October 17.

O'Driscoll, Patrick. 1995a. Webb allegedly tried to bend DIA hiring rules. *Denver Post.* May 26.

———. 1995b. Mayoral cronyism at DIA alleged. *Denver Post.* May 26.

————. 1995c. Ex-airport official files bias complaint. *Denver Post.* June 10.

Paulson, Steven. 1994. Inspector says photos are proof of poor work at new Denver airport. *Philadelphia Inquirer.* October 17.

Peat Marwick Mitchell & Co. 1980. Assessment of General Financial Requirements. Prepared for Denver Regional Council of Governments.

Phillips, Don. 1995. Airport funding under investigation. *Washington Post.* March 13.

Purdy, Penelope. 1995. Webb won, but serious questions remain. *Denver Post.* June 11.

Rebchook, John. 1995. Investor sues Denver. *Rocky Mountain News.* February 28.

Reid, T. R. 1989. Denver votes to construct world's largest airport. *Washington Post.* May 17.

Rocky Mountain News. 1991. Chronology. March 17.

————. 1995a. DIA fees 2nd highest in U.S. January 29.

————. 1995b. A second term for Webb. June 7.

————. 1995c. Bill bans funds for DIA runway. July 1.

————. 1995d. DIA we have a problem. July 13.

————. 1995e. Sen. Brown seeks DIA moving-expense probe. August 6.

Russell, James. 1994. Is this any way to build an airport? *Architectural Record.* November 1994:36.

Sahagun, Louis. 1994. Denver airport ready for takeoff—and a bumpy ride. *Los Angeles Times.* February 1.

————. 1995. Denver airport woes add up to distressing start. *Los Angeles Times.* February 24.

Schwartzkopf, Emerson. 1986. Making it fly. *Rocky Mountain Business Journal.* September 1:1.

S. G. West & Associates. 1992. DIA Executive Assessment.

Sleeth, Peter. 1992. CenterPort's key financier dodges limelight. *Denver Post.* January 26.

Spencer, Gil. 1995. Mr. Peña and a pair of probes. *Denver Post.* March 26.

U.S. General Accounting Office (GAO). 1992. Letter by Ken Mead to Representatives W. Lehman, L. Coughlin, B. Carr, F. Wolf. September 14.

————. 1994. Denver International Airport: Baggage Handling, Contracts and Other Issues. August 9.

————. 1995. Denver International Airport 1. Testimony of Michael Gryszkowiec, May 11.

U.S. News & World Report. 1995a. A Taj Mahal in the Rockies. February 13:48.

————. 1995b. A building, a tragedy. February 13.

Veazey, Dick. 1995. Interview. October 18:53.

Wall Street Journal 1995a. Peña's folly. February 23.

————. 1995b. Federal programs that could be affected. June 13.

Weber, Brian. 1995a. Mayor candidates question minority contracts. *Rocky Mountain News.* March 8.

————. 1995b. Webb. *Rocky Mountain News.* April 16.

Westword. 1993. High fliers, low behavior. February 17.

Wilmsen, Steven. 1991. *Silverado.* Bethesda, MD: National Press Books, Inc.

World Airport Week. 1995. NSIA considers opening with extra runway. July 1.

10

Airport planning theory in perspective

Since master planning for airports is flawed at the core, is logically indefensible, and produces unsatisfactory results, it must be replaced—RICHARD DE NEUFVILLE[1]

In Chapter 1, we discussed the problems that megaprojects commonly encounter. Subsequent chapters amply demonstrated that airports are megaprojects par excellence. We suggested that part of the problem in regards to airport planning is the universal reliance on a theoretical model—the rational comprehensive model—that itself possesses many shortcomings. The critical question this chapter addresses, therefore, is whether alternative models exist that can produce better results.[2]

Models can possess both explanatory and prescriptive elements. On the one hand they seek to explain the ways in which decisions are made and on the other they offer guidance on how to make better decisions. To clearly assess the utility of the rational comprehensive and other models, it is necessary to distinguish between these elements. Certain models may possess considerable explanatory power concerning the ways in which decisions are made but are not designed for prescriptive purposes. Other models serve much better as blueprints but may not be effective in describing how planning and decision-making actually occur. In the sections that follow we shall assess the relevance of this group of models in both these senses—to analyze the extent to which they fit the DIA case and also to consider the extent to which they can offer future guidance to planners and policy-makers.

1. From de Neufville 1991.
2. This chapter draws heavily on Szyliowicz and Goetz (1995). An expanded discussion will appear in a forthcoming article by Goetz and Szyliowicz (1997).

Models of decision-making

At best, the rational comprehensive model (discussed in Chapter 1) represents *only* an ideal form of how planning and decision-making is supposed to occur. Scholars have, over the years, noted its deficiencies and attempted to develop alternative models that better describe the actual process and that could serve as guidelines for planners and decision-makers. One of the first was Herbert Simon (1955), who developed an *organizational approach* in which decisions are based on organizational procedures. Planners and decision-makers are "satisficers" rather than "optimizers." Plans are chosen and decisions are made on the basis of whether an alternative is judged to solve the problem satisfactorily, not necessarily whether it is the best possible solution conceivable. In the organizational perspective, much more attention is paid to the psychology of planning and decision-making, in the form of individual and group perceptions of risks, priorities, and consequences (March and Simon 1958).

Another criticism of the rational model was put forward by Charles Lindblom (1959) who posited that planning and decision-making occurred according to an *incremental approach*. Suggesting that the rational approach was "fact greedy" and consumed too much time and resources to be practicable, Lindblom argued that planners and decision-makers actually made only incremental, or marginal, changes from existing policies or solutions. Because of budgetary constraints, planners and decision-makers cannot consider all possible alternatives in the process and instead engage in making "successive limited comparisons" by a branch method based on previous related experiences.

Arguing that the criticisms of the rational model were valid but that the incremental approach possessed major drawbacks, Amitai Etzioni (1967) blended together elements of both to suggest a *mixed scanning approach* to planning and decision-making. He argued that incrementalism placed too great a reliance on precedent and that its approach was inherently conservative. The likelihood of new and bold initiatives emanating from an incremental approach was quite low. New problems for which there are no previous policy solutions were especially difficult to address using an incremental approach.

Accordingly, Etzioni proposed a two-stage process in which the entire field of potential issues and problems is broadly scanned to identify specific elements that require more consideration as part of a detailed second-stage analysis in which a rational approach may be more applicable. Etzioni used the analogy of weather satellites, which employ less-discriminating telescopic tools to cover the large-scale field and more detail-enhancing microscopic tools to focus on areas of detectable disturbance that require further scrutiny. Furthermore, these tools can be used in any order, according to circumstances and needs.

Mixed scanning can thus be regarded as an improvement over the rational model in that most issues would not require an in-depth analysis and would continue to be monitored periodically. Only the most pressing issues are selected for more exhaustive analyses. Mixed scanning is regarded as an improvement on incrementalism because that approach may miss obvious trouble spots in unfamiliar areas due to its excessive reliance on precedence and past experience (Etzioni 1967, 1986; Levy 1994).

In an effort to test the efficacy of various models, Graham Allison (1971) assessed the process of decision-making during the Cuban missile crisis. He applied the rational-actor, *organizational behavior* and *bureaucratic politics* models to this case. The organizational behavior model, which draws upon earlier work in organizational theory, suggests that decisions are made on the basis of established procedures within an organization. According to the bureaucratic-politics model, plans and decisions are the result of political bargaining involving both governmental and nongovernmental players. The personalities and abilities of individuals to persuade and cajole others regarding a certain course of action is explicitly recognized in this model. Often, previously determined plans undergo substantial modification in response to changing political circumstances.

More recently, Harold Linstone (1984, 1994) has attempted to provide planners with an approach that would avoid the shortcomings of existing models. He suggests that it is necessary to analyze any project through three principal perspectives: a technical perspective (essentially the rational model), an organizational perspective (includes organizational approaches and models, incrementalism, and the bureaucratic politics model), and a personal perspective (which addresses the realm of *cognitive* models developed by such scholars as Axelrod 1976; Jervis 1976; and Steinbruner 1974). The cognitive models on which Linstone's personal perspective is based attempt to explain planning and decision-making on the basis of the values, beliefs, and mind-sets of the key actors. Linstone also includes such variables as personality, charisma, intuition, self-interest, and leadership in this category.

In more recent years, growing concern with the risks posed by new technologies has led to the development of a rich body of literature that focuses on how decisions should be made so as to minimize the impacts of technological failures. Scholars have identified both the nature of the problem and suggested various strategies to help guide the planners. At the core of these strategies is a need to anticipate risks, to reduce uncertainty, and to maintain flexibility (Collingridge 1992; Lowrance 1976; Morone and Woodhouse 1986; Perrow 1984).

Although all these alternative approaches vary widely, they have been grouped into a general category called "strategic" (Morone and Woodhouse 1986) or "adaptive" (Szyliowicz and Goetz 1995) because they share two common features. First, they differ from the rational model in terms of the assumptions that they make in regard to political variables. The rational model has as its core the concept of a single, all-powerful actor (the state, organization, or leader), who makes decisions according to a set of specified rules and implements them without difficulty. The adaptive approach tends not to make such assumptions, although the role of power is treated differently by various scholars. Incrementalism, for example, suggests that the process of "mutual adjustment" between groups leads to better decisions but fails to give adequate recognition to power inequalities. In the case of the bureaucratic politics model, decisions are explicitly the outcome of a political process of bargaining and negotiations and the cognitive models focus on the ideological orientations, values, and beliefs of key actors.

Second, the "adaptive" models all agree on the advantages of maintaining flexibility, an important element when dealing with large projects, since these frequently cannot adapt to changing conditions and often turn out to be white elephants (Collingridge 1992; Collingridge and James 1991; Morris and Hough 1987). Flexibility, however, is an elusive concept, and it is necessary to specify more precisely its content. A recent study by Evans (1991) has attempted to differentiate analytically among the many concepts that are often grouped with flexibility or are considered dimensions thereof.

To begin, Evans builds on work by Stigler (1939) by differentiating between adaptability and flexibility. Adaptability represents a one-time change within an organization that permits it to function more effectively in new conditions. In other words, an organization either by luck or conscious decision develops those attributes that are required in order to be successful in a changed environment. Flexibility is a more dynamic concept. It represents the ability to make continuous adjustments in constantly changing conditions. In other words, it involves not only adaptation, but also re-adaptation. Flexibility thus becomes the broad category under which numerous related concepts are incorporated. Particularly relevant are robustness, hedging, resilience, and corrigibility.

Robustness is the degree to which an organization is prepared to function after having been subjected to unanticipated events. Resiliency, on the other hand, is the ability of an organization to continue to function after experiencing unexpected changes. Hedging is a defensive strategy that minimizes negative impacts from the environment by building in redundancy and backup systems. Corrigibility is the ability to learn from and adapt to new conditions. In short, hedging and corrigibility are strategies; robustness and resilience are qualities of an organization. Not all scholars agree, however, with the view of hedging just identified. Collingridge (1983) has argued that hedging leads to worse outcomes than "flexing." By hedging, however, he means a strategy that seeks to avoid the worst possible outcome, rather than anticipating unforeseen impacts; flexing, on the other hand, seems to correspond closely to "corrigibility," because it entails the ability to alter course as new information becomes available.

These concepts all have a common thread—they represent elements of flexibility that guard against potential negative consequences of changes in uncertain environments. They can, however, be differentiated on the basis of the time dimension involved, i.e., whether they are anticipatory or reactive in character. The former incorporates robustness and hedging, while the latter includes corrigibility and resilience (Evans 1991).

Related research suggests that organizations should function in a manner that minimizes risks and errors. The work of several scholars (Morone and Woodhouse 1986; La Porte and Consolini 1991), which has been labeled "high reliability organization theory" (Sagan 1993), is especially relevant. This literature emphasizes such organizational variables as committed political elites and organizational leaders, high levels of redundancy, an emphasis on training, decentralization, an appropriate organizational culture, and an emphasis on organizational learning (Sagan 1993). Through such measures an organization can achieve the qualities of resilience and robustness required to function in an uncertain environment.

Such concepts and approaches are especially relevant to cases like DIA because of their inherent inflexibility, which stems from both technical and social considerations, such as high capital costs, long lead times, centralization, technical orientation, and alignments of interested coalitions. Under these circumstances it becomes essential to consider whether it is possible to structure a planning process that enhances flexibility. We return to this critical point in our last section. First, however, we discuss the relevance of the alternative approaches to the planning of Denver International Airport.

The character of the Denver airport planning process

The previous chapters have highlighted the complexities involved in planning and implementing a large project. Many organizations played a prominent role—the FAA/DOT, Adams County, the City of Denver, DRCOG, the tenant airlines, and Congress, among others. Personalities and individuals also greatly influenced the process—Mayor Federico Peña's decision to reconsider existing plans, DOT Secretary Sam Skinner's steadfast support of the new airport, the willingness of Adams County commissioners to compromise, Governor Roy Romer's commitment to the project, Richard Fleming's (Greater Denver Chamber of Commerce) strong backing of the new airport, and Mayor Webb's many decisions.

Each of these actors not only enjoyed individual power positions, but it should be obvious that the strategic decision to build the airport was implemented because a powerful coalition was assembled. This coalition included the leading political figures in the city and the state, the leading business organizations, the trade unions, banks, developers, real estate firms, bond houses, and the like. Opposing this coalition was a group of concerned citizens with limited financial and human resources, some hotel companies, and the airlines, but these groups never coalesced into a single powerful coalition.

Furthermore, many specific events influenced the process. These include the election of Mayor Peña, the Memorandum of Understanding between Denver and Adams County, the two airport referenda in Adams County and in Denver, federal funding and the successful bond sales to finance the project, and the concessions that the airlines were able to obtain in return for agreeing to use DIA.

Clearly, this project, like any other, went through a series of stages. Although there is no universal agreement among scholars as to how stages should be delimited, or how many there should be, all schemes incorporate such elements as project identification, design, appraisal, and implementation. The actual relationship between a project and any such scheme, however, is more complex than the idealized schemes suggest.

Accordingly, on the basis of the actual historical events related to the new Denver airport, we can identify the following five stages as most relevant for analysis:

1. Problem conceptualization and definition (1974–1983)
2. Project selection, negotiation, and decision (1983–1985)
3. Project formulation and design (1985–1989)
4. Public approval and ratification (1988–1989)
5. Project implementation (1989–1995)

Although similar stages may be found in many manuals, including those designed for airport planning, it is appropriate to emphasize three points. First, these stages emerge from our analysis of the actual events. Second, we highlight the role of public participation throughout the process and accord a specific stage to public approval and ratification. The role of public participation has been largely ignored until quite recently and even now is often conceptualized quite narrowly. This is, as we shall see, the case for the FAA planning process.

Finally, we must note important distinctions between the actual planning process and the rational model. Despite the appearance of "rationality" that the model projects, the entire process was subject to a range of influences that are not accounted for by the type of activities called for by the rational model. An obvious example is Governor Romer's attempt to influence voters in Adams County to support the new Denver airport, an effort that proved crucial to a favorable election outcome, and the later baggage system decision with its disastrous consequences. This is not to suggest that no aspect of the process was carried out on the basis of rationality, but that it is impossible to characterize any specific stage according to only one of the decision-making models cited earlier. Rather, as we show below, various actors were involved in each stage, all of whom could, and actually did, conform to a different model.

In Stage I, Problem conceptualization and definition (1974–1983), the major actors were the City of Denver, DRCOG, the FAA, and Adams County. Since 1974 DRCOG and the FAA were addressing the future problem of limited capacity at Stapleton by conducting a planning process according to the rational model. Denver, however, was behaving incrementally, identifying ways of increasing capacity at its existing facility. By 1979 Denver was preparing to expand Stapleton onto Rocky Mountain Arsenal (RMA) land, a move strongly opposed by Adams County. To understand Adams County's actions, it is necessary to remember the legacy of relations between the two communities, as well as its own plans for the RMA. It could thus best be described, at this stage, as an external force in opposition to any change in the status quo.

From an overall perspective, the decision process underwent a marked transformation during this first stage. At first, the process seemed to follow the rational model, but as the issue gained in saliency, the City of Denver emerged, quite naturally, as the leading actor while Adams County moved into opposition. At this point the process was conforming increasingly to the bureaucratic politics model, wherein questions of power, personalities, and leadership became paramount.

Stage II, Selection, negotiation, and decision (1983–1985), began with the election of Federico Peña in May 1983. It is not at all evident that if the incumbent mayor, Bill McNichols, had been reelected, the decision to build a new airport would have been made. Indeed, the new administration reviewed the issue and decided that a new airport was preferable to expansion onto the Rocky Mountain Arsenal. The process by which this decision was reached may have been "rational," but the only change that had occurred since the earlier decision was new political leadership in Denver. Hence, one must emphasize the willingness of the new mayor to seek an accommodation with Adams County over the expansion issue and to address the concerns of neighborhoods that opposed expansion due to noise concerns..

To implement such a decision, however, necessitated building a powerful coalition and establishing cooperation between two hitherto warring entities, Denver and Adams County. The Memorandum of Understanding (MOU) signed in January 1985 symbolized the new entente. Accordingly, this second stage is characterized by extensive negotiations and bargaining between Denver and Adams County and a powerful political effort to obtain voter support—activities that are not recognized by the orthodox rational model.

Stage III, Plan formulation and design (1985–1989), witnessed the selection of the exact site and the preparation of specific plans for each aspect of the new airport. These included the terminal, the concourses, the baggage-handling systems, the runway layout, the air traffic control system, the air cargo facilities, the environmental impacts, and the financing of the project. Each of these received detailed attention and involved numerous consultants, architects, designers, and other technical experts.

The key actors were the city, which was organizing the project; the FAA, whose specific rules, regulations, and guidelines influenced and often determined the nature of the plans; and the airlines, which assumed a more critical role because of changes in the operating environment. As a result of deregulation starting in 1978, the airlines were granted the freedom to determine their city service patterns and, as a result, gained the power to determine which airports were to thrive and which were to stagnate. Airlines have traditionally been key players in airport planning, especially regarding design and marketing elements, but the fundamental decision to build a new airport has always rested with the local government and national governmental entities. This remains true today, but local governments now have to negotiate with airlines, which have the option of abandoning cities should the costs be deemed too onerous. This fact gives the airlines enormous power and the ability to extract major concessions from any community that wishes to alter its airport facilities. Airport planning has been transformed into a more complex bargaining game in which the airlines can and do use their increased leverage to shape outcomes. The new Denver airport, the first to be built in this post-deregulation environment, marks the first time in U.S. history that the airlines played such a pivotal role.

The City of Denver found itself attempting to proceed on the basis of the rational approach, carefully formulating the plan and designing every aspect of the airport according to accepted practice. However, underlying this planning was a major (and unrealistic) assumption—that the airlines (passenger and cargo) would eventually sign on and approve the planned design.

Next, let's turn to Stage IV, Public approval (1988–1989). While it has become commonplace to talk of public participation in large technological projects, the pattern of such participation is often merely formal and symbolic, serving to ratify decisions made earlier. In this case, however, the public was explicitly involved in the fundamental decisions through two separate referenda. But its role was circumscribed by the nature of the political environment, since, as noted earlier, the "pro" forces comprised a powerful coalition with access to extensive financial and political resources (including the strong support of the media).

The concept of coalitions involved in such activities are not included in the classic rational model. In the FAA planning approach, for example, public participation is limited both in form and function. Public hearings appear as the only channel, and these are integrated into the planning process at only two points:

1. After the plans have been developed, in the case of expansion.
2. Prior to site approval, when a new airport is planned (see Fig. 1-4 in Chapter 1).

Public participation, however, can take many forms and occur at different stages. The standard airport planning approach does not seem to recognize this point. The public's role appears to be limited to what skeptics might call legitimization; the public is apparently not expected to participate in making as fundamental a decision as whether a new airport is needed. As we noted, this was not the case for the planning of the new Denver airport, although the process did essentially legitimate a decision that had been made by the political and business elites.

Although the outcome was favorable to their interests, the policymakers did not anticipate such extensive public participation. The first referendum, the Adams County vote to allow Denver to annex the land on which to build the facility, was mandated by an existing state amendment, without which it is not at all clear that this election would have been held. The second referendum, the Denver vote to ratify the project, was held because the Peña administration feared that in the absence of an affirmative vote, opposition to the project would continue to grow and undermine its legitimacy. Thus, although the policy-makers preferred the traditional planning approach, the political structure and the political interests of the Peña administration mitigated against it. Accordingly, this stage highlights an important point—the ways in which the structure influences the decision process generally and the pattern of public participation specifically.

In Stage V, Implementation (1989–1995), Denver had been actively designing the airport on the basis of assumptions about traffic demand and the airlines. These assumptions proved to be only partly right. Since passenger traffic had been declining from 1986 to 1990, the airport had to be downsized significantly. This was a relatively simple matter since the original design provided for such exigencies, although the change raised questions about the project's financial viability.

More serious were the consequences of the assumptions that had been made about the airlines. Although both Continental and United eventually came to terms, Denver decided to start construction before the airlines had approved the plan. The airlines later demanded, and were granted, numerous large and small changes in many aspects of the design. Hence, what was assumed to be a relatively rational process changed dramatically as modifications had to be accommodated within an unrealistic schedule, with, as was demonstrated in Chapter 9, disastrous consequences for the budget, quality control, employee morale, and project management (Russell 1994).

Perhaps the most serious assumption involved the totally automated baggage handling system demanded by United, which forced additional modifications and significantly delayed the airport's opening. The planners apparently never considered the problems involved in greatly increasing the scale of a complex technology, especially one so dependent on the development of a large-scale software system of a type in which failures are commonplace (Gibbs 1994).

Thus this stage also turned out to be permeated by political considerations. Developments in this stage were greatly influenced by the increased power of the airlines and their ability to leverage the city into accepting their demands. The project managers found themselves in a situation of implementing changes, the consequences of which, for whatever reason, were not foreseen, while simultaneously pressing to maintain a schedule that had been drawn up prior to the myriad modifications.

Process and models

To what degree does the decision to build the new airport approximate the rational comprehensive model? The project sponsors point out that it was a rational process because several alternatives had been identified, their merits and demerits weighed, and the best one selected. On the other hand, skeptics have pointed to the fact that, although the key actors responsible for the design and implementation of the project—the city and the FAA—may have thought that they were following the rational model, the other key actors, especially the airlines, pursued their own agendas, which were not consistent with those of the city and the FAA. The airlines played a role that greatly exceeded that suggested by the FAA's master planning process (which merely calls for "coordination with users"). Similarly, the pattern of public participation also did not conform to the model. The multiplicity of actors and their differing strategies and interests ensured that the decision would actually be determined primarily by political and cognitive factors.

What is particularly striking is the critical role of power. The key turning points in the project—the Memorandum of Understanding with Adams County (Stage II), the ratification by the voters in the two referenda (Stage IV), DOT's ultimate endorsement in the form of the $500 million commitment (Stage IV), and the Continental and United agreements (Stage V)—can only be explained by the nature of the political process. In each case, bargaining and negotiations between organizations and groups were involved, and the outcome was influenced by power distributions and the perceptions and belief systems of key actors. These conclusions corroborate the importance of political factors and further validate the criticisms that have been levied against the rational actor model.

On the other hand, some of the models that we grouped under the "adaptive" category seem to possess more explanatory power. Linstone's model does apply in explaining the process, since it explicitly recognizes the role of organizations and individual actors in an interactive system. The bureaucratic politics model provides further insights because it highlights the critical role of political factors. The cognitive models are also relevant, because the perceptions of the key actors, especially those of Mayor Federico Peña, played a key role.

This permits us to draw an important conclusion. The degree to which the rational model is approximated depends on the power of the agency that designs and implements the project and the orientations of its elite toward such an approach. In the case of DIA, the FAA's jurisdiction was limited; most key decisions were made by the city interacting with the airlines, and each had its own point of view or interests that it wished to maximize. Thus, at the same time that the FAA imposed its "rationality" upon the process, the City of Denver and the airlines were also making what they perceived to be rational planning efforts. Sometimes these coincided, but more often they conflicted. However, none of these actors was in a position to control the process. Hence, the City of Denver attempted to initiate a rational process but proved unable to sustain it because it could not control the process and had to engage in a complex process of bargaining and negotiation.

The implications of this development deserve to be emphasized. They can be generalized as follows: The power of public agencies to make decisions unilaterally has decreased, thus making the rational-actor model less relevant than ever because the airlines are now in a position to extract significant concessions that skew the process. This obviously occurred in the DIA case, in which Continental successfully forced a fundamental change in the design, and United's desire for an automated baggage system and other concessions led to significant cost overruns and delays.

Not only does the rational comprehensive model fail to explain how the strategic decision was made, the attempt to apply it led neither to a trouble-free design nor effective implementation. One must stress, however, that many specific design elements were treated according to standard engineering rationality and that many of the problems that arose were not due to the efforts of the technical experts to behave in a rational manner but to the inevitable intrusion of political and other factors. Planners cannot (unless they are working for an all-powerful ruler) impose their version of rationality upon the process. At the explanatory level, therefore, it is obvious that the rational comprehensive model has relevance only to the specific technical aspects of the project that technical experts control.

Toward flexibility

The foregoing conclusions have important implications for megaproject planning and design. The rational approach clearly had limited power to explain both the decision and the implementation process in the case of DIA. It is at best an ideal that realistically cannot be achieved because of the two fundamental reasons identified above. First, decision processes are likely to involve multiple actors, all pursuing their own agendas. Second, given the nature of human affairs, unexpected events will inevitably arise, creating unanticipated difficulties. For both these reasons, we suggest that the planning of airports and all large transportation projects be approached from a paradigm that recognizes the great uncertainties that are inherent in such endeavors.

This is not to suggest that decision-making should completely ignore the rational model. It retains its importance as an ideal toward which it is necessary to strive. Its strength lies in its ability to provide a formal framework within which analyses (based on its premises) can be carried out. In this chapter, for example, we utilized it to good effect in describing how the process unfolded. But it is no more than a framework, and the process which takes place is inevitably influenced by political factors that are driven by a different rationality.

A more flexible approach to airport planning must begin with a recognition that the very nature of these projects involves increased risk because of their inflexibility. Essentially, the larger the scale and complexity of the project, the greater the risks. Analytically, however, one must distinguish between risk and uncertainty. The former involves "known" uncertainties, those which can be anticipated and evaluated; the latter, those which are totally unexpected.

The airport illustrates this point well because both risks and uncertainties were involved. Being a large public development project that depended on financial outcomes involving entrenched political mind-sets, difficulties could be anticipated. These could be dealt with only if the political elite created a management structure that was sensitive to the inherent problems and empowered to address them. This should be an integral part of any project. Potential risks posed by uncertain forecasts and specific technologies should be identified, analyzed, and measures developed to minimize their impact. In addition, one must be prepared to deal with the totally unexpected.

The most obvious problem in airport planning specifically, and transportation planning generally, is created by the issue of forecasting. Projections of future demand are notorious for the large margins of error commonly involved. This was demonstrated once again by the forecasts of passenger demand by the FAA and the City of Denver's consultants. They proved to be far too optimistic and had to be downsized significantly over time. The reasons for this phenomenon are many and complex. First, one cannot ignore the cognitive, political, and ethical dimensions involved in the client-consultant relationship. Perhaps more important, in this case, however, are the inherent difficulties in forecasting passenger enplanements in the highly volatile aviation industry.

Although any forecast can prove erroneous and estimates of passenger demand have traditionally been subject to significant errors, the new environment has greatly exacerbated this tendency. The critical nature of aviation forecasts for airport planning and design and of demand forecasts for any type of project requires that ways of generating more credible forecasts be developed and used. Furthermore, the planning process must be structured in a manner that permits change in the face of significant errors. Since forecasting is likely to remain more of an art than a science, the need for such a planning process will not disappear.

A second major factor that always creates difficulties involves technology; the newer and more untested the technology, the more likely one is to encounter problems. This was obviously true of the baggage system, which was similar to a technology that had previously required considerable debugging in an earlier application at Frankfurt Airport. It was now expected to operate flawlessly at increased levels of scale and complexity and to do so without allowing more time in the construction schedule for unforeseen difficulties that could have been readily anticipated. Driven by the need to open the airport quickly to defray expenses, the decision-makers created a disastrous situation.

Since the baggage system is a critical component of any airport, the decision not to hedge represented a remarkable gamble. If a reasonable time frame (an extension of two to three years has been mentioned) could not be accommodated, then a backup system should have been built. Ironically, that is the situation today, albeit at much higher cost. If the planners had been more attuned to the very high risk associated with the complex technology and massive scale of the proposed automated baggage system, the airport could have opened on time at considerable financial and other savings.

The problem of anticipating risk goes far beyond cutting-edge technologies. Aside from the baggage system, the new Denver airport was not a project that depended on high-risk technology (such as, for example, a nuclear power plant). Most of the technologies were proven; the biggest risks were financial and political. Hence, one should have expected the traditional approach to large project planning to have produced a successful outcome. That it did not produce good results even under what were reasonably favorable circumstances underlines the importance of developing new planning orientations that can minimize risk.

The planners of DIA actually attempted to do so, to some degree. The physical design of the airport facilitated future expansion and, as proved necessary, contraction. Furthermore, contingency funds were built into the budget in anticipation of possible revenue shortfalls, but these were overwhelmed by the scale of the problems encountered. Interest on the bonds amounting to $1 million a day had to be paid for 16 months while no revenues were generated.

Despite these modest instances of attempts to achieve flexibility, the overall approach remained relatively rigid. The planners and decisionmakers really did not want to consider, and thus never provided for, the possibility that the project could be delayed or even canceled. The great contribution of a truly flexible approach would be to permit adaptation to changing conditions. In short, when discrepant information begins to accumulate that challenges the assumptions on which the original project was based, the project should be re-evaluated and new decisions reached about its critical elements.

Preparing for the unexpected might involve a series of "go/no go" checkpoints, whereby at specific decision stages the situation would be evaluated anew and decisions made on the basis of existing conditions and new information. This strategy is inherent in much of the literature that advocates flexibility and incrementalism (Collingridge 1992; Weiss and Woodhouse 1992). Richard de Neufville has long advocated such an approach to airport planning specifically, and there is some evidence that it has been successfully applied, most notably in Sydney, Australia, where land for a new airport site was acquired and "banked" (de Neufville 1991).

The appeal of such an approach is obvious. There is much to be gained from creating a more modular planning, design, and construction format that allows for a series of checkpoints at key stages in the process. If such a structure could be laid out by the FAA, then all airport planning would be subject to a process in which constant feedback from a highly uncertain external environment could alter decisions about how to proceed. It would have to be mandated via the FAA or a similar federal level agency to standardize decision-making, thereby relieving pressure on local politicians or financial bond brokers and holders.

Despite its theoretical appeal, however, this option encounters serious practical obstacles that render it extremely difficult to implement, especially now given the current trend of government decentralization and privatization. In an environment in which more responsibility is being given to local government and local entrepreneurs, the most that the FAA would probably be able to do is to suggest such a course of action, but without some relatively independent group overseeing the feedback, the process would once again be controlled by the same group that has a vested interest in seeing it go forth unimpeded.

This difficulty highlights the ways in which decision-making structures shape the content of decisions. Large projects are usually initiated because a powerful coalition has formed to support it, and few policy-makers would be willing to run the risk of alienating some or all of its members. For their part, politicians would be unlikely to change their decisions on projects of this magnitude because they would fear that voters would view their original decision as incorrect and perceive them as indecisive "straddlers." Nor can one ignore the problem of regaining the initial momentum if the project were delayed. It is no coincidence, after all, that many projects are deliberately put on a fast track and front-loaded. Finally, substantial negative fallout would ensue if the project was halted and financing arrangements and contracts were breached.

In this case, the new Denver airport involved a policy decision that could not be reversed easily even if early feedback mechanisms that provided new information were in place. It was an extremely large undertaking that had been made on an all or nothing (build or no-build) basis, i.e., a lumpy decision. Once the strategic decision to build was made, the project's leaders never seriously contemplated the possibility that it should be slowed or stopped, regardless of external feedback; thus, retaining flexibility was virtually impossible. The project also developed its own momentum; it was locked in financially (because of the bond sales and the necessity to pay them off through the new airport) and politically (because of the agreement with Adams County, the two referenda, and the clear decisions made by Peña, Romer, Webb, and others whose political fortunes became tied to following through on the airport). In other words, even though conditions changed and passenger traffic decreased, forces were at work to ensure that the new facility would not be halted.

Despite such powerful obstacles, however, an approach that permits change when new conditions arise retains its validity and must be given thoughtful consideration if disastrous consequences are to be avoided. When decisions with tremendous implications for a community are made in a highly uncertain environment, adherence to an inflexible plan is not advisable. The challenge that decision-makers and planners face is how to structure such projects in a manner that protects them from political damage and permits adaptation in response to changed circumstances. This challenge requires the creation of new structures that are sensitive to the inherent problems of large projects and are empowered to address them. Rigorous risk analyses should be undertaken whereby potential risks can be identified and analyzed and measures developed to minimize negative impacts. It also requires the emergence of new values among the public and decision-making elites.

Culture change of this type is a slow process. In the meantime, airports are still being designed and built. Accordingly, one should focus on possible structural arrangements that can enhance the chances that the basic decision is a sound one and that its implementation can be carried out in a way that minimizes the negative impacts of the unexpected.

Public participation (discussed in detail in Chapter 3) is one key element. Airports and other megaprojects have major consequences for communities, and though public participation is now often required by statute (such as in the case of environmental impact statements), participation is often not structured in a manner that permits the public to make a genuine contribution to the basic decisions involved. In the case of DIA, for example, although two elections were held, many observers believe that the pro-airport coalition's resources (including strong media support) did not permit a genuine debate, and, indeed, considerable pressure was brought to bear upon its opponents.

Achieving meaningful public participation is no easy matter and entails various costs, but these costs are outweighed by numerous benefits—if implemented wisely. It legitimizes the decision, permits all issues to be debated, and facilitates the generation and evaluation of different solutions. Furthermore, if the need for flexible implementation is accepted, then the political leadership would find it easier to modify the project to meet unexpected developments.

Enhanced public participation does not mean that the implementation of the project should be highly politicized. Indeed, it is appropriate to consider the utility of "buffer" organizations that will shield the project from politics. Semi-autonomous bodies that are appointed or elected, such as the Port Authority of New York, or other regional organizations may well provide models that are worthy of consideration. Whatever the structure, it is obviously desirable that the actual implementation be left in the hands of professionals—turnkey projects have obvious appeal—and that the municipality not yield to the temptation to micromanage the project.

Even with "meaningful" public participation, however, the strategic decision to design a transportation project may still be based on questionable assumptions. This problem extends to all public policy decisions, and scholars have identified various mechanisms to identify such sources of future problems. One of the most promising involves the "devil's advocate"—one or more individuals who are explicitly assigned the task of opposing the policy. Such an approach is certainly not foolproof—the opponent may play a merely symbolic role or be marginalized (Jervis 1976). The fundamental idea, however, deserves more detailed attention. Not only might it contribute to better decisions within the planning organization, but publication of the analyses would serve an important public-education function that would enhance the possibility of meaningful participation by the community and other potential users.

Another useful technique that might involve the community in a way that promotes consensus on difficult choices has been suggested by Lewis (1995). He argues that since forecasts are based on assumptions made by planners in a closed environment, genuine public involvement requires that these assumptions be examined by the community and other actors within a risk-analysis framework that assigns probabilities to various options. This approach is consonant with our general philosophy, because it recognizes the inherent fallibility of most forecasts and the need to develop a consensus within the community concerning what are usually highly controversial and politicized projects. Whether this goal can in fact be achieved depends on many sociopolitical factors that may well mitigate against such an outcome. Nevertheless, even if no agreement emerges—and it may well be no easy matter to agree upon the probabilities—the application of such a procedure, which involves meetings with facilitators, permitting everyone's views to be carefully and fully considered, should minimize the danger that a group would remain highly hostile and polarize the community over the issue.

Such an orientation clearly fits the second category of planning and decision models. But we do not advocate that the analytic (rational) approach to transportation planning be completely abandoned; rather it must be improved and made more realistic in terms of how planning actually occurs. The rational model can be modified so as to yield improved outcomes in the following ways:

1. Incorporate a formal recognition of the enlarged roles of the public and other users of transportation projects
2. Make the rational process more flexible along the lines outlined above.

It is important to recognize that in the case of large projects, a focus on optimal efficiency represents false economies; there are considerable advantages to building in redundancy, increasing margins, laying aside reserves, and developing mechanisms that ensure prompt and precise feedback and the development of an organizational culture that emphasizes learning and adaptation.

Conclusions and recommendations

Criticisms of the rational planning model, especially as it pertains to airports, can be divided into two categories: internal and external to the model. Accordingly, our recommendations are similarly grouped, and we start by offering "internal" recommendations that seek to improve the process by working within the traditional rational theoretical framework.

First, the changed policy environment affecting the airline industry requires that all the stake holders become more serious participants in the process. Deregulation has fundamentally changed the roles of the airlines, local governments or airport authorities, and the FAA. This is particularly true of the more pivotal, and perhaps more powerful, role that airlines now assume. Furthermore, deregulation has greatly increased the level of uncertainty in aviation forecasting. Second, since the public is also a major stake holder, it too must be incorporated into planning and decision-making in a more meaningful way. Third, there needs to be more standardized responsibility for airport planning at the local level. The current institutional pattern of airport ownership and control is a mixture of municipalities, regional authorities, or states, with the FAA overseeing the system.

This situation has inherent weaknesses. The DIA experience strongly suggests that ownership and control of large regional airports should be moved from individual municipalities to regional authorities or state governments. Airports of this scale serve a much larger relevant community who deserve to have a voice in planning and decision-making. Decisions and mistakes made by the City of Denver are affecting the entire region because it is the local origin and destination (O&D) passengers who must ultimately bear the costs of the new facility. How this can be accomplished, given the numerous benefits that cities like Denver enjoy from their airports, is difficult to specify. It may be possible, however, to develop arrangements that would compensate the city for relinquishing control.

The above recommendations, however, do not deal with fundamental criticisms that have been leveled at the rational model. They are, in a sense, only Band-Aids. Accordingly, it is necessary to identify recommendations that deal with the criticisms that are external to the rational model. Thus we argue that continued reliance purely on the rational approach is futile. New models based on the more flexible adaptive approaches are more appropriate to the complexities inherent in contemporary transportation planning. A structural and decisional framework that emphasizes flexibility is necessary.

Structural elements of flexibility refer to the development of organizations that are robust and resilient. Although this is an important goal, we recognize that there are limits to which organizations can become so flexible that they are able to adjust to any condition or become practically error-free. Hence, efforts at improving organizational structures and management procedures, preparing in advance to deal with risks and uncertainties, are a necessary but not a sufficient condition; they must be supplemented by a decisional approach that is oriented toward the need for rapid adaptation.

Above all, the decision process must be one that emphasizes the other two elements of flexibility—hedging and corrigibility. And, once the decision has been made, it should be constantly monitored and an active and ongoing search for new information undertaken. It is also important to note that in the case of large projects, a focus on optimal efficiency represents false economies; there are considerable advantages to building in redundancy, increasing margins, developing mechanisms that ensure prompt and precise feedback, and creating an organizational culture that emphasizes learning and adaptation.

To be flexible does not mean total abandonment of the rational model. In fact, some aspects of hedging and corrigibility (although not robustness and resiliency) can be incorporated into a rational framework. But in their totality and in the orientation that underlies these elements, the adaptive approach specified represents a significant departure from the fundamental rigidity of the rational model.

There are encouraging signs that planners in the FAA and elsewhere are cognizant of the new environment. New approaches, such as the ones identified here, are gaining increasing acceptance. The FAA has a vital role to play in two regards. First, it is the only national organization that can promote new approaches, given the nature of U.S. airport planning. Second, the FAA itself is functioning in a rapidly changing environment. Accordingly, the degree to which the FAA is able to operate in a manner consonant with the requirements of an adaptive approach deserves detailed attention.

In the case of airports, this requires that the FAA develop and mandate a process that is both more realistic and produces better outcomes. Unfortunately, the FAA's airport planning model only provides an idealized approach that cannot produce optimal results. To be effective, the planning process must also be realistic in terms of how projects actually proceed. Specifically, as we have emphasized, any model must recognize the reality of political variables, notably the new roles of the public and the airlines.

Also relevant is the responsibility of the FAA to the local communities and other stake holders, because the present situation possesses an inherent conflict. On the one hand, the FAA is expected to control the national airspace system, but it has limited influence over the airport infrastructure. Current political trends toward decentralization and privatization are likely to further aggravate the problem. Yet, the FAA and DOT could improve the present situation by developing general policy standards in the following areas:

1. New airports should be created by regional authorities so that neighboring communities not be politically disenfranchised.
2. More stringent oversight with respect to the ways in which federal funds are disbursed.
3. Intermodal access, especially regional commuter rail systems.

The FAA, as an organization, needs to demonstrate that it possesses the quality of adaptability. Earlier, we identified adaptability as one of the key elements that permits an organization to deal with a changing environment. Such a characteristic is clearly needed, given the changed environment in airport planning. In the more technical terms used earlier, the FAA, as an organization, should concern itself explicitly with the degree to which it has developed an effective learning curve, adaptability, and resilience.

However, planners tend not to be principal decision-makers. Thus, if projects are to be designed and implemented successfully, the political elite must possess an appropriate set of values—a commitment to analytical approaches coupled with a sensitivity to the need for flexibility, a commitment to genuine public participation, and a willingness not to impose narrow political objectives on the project's design and implementation.

Clearly, the planning of airports and other large projects has entered a new era. The serious problems of infrastructure that plague the United States and many other countries cannot be resolved effectively or efficiently without new approaches that incorporate the principles outlined here. The first step in addressing these challenges is to recognize the shortcomings of existing practices and to develop structures and tools that will enhance the ability of practitioners and decision-makers to produce better outcomes.

Lessons learned

- *Efficiency should not be defined in narrow economic terms.* As DIA's experience with its baggage system demonstrates, redundancy in the form of backup systems may enhance efficiency.
- *Develop mechanisms to promote community consensus on critical choices.* Since all choices involve costs and benefits, it is essential to make decisions in a manner that does not lead to parts of the community enjoying most of the benefits and another most of the costs.
- *Develop arrangements to force careful public discussions of the project's assumptions.* Assumptions drive decisions, but they are often proven erroneous and are seldom debated fully and openly.
- *Develop a new approach to airport planning that recognizes the importance of maintaining flexibility throughout the process.* Such an approach should be mandated by the FAA.
- *Promote the development of organizations that are prepared and equipped to deal with risks and uncertainties.* The organizations that are responsible for planning and implementing the project should be as robust and resilient as possible, prepared to monitor progress, and constantly seeking new information.

References

Allison, G. T. 1971. *Essence of Decision: Explaining the Cuban Missile Crisis.* Boston: Little, Brown, and Company.

Ashford, N. and P. Wright. 1992. *Airport Engineering,* 3d ed. New York: John Wiley & Sons.

Axelrod, R. 1976. *Structure of Decision.* Princeton, NJ: Princeton University Press.

Booz-Allen and Hamilton, Inc. 1986. The Regional Economic Impact of Stapleton International Airport and Future Airport Development. Prepared for the City and County of Denver. Denver, Colorado.

Brecher, M. 1975. *Decisions in Israel's Foreign Policy.* New Haven: Yale University Press.

Browne, Bortz, and Coddington, Inc. 1986. A Land Use Analysis of the Environs of Stapleton International Airport. Prepared for the City and County of Denver, Denver, Colorado.

Collingridge, D. 1983. Hedging and flexing: Two ways of choosing under ignorance. *Technological Forecasting and Social Change.* 23:161–172.

__________. 1992. *The Management of Scale: Big Organizations, Big Decisions, Big Mistakes.* London: Routledge.

Collingridge, D. and P. James. 1991. Inflexible energy policies in a rapidly changing market. *Long Range Planning.* 24:101–108.

Colorado National Banks. 1989. Ready for Takeoff: The Business Impact of Three Recent Airport Developments in the U.S. Prepared by Coley/Forrest, Inc., Denver, Colorado.

de Neufville, R. 1976. *Airport Systems Planning.* Cambridge, MA: MIT Press.

__________. 1991. Strategic planning for airport capacity: An appreciation of Australia's process for Sydney. *Australian Planner.* 24:174–180.

__________. 1994. The baggage system at Denver: Prospects and lessons. *Journal of Air Transport Management.* 1:229.

de Neufville, R. and J. Barber. 1991. Deregulation induced volatility in airport traffic. *Transportation Planning and Technology.* 16:117–128.

Etzioni, A. 1967. Mixed-scanning: A third approach to decision-making. *Public Administration Review.* 27:385–392.

__________. 1986. Mixed-Scanning Revisited. *Public Administration Review.* 46:8–14.

Evans, J. S. 1991. Strategic flexibility for high technology manoeuvres: A conceptual framework. *Journal of Management Studies.* 28:69–89.

Federal Aviation Administration (FAA). 1985. Airport master plans. Advisory Circular AC 150/5070-6A.

Gibbs, W. W. 1994. Software's chronic crisis. *Scientific American.* September.

Goetz, A. R. and Szyliowicz, J. S. 1997. Revisiting transportation planning and decision-making theory: The case of Denver International Airport. *Transportation Research*, pt. A.

Hall, P. 1982. *Great Planning Disasters.* Berkeley, CA: University of California Press.

Hamer, A. M. 1976. *The Selling of Rail Rapid Transit: A Critical Look at Urban Transportation Planning.* Lexington, MA: D.C. Heath.

Hayes, M. T. 1992. *Incrementalism and Public Policy.* New York: Longman.

Hirschman, A. O. 1967. *Development Projects Observed.* Washington, DC: Brookings Institution.

Jervis, R. 1976. *Perception and Misperception in International Politics.* Princeton, NJ: Princeton University Press.

Johnston, R. A., D. Sperling, M. A. DeLuchi, and S. Tracy. 1988. Politics and technical uncertainty in transportation investment analysis. *Transportation Research.* 21A:459–475.

La Porte, T. R. and P. M. Consolini. 1991. Working in practice but not in theory: Theoretical challenges of "high reliability organizations." *Journal of Public Administration Research and Theory.* 1:19–47.

Leib, J. 1995. Another runway on itinerary? *Denver Post.* November 6.

Levy, J. M. 1994. *Contemporary Urban Planning,* 3d ed. Englewood Cliffs, NJ: Prentice-Hall.

Lewis, D. 1995. The future of forecasting: Risk analysis as a philosophy of transportation planning. *TR News.* 177(March-April):3–9.

Lindblom, C. E. 1959. The science of muddling through. *Public Administration Review.* 19:79–88.

Linstone, H. A. 1984. *Multiple Perspectives for Decision-Making: Bridging the Gap Between Analysis and Action.* New York: North-Holland.

__________. 1994. *The Challenge of the 21st Century: Managing Technology and Ourselves in a Shrinking World.* With I. I. Mitroff. Albany, NY: State University of New York Press.

Lowrance, W. W. 1976. *Of Acceptable Risk: Science and the Determination of Safety.* Los Altos, CA: W. Kaufmann.

March, J. G. and H. A. Simon. 1958. *Organizations.* New York: John Wiley and Sons.

Morone, J. G. and E. J. Woodhouse. 1986. *Averting Catastrophe: Strategies for Regulating Risky Technologies.* Berkeley: University of California Press.

Morris, P. and G. Hough. 1987. *The Anatomy of Major Projects.* New York: John Wiley and Sons.

Perrow, C. 1984. *Normal Accidents: Living with High-Risk Technologies.* New York: Basic Books.

Rifkin, G. 1994. What really happened at Denver's airport. *Forbes.* April 29.

Rondinelli, D., ed. 1977. *Planning Development Projects.* Stroudsburg, PA: Dowden, Hutchinson and Ross.

Russell, James S. 1994. Is this any way to build an airport? *Architectural Record.* November.

Rycroft, R. W. and J. S. Szyliowicz. 1980. The technological dimension of decision-making: The case of the Aswan High Dam. *World Politics.* 32:36–61.

Sagan, S. D. 1993. *The Limits of Safety: Organizations, Accidents, and Nuclear Weapons.* Princeton, NJ: Princeton University Press.

Simon, H. 1955. *Administrative Behavior.* New York: MacMillan & Co.

Steinbruner, J. D. 1974. *Cybernetic Theory of Decision.* Princeton, NJ: Princeton University Press.

Stigler, G. J. 1939. Production and distribution in the short run. *Journal of Political Economy*. 47:305–327.

Sussman, G. and F. Gray. 1986. Implications of the construction of major new airport facilities for economic development in the metro Denver region. Prepared for the City and County of Denver. Denver, Colorado.

Szyliowicz, J. S. and A. R. Goetz. 1995. Getting realistic about megaproject planning: The case of the new Denver International Airport. *Policy Sciences*. 28:347–367.

U.S. General Accounting Office (GAO). 1994. New Denver airport: Impact of the delayed baggage system. Washington, DC: GAO.

Wachs, M. 1985. Planning, organizations and decision-making: A research agenda. *Transportation Research*. 19A:521–531.

Webber, Melvin M. 1976. The BART experience—what have we learned? *The Public Interest*. 45:76–108.

Weiss, A. and E. Woodhouse. 1992. Reframing incrementalism: A constructive response to the critics. *Policy Sciences*. 25:255–273.

11

What have we learned?

Mike Keefe, *Denver Post.* Reprinted by permission.

> *"[W]hen our children's offspring look back into history, 40 or 50 years from now, they will see an airport whose design, cost and purpose will seem ridiculously obvious."*—CHUCK GREEN, EDITOR, DENVER POST[1]

In this concluding chapter, we address the three fundamental issues raised by this project:

1. What were the costs and benefits of DIA?
2. How was Denver able to plan, design, finance, and build a major airport when so many other cities had failed?
3. What lessons emerge?

In doing so, we necessarily summarize, and in some cases develop, points that were made in previous chapters.

1. Green (1995).

Costs and benefits of DIA

Ultimately, the success or failure of DIA must be measured by what was promised versus what was delivered. Many endorsed DIA on the basis of its promise, expressed in the 1989 city referendum, "to create more jobs, stimulate the local economy, and meet future air transport needs . . . using no city taxes." The jury is still out as to whether those objectives will be fulfilled, but it is possible to draw some preliminary conclusions.

Why did DIA cost so much?

The most impressive failures surround DIA's embarrassing string of delayed openings and its massive cost overruns. DIA suffered the following delays of its scheduled opening dates:

October 29, 1993

December 19, 1993

March 9, 1994

May 15, 1994

February 28, 1995

DIA suffered the following revisions in projected costs:

1988	$1.5 billion
1989	$1.7 billion
1991	$2.7 billion
1992	$3.8 billion
1993	$4.4 billion
1995	$4.8 billion
1995	$5.3 billion

The authors have concluded that the airport cost considerably more than it should have and that any future airport project should attempt to constrain costs better than Denver did in building DIA. The primary causes of the delays and cost overruns at DIA were the following:

- Mayor Peña's decision to break the airport into a large number of small contracts to distribute the largesse to the largest number of recipients.

- The excessive time consumed between the decision to build a new airport and the acquisition of the land, allowing land owners to develop specious development plans and water rights to drive up the cost.
- Mayor Peña's decision to sign scores of contracts on the eve of his departure as mayor to put the project beyond the "point of no return," so that mayoral candidate Don Bain could not, if elected, mothball the site.
- Mayor Webb's capitulation to the airlines' belated demands for massive scope changes and additions after construction had begun.
- Mayor Webb's decision to expand the automated baggage system beyond United's Concourse B to cover all three concourses.
- Mayor Webb's focus on securing contracts and concessions for his friends and supporters while failing to focus on more crucial airport issues.
- The failure of the federal government to provide appropriate oversight.

Rebuttal: Relative cost

DIA's $4.8 to $5.3 billion is a staggering figure, standing alone. However, as Albert Einstein proved, everything in the universe is relative to everything else. How does DIA stand up against what is being spent on airport infrastructure elsewhere?

First, let's examine DIA's relative size. At 53 square miles, DIA is twice the size of Manhattan Island, New York, or about the size of the District of Columbia. Figure 11-1 reveals DIA's size relative to other major U.S. airports.

Denver's longest runway is 12,000 feet. Figure 11-2 reveals DIA's size as measured in length of runways vis-à-vis several other major U.S. airports.

Another measuring stick is cost per gate. In a crude sense, with a cost of $4.8 billion and 88 gates, DIA works out to a cost of more than $50 million per gate. Pittsburgh's new terminal cost $690 million, or $9 million per gate. The industry average was $5 million per gate in the 1980s (Chandler 1993a).

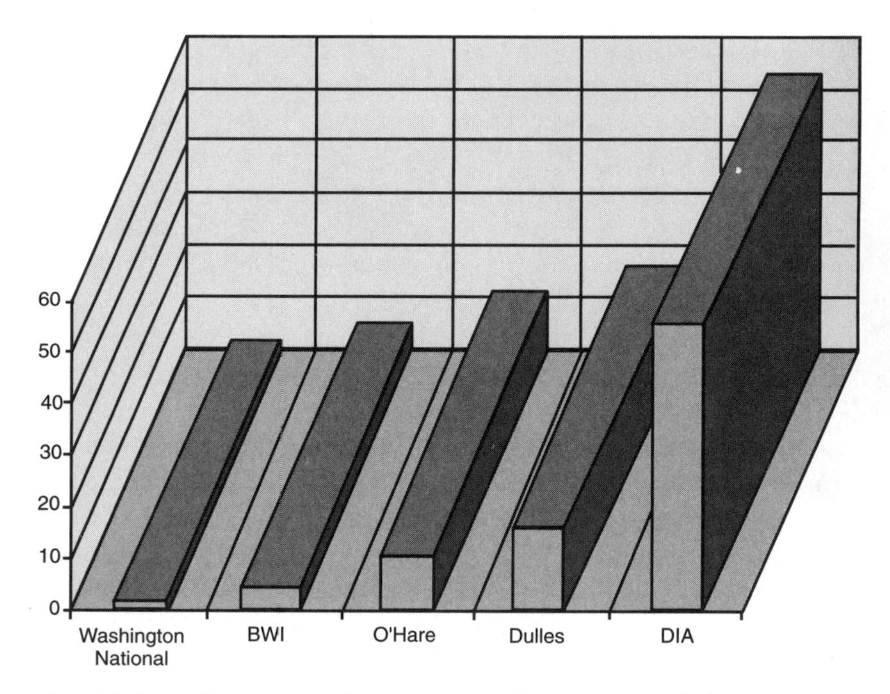

Fig. 11-1. *Relative size of U.S. airports (in square miles).*

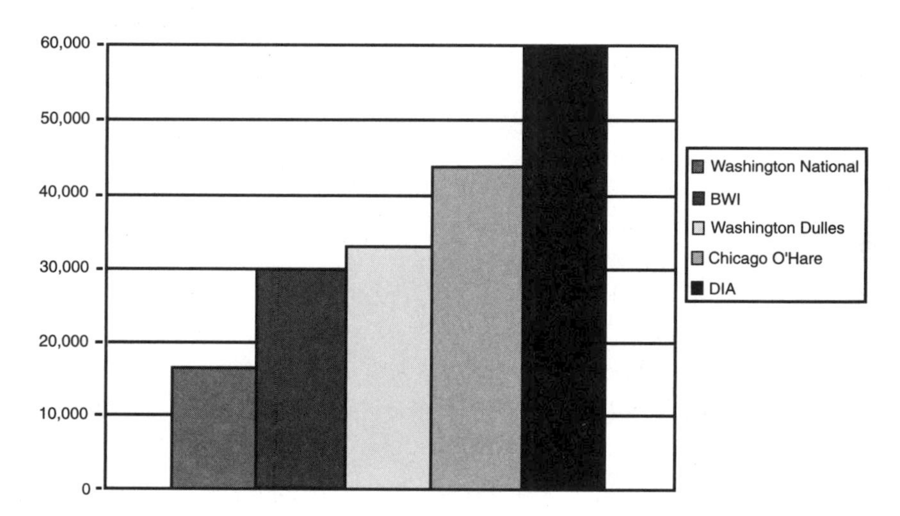

Fig. 11-2. *Total runway length (in feet) at major U.S. airports.*

Let's examine relative costs for expansion and new airport construction at several of the major projects around the world (Fig. 11-3):

AIRPORT	COST PER PASSENGER (US$)	TOTAL COST IN BILLIONS (US$)	ANNUAL PASSENGERS IN MILLIONS	ANNUAL OPERATIONS IN THOUSANDS	NUMBER OF RUNWAYS
Bangkok	114	4.0	35	203	2
Denver	118	3.9	33	530	5
Seoul	155	4.2	27	285	1
Kuala Lumpur	200	5.0	25	250	2
Hong Kong	346	9.0	26	165	1
Munich	407	6.1	15	185	2
Kansai	609	14.0	23	160	1

Fig. 11-3. *Cost comparisons of the major airports being built around the world in the 1990s.*

- Pittsburgh completed a new terminal facility in 1988 for $690 million.
- Chicago needs a third airport (to complement O'Hare and Midway). At the city's preferred site, the clean-up costs alone will be $10 billion.
- Miami International Airport will spend more than $2.5 billion for terminal, concourse, and parking garage expansion or rehabilitation, and 11 new cargo buildings (Fig. 11-4).
- The Port Authority of New York and New Jersey will overhaul JFK International Airport for $4.3 billion and would like to spend $2.5 billion on a rail line linking Manhattan with JFK and LaGuardia airports.
- Boston will spend $4 billion merely on a tunnel to link downtown to Logan Airport.
- San Francisco is spending $1 billion on a terminal.
- Metropolitan Washington Airports Authority will spend $2 billion on construction of new and expanded terminals at Washington National and Dulles Airports (Fig. 11-5).
- Philadelphia will spend $1.3 billion to purchase 30 acres, rebuild aprons and taxiways, remodel the terminal building,

and construct one commuter runway. The 5000-foot runway will cost $210 million, compared to the $75 million it took to build each of DIA's 12,000-foot runways (Fig. 11-6).

- Dallas/Fort Worth is spending $350 million to build a new runway.
- If it can overcome the environmental hurdles, Seattle will spend more than $350 million to build a new runway.
- Hong Kong is spending approximately $20 billion for a new airport and rail link.
- Kansai International Airport in Osaka opened with just one runway at a cost of $14 billion (Fig. 11-7).
- Tokyo spent $10 billion on airport construction.
- Japanese officials are funding a study for an offshore floating airport that will cost more than $20 billion.
- The New Seoul International Airport will cost $7 billion.
- Kuala Lumpur International Airport in Malaysia will cost $3.5 billion.
- Munich, Germany's Franz Joseph Strauss Airport, which opened in May 1992, with two runways, cost $5 billion (DIA 1995b; *Denver Post* 1995a; *Wall Street Journal* 1995).

Miami

$ 2.5 Billion in Capital Improvements

- **Double size of Terminal**
- **New Concourse A**
- **Refurbish Concourse G & H**
- **Widen Concourse D, E & F**
- **Increase gates from 110 to 140**
- **New roadway system to increase curbside capacity**
- **New parking garage**
- **$ 500 Million for Cargo expansion (11 new buildings; new roadway system for cargo)**

Source: Metro Dade County, Aviation Department

Fig. 11-4. *Miami is spending $2.5 billion for incremental airport improvements and expansions.*

Washington National

$932.6 Million Capital Development Program

- New North Terminal - 35 Gates
- Main Terminal Rehabilitation - 9 gates
- **44 gates total - *NO INCREASE***
- 2 new parking structures
- Net gain of 3500 parking spaces

Source: Washington Airports Auth

Fig. 11-5. *Nearly a billion dollars is being spent at Washington National Airport for these improvements.*

Philadelphia

Completed Capital Project

- $ 130 Million for new International Terminal

$ 902 Million in Capital Improvements

- $ 210 Million for a new 5000 ft. Runway
- $ 200 Million for new Terminal F
- $ 71 Million for Baggage & Passenger Circulation Improvements
- $ 70 Million for new Terminal A Prime
- $ 50 Million for Airfield & Apron Rehabilitation
- $ 36 Million for improvements to Existing Facilities
- $ 35 Million for People Mover System (between Terminals & Remote Parking)
- $ 25 Million for East Cargo City Development
- $ 20 Million for new Commuter Terminal
- $ 20 Million for Airport Roadway Improvements
- $ 20 Million for Relocation of Hwy. 291 (gains 44 acres of land)
- $ 20 Million for Land Acquisition
- $ 15 Million for Central Plant Rehabilitation

Source: City of Philadelphia, Department of Commerce, Aviation Division

Fig. 11-6. *Philadelphia is spending nearly a billion dollars on improvements.*

Kansai, Japan

$14.4 Billion for New Airport

- **1 Runway - 11,483 Ft.**
- **First year - 21.8 Million Projected Passengers**
- **Capacity - 30.7 Million Passengers**
- **Landing Fee - $25.10 per metric ton**
- **Cost to land 747-400 - $10,000**
- **Cost for 2 additional runways - $21.4 Billion**

Source: *Aviation Week & Space Technology*

Fig. 11-7. *Osaka's new airport cost more than $14 billion.*

Of course, comparing these projects to DIA is like comparing apples and oranges. At Osaka, Hong Kong, and Seoul, dry land had to be created from the seabed up, an enormously expensive engineering feat, much more expensive even than the overpriced near-desert farmland that Denver had to purchase. And many of the airport expansions listed are tremendously expensive because the airports are hemmed in on all sides by development and hostile neighbors.

Nonetheless, just raw dollar comparisons suggest that DIA, with five runways and room for 12, three concourses and room for five, and a huge landside terminal that can be doubled in size, looks like a bargain, even at $5.3 billion.

The passenger cost of air travel at DIA

An airport that should have cost a little more than $3 billion cost about $5 billion. Ultimately, people flying from or to Denver pay the piper for the enormous cost overruns at DIA.

Several studies by the U.S. General Accounting Office of pricing practices by dominant airlines at concentrated hubs reveals that passengers who begin or end their trips at such hubs pay prices that are between 18 and 27 percent higher than in competitive markets (Dempsey and Goetz 1992, p. 252). Denver's business traveler's air fares rose an astounding 46 percent over the previous year by the summer of 1995 (Leib 1995b; Luzzadder 1995). United Airlines insisted that it raised prices only 15 to 17 percent, but these data apparently include frequent-flyer mileage tickets, which cost zero dollars (Imse 1995c). The U.S. General Accounting Office found that Denver's air fares had increased 38 percent (GAO 1996).

The city's revenue needs at DIA are about $450 million a year (compared with DFW's $218 million), most of it for debt service (Chandler 1993b). On top of that is added the monopoly premium once Denver became a fortress hub. Thus, more than half a billion dollars a year are being sucked out of the Colorado economy (Figs. 11-8, 11-9, and 11-10), and that for an airport that was launched to stimulate economic growth. Some observers noted that if Denver is perceived to be an expensive air destination, then surely Colorado's $8 billion tourist economy, as well as its convention and hotel business, would decline. Even Fortune 1000 companies might not locate in a city where the cost of travel was among the highest in the nation. In 1989, Continental Airlines' Frank Lorenzo warned that in building DIA, "Denver will price itself out of the market" (*High Country News* 1995b).

Gordon Yale, a Denver securities expert, observed:

> *While I do not think it is inevitable that this project will fail, it will certainly not, in the short-term, produce what it was supposed to. . . . Instead of serving as a magnet for traffic, it has driven traffic away. Instead of three hubs and competitive air fares, it will have one hub and among the highest fares in the nation. Nor has it made us an international gateway, or sparked a real estate boom (Sahagun 1995).*

The causes of the sharp increase in the cost of air travel to Denver were, as we have seen, the massive cost overruns. In 1991, then-mayor Federico Peña assured Denver that DIA's cost to the airlines per enplaning passenger would be between $10 and $11, excluding an efficiency credit of $2.25 (Peña 1991). By 1994, this estimate had grown to $15.19 (*Denver Post* 1994). But the cost of DIA at opening was $18.80 per passenger, or more than three times the per-passenger cost ($5.85) at Stapleton. Larger than expected revenues from concessions and parking have since allowed DIA to lower the per-passenger cost to $16.85 (DIA 1996). But that is in addition to the $3 passenger facility charge (a federal tax that is essentially rebated to local airport authorities) levied on each ticket

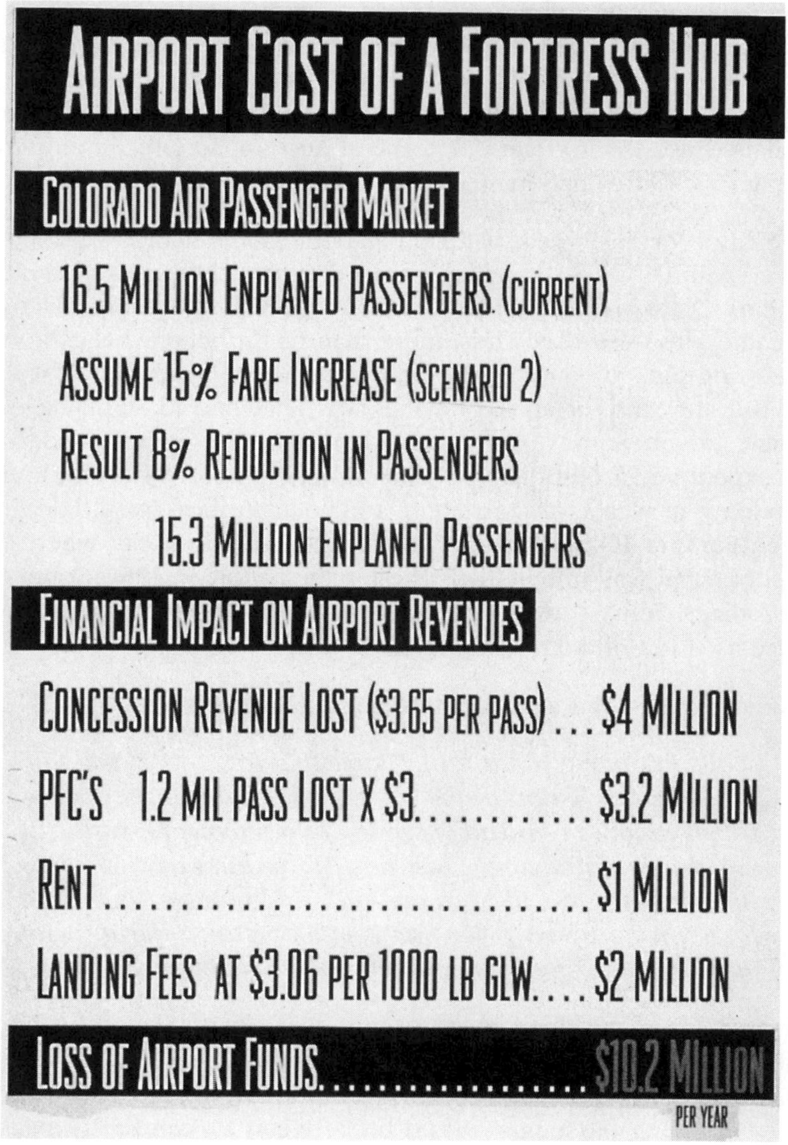

AIRPORT COST OF A FORTRESS HUB

COLORADO AIR PASSENGER MARKET

16.5 MILLION ENPLANED PASSENGERS (CURRENT)

ASSUME 15% FARE INCREASE (SCENARIO 2)

RESULT 8% REDUCTION IN PASSENGERS

15.3 MILLION ENPLANED PASSENGERS

FINANCIAL IMPACT ON AIRPORT REVENUES

CONCESSION REVENUE LOST ($3.65 PER PASS) . . . $4 MILLION

PFC'S 1.2 MIL PASS LOST X $3. . . . . $3.2 MILLION

RENT . . . . $1 MILLION

LANDING FEES AT $3.06 PER 1000 LB GLW. . . . $2 MILLION

LOSS OF AIRPORT FUNDS. . . . . $10.2 MILLION

PER YEAR

Fig. 11-8. *With Continental Airlines' decision to abandon Denver as a hub, United Airlines enjoyed a fortress hub monopoly. This chart reveals the potential airport revenue loss attributable to monopoly fares.*

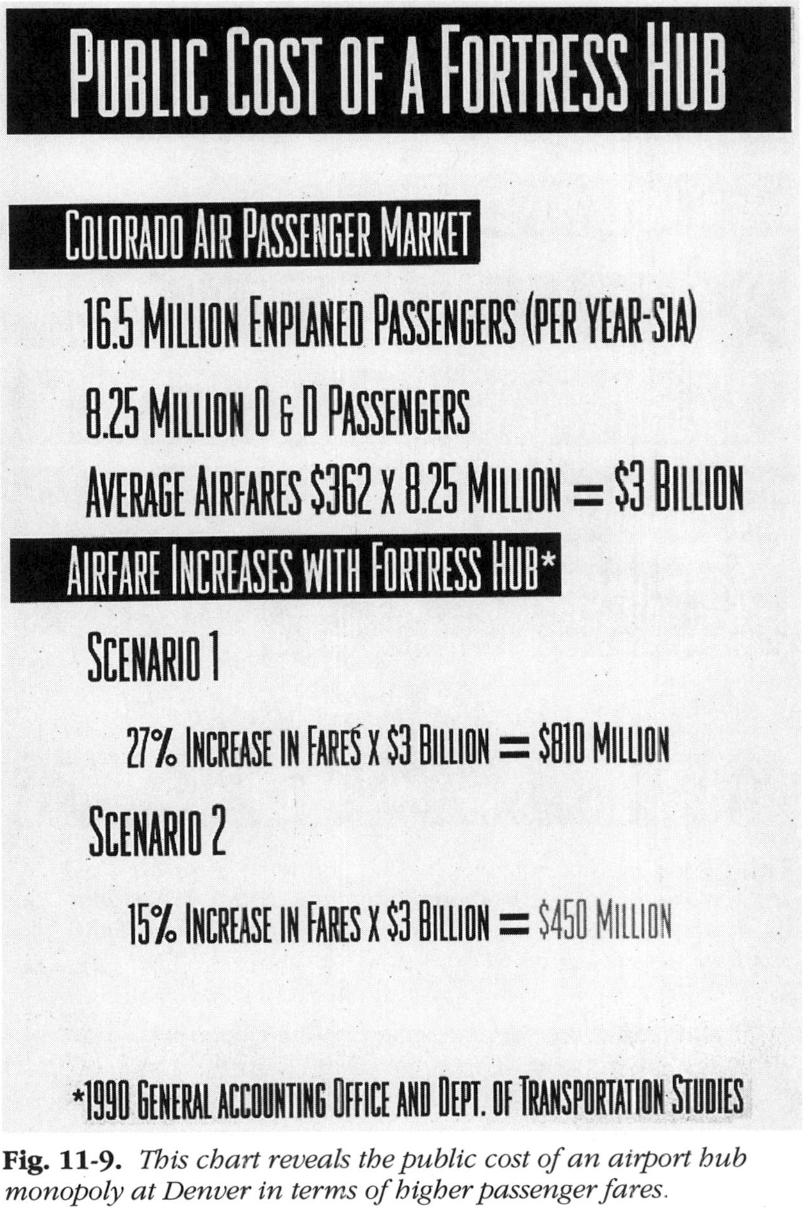

PUBLIC COST OF A FORTRESS HUB

COLORADO AIR PASSENGER MARKET

16.5 MILLION ENPLANED PASSENGERS (PER YEAR-SIA)

8.25 MILLION O & D PASSENGERS

AVERAGE AIRFARES $362 X 8.25 MILLION = $3 BILLION

AIRFARE INCREASES WITH FORTRESS HUB*

SCENARIO 1

27% INCREASE IN FARES X $3 BILLION = $810 MILLION

SCENARIO 2

15% INCREASE IN FARES X $3 BILLION = $450 MILLION

*1990 GENERAL ACCOUNTING OFFICE AND DEPT. OF TRANSPORTATION STUDIES

Fig. 11-9. *This chart reveals the public cost of an airport hub monopoly at Denver in terms of higher passenger fares.*

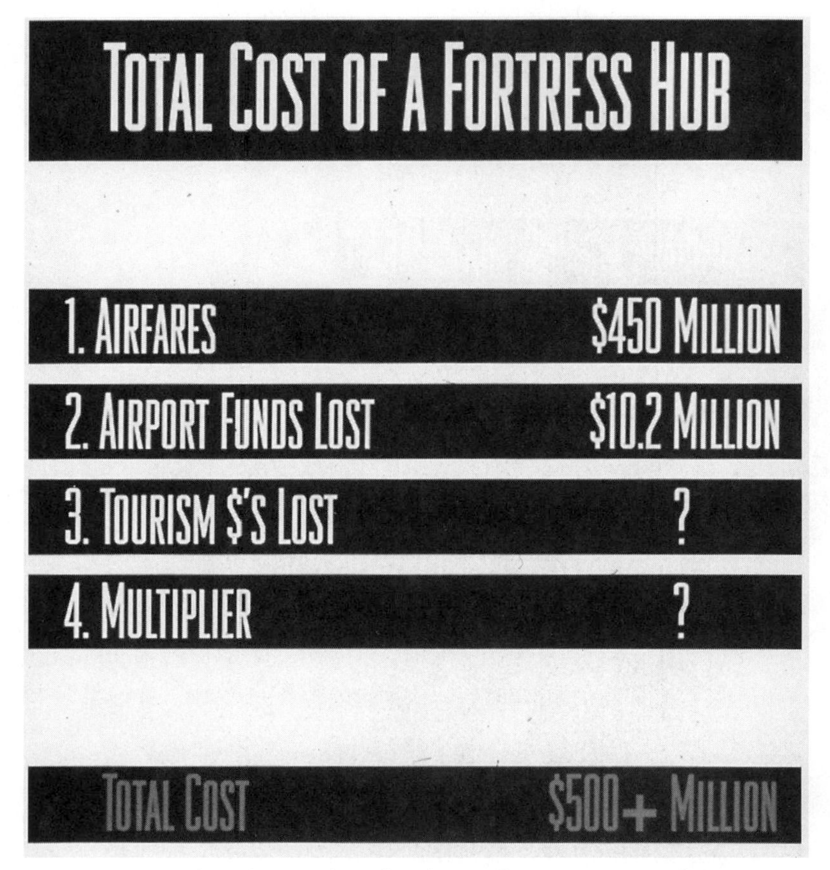

Fig. 11-10. *This chart takes the data of Figs. 11-8 and 11-9 and adds them together, showing the total cost to the airport and its passengers of the hub monopoly enjoyed by United Airlines. This is a highly conservative estimate.*

sold at DIA and which for several years had been levied at Stapleton to help pay for DIA—raising $39 million in 1993 alone (Chandler 1993a). DIA's cost to airlines was nearly 1 cent per available seat-mile, a very significant number when one remembers that costs range between 7 cents (for the nonunion low-cost entrants) and about 12 cents (for the established high-cost megacarriers) per available seat-mile. The table compares DIA's per-passenger cost to those of several major airports.

Per-passenger cost at selected airports

Salt Lake City	$1.48
St. Louis	$2.79
Dallas/Fort Worth	$4.11
Chicago O'Hare	$5.22
Denver Stapleton	$5.85
New York JFK	$9.16
Washington Dulles	$9.82
New York LaGuardia	$12.89
Newark	$15.29
Denver International Airport	$16.85

Rocky Mountain News 1995a.

The city insisted that the $18.80 figure is bloated because at Denver, many traditional tenant improvements (e.g., loading bridges, commuter facilities, aircraft docking systems, communications equipment, baggage and mail sorting equipment, flight display monitors, and ticket and service counters) were financed by the city. The city could borrow money at lower interest rates than most airlines, whose debt was rated junk by Wall Street. The city "conservatively estimates the value of DIA's in-place equipment and tenant finish items as $5.10 per passenger" (DIA 1995a). The city also lauded DIA's airfield operational savings:

> *With an efficient runway spacing and layout, with room between concourses for continuous air traffic flow, engineering and financial consultants estimate that DIA will save $70 million annually for the airlines. The airlines estimate the savings to be $27 million. Using the conservative figure of $27 million, the savings translate to $1.68 per passenger. Subtracting $5.10 and $1.68 from $18.80 results in a net figure of $12.02 (DIA 1995a).*

Either number, $18.80 or $12.02, is an enplaned passenger number, meaning that you double it to determine the impact on a consumer's round-trip air fare. Assume the city is right—that at its opening DIA cost the airlines only about $24 round-trip. The airlines raised round-trip ticket prices on Denver origin and destination passengers by $40. Wellington Webb blasted United Airlines for the decision, saying, "It is no secret that when one air carrier becomes dominant in a market, that they develop as a fortress hub, and as a result, fares have always increased" (Flynn and Hubbard 1995).

Two reasons account for the sharp increase in air fares well beyond the amount estimated by the city. First, connecting traffic was immune from the higher costs, and second, Denver became a monopoly hub airport.

Connecting passengers were immune from these high costs because of competitive considerations. A passenger seeking to travel, say, from the Midwest to California would have the option of flying American Airlines (via Chicago or Dallas), Northwest (via Detroit or Minneapolis), TWA (via St. Louis), Delta (via Cincinnati or Salt Lake City), America West (via Phoenix), or United (via Denver). Hence, the higher per-passenger burden reflected by DIA's costs could not be passed on to connecting passengers who could just as easily route themselves over an alternative hub airport. That meant that the origin and destination (in industry jargon, O&D) passenger would bear the brunt of DIA's high costs. Fifty-eight percent of DIA's traffic is O&D (GAO 1996).

Thus, shortly before DIA opened, United Airlines announced it would raise fares $20 per ticket (a queer way of saying the round-trip price just went up $40). Many other airlines followed suit.

But that's not the whole story. As Continental Airlines reduced its frequencies at Denver, peeling off aircraft to start ill-fated CALite east of the Mississippi, United raised fares in every market in which competition subsided—and substantially. Thus, the $40 DIA premium was on top of fares that had already been increased sharply in every market Continental had exited. By mid-1995, Continental Airlines was flying between Denver and only Houston, Cleveland, and Newark.

Other low-fare carriers had retreated from Denver's high costs as well. Morris Air, consumed by Herb Kelleher's Southwest Airlines, dropped its flights from Salt Lake City. Kelleher insisted a low-fare, low-cost carrier could not maintain its niche flying from a high-cost airport. Midway Airlines (no relation to the first Midway Airlines, which was liquidated), ceased serving Denver from Chicago, moving its planes east to feed American Airlines out of Raleigh/Durham, North Carolina. GP Express (formerly a Continental Express feeder) folded its tent, depriving Denver of competitive air service to Rocky

Mountain ski destinations (Berenson 1995a). MarkAir, which had moved its operations from Alaska to Denver while in Chapter 11 bankruptcy, was shut down by the FAA for alleged safety violations. MarkAir's six 737s served such major cities as Dallas, Phoenix, Las Vegas, Chicago, Kansas City, Los Angeles, Seattle, and San Diego. Although MarkAir was much maligned for its poor service, as the business editor of the *Rocky Mountain News* observed, "Love it or hate it, MarkAir kept airline prices lower here" (Knox 1995). But MarkAir was liquidated in the fall of 1995, owing DIA several million dollars in overdue rent.

United Airlines, which now dominates the Denver fortress hub, had met the low fares on a capacity-controlled basis. With the departure of the competitive stimulus, United reduced the number of seats for which the lowest fares were available and in some instances eliminated them altogether. United was hardly a benign monopolist. For example, when MarkAir exited the Denver-San Francisco market, United raised its lowest weekday nonstop fare from $350 to $1064 overnight (Berenson 1995b).

By mid-1995, United Airlines was carrying about 70 percent of the passengers at Denver. Although Denver's air fares were rising more sharply than in any other major American city (business travelers were paying 46 percent more), United denied that it was gouging its customers (Leib 1995b). By fall 1996, United and its code-sharing affiliates controlled 78 percent of DIA's traffic.

Of course, the total cost to passengers is more than ticket prices. It includes the price of parking or the cost of a taxi ride or car rental from the airport to one's business or residential destination. DIA is significantly more expensive than Stapleton in both categories. Taxi fare grew from less than $10 for the ride from downtown Denver to Stapleton International Airport to about $40 from downtown Denver to DIA (Baca 1995). As one might expect, van and shuttle traffic, as well as car rentals, began to grow, while taxi demand declined. DIA per-passenger costs and distances from downtown are among the highest in the United States. These factors may well siphon tens of billions of dollars of revenue out of Denver in the long term, dampening discretionary traffic demand in an economy strongly dependent on tourism.

Some of it may flow to neighboring communities, since there is an economic principle of price elasticity of demand in airline economics which posits that as the price of air transportation declines, the geographic radius of the airport expands. That is to say, price-sensitive travelers are willing to pack the kids in the station wagon and drive to a distant airport to save a few hundred dollars. Colorado Springs is about 75 miles south of Denver. Although the Springs (as it is affectionately referred to by Coloradans) has a small population base, it can draw traffic from Pueblo, to the south, and south suburban Denver, to the north. A study commissioned by the Greater Colorado Springs Economic Development Council found that an airline could offer fares of $49 one-way to 40 major cities and still turn a profit (Hellman 1994). In 1994, Colorado Springs opened a new 270,000-square-foot terminal facility with 12 gates for a cost of $39.5 million—allegedly on time and under budget (Pavlik 1994).

One of its major tenants was Western Pacific, a new discount airline that inaugurated service from Colorado Springs to Los Angeles, San Francisco, Seattle, Phoenix, Chicago, Dallas, Las Vegas, Kansas City, and Oklahoma City, with one-way fares as low as $99 (Leib 1994). Its lowest walk-up fare to San Francisco was $417, while United's fare from Denver to San Francisco was $1080 (*Denver Post* 1995b).

While passenger traffic dropped 6 percent at DIA in May 1995, it rose 67 percent at Colorado Springs (Knight 1995). Denver estimated traffic at DIA would be down 2.6 percent in 1995 but that the city would enjoy 2.4 percent annual growth during the years 1996 to 2000 (Leib 1995a). DIA critic Michael Boyd estimated Denver traffic would drop to fewer than 32 million passengers in 1995, about 600,000 less than the preceding year (Imse 1995b); it actually fell to a little more than $31 million. The 100,000 El Paso County residents who formerly flew from Denver now had a convenient and low-cost alternative, and for the price conscious, southbound Interstate 25 was the highway to heaven (Leib 1995d). Colorado Springs Airport estimated that about 20 to 30 percent of its traffic was coming from the Denver area.

One new low-cost, low-fare, full-service airline emerged at DIA to challenge United's monopoly grip—Frontier Airlines, a Denver-based carrier founded and staffed by many of the individuals from the original Frontier Airlines (which was folded into Continental in 1986). By 1996, Frontier's presence had lowered fares significantly on more than a dozen routes radiating from Denver, although after it became profitable, United engaged in various predatory practices to drive it from the market.

The national picture

One of DIA's major benefits is its contribution to the national airspace system. Its tremendous efficiencies will lead to fewer delays and cost savings, though the exact figure is hard to quantify. True, growth in U.S. domestic air traffic began to slow in 1987, but international traffic continues to grow robustly. Although domestic demand will not likely reach the levels of the faster-growing international market, domestic demand will likely continue to march steadily upward. Across the national air traffic landscape, we see airports bursting at the seams, unable to meet demand without enormous delays that impose significant costs on airlines and their passengers. DIA will assist in relieving national air gridlock, particularly on United Airlines' vast domestic and international route network (had Continental not abandoned the Denver hub, the national efficiencies would have been spread more ubiquitously). And DIA will be a significant improvement over the airport it replaced, since Stapleton was too often shut down because of inclement weather, causing national gridlock. In 1990, Denver's Stapleton Airport was responsible for 46,000 hours of delay, which rippled across the entire national air traffic system. The new Frontier Airlines, which inaugurated operations in 1994, announced it was inspired to begin operations focused on Denver because of the economies and efficiencies a new airport would provide. Many of Frontier's officers were with the original Frontier Airlines and were well acquainted with the bottleneck that Stapleton had become.

Over time, DIA can be expanded at a relatively low cost to accommodate future growth in air traffic. Its positive contribution to air traffic congestion will become more and more evident in the future.

A city with room to grow

One of the motivations for the airport was that it provided the first opportunity in two decades for Denver to expand its city limits and take advantage of suburban-style growth and corresponding improvement in the city's tax base. Stapleton Airport's 7 square miles would be freed up for development, as would the strip along Peña Boulevard. Federico Peña described this as among DIA's greatest accomplishments. Responding to criticism of his role in promoting DIA, Peña says, "I still believe today that most people in Denver don't understand the impact of annexing 50 square miles to the city. It is a tremendous economic opportunity. It expanded the city and county of Denver, which was before that 110 square miles. That had never been accomplished by any mayor in 30 or 40 years." Further, Peña said, "I strongly believe that five, 10, 15, 20 years from now, people will look at the airport, will look at the land that was annexed, will look at the redevelopment opportunities of Stapleton and say this is one of the best investments and economic decisions for the city that was made, many many years ago" (*Rocky Mountain News* 1995b).

Of course, economic growth is a mixed blessing. Denver already has a significant air pollution problem, with high carbon monoxide, particulate, and ozone levels, the principal cause of which is the internal combustion engine. Denver also suffers from growing highway congestion and sprawl, with two million metropolitan inhabitants spread over six Colorado Front Range counties. Growth in the northeast metropolitan quadrant, stimulated by the new airport, will further spread the megalopolis across the plains. In the 21st century, the Denver urbanized region will engulf an area stretching from Ft. Collins to Colorado Springs.

Colorado is among the nation's fastest growing states. Californians represent the largest influx of new residents. Polls show that people who have resided in Colorado less than two years are far more concerned about the negative implications of unconstrained growth than are people who have lived here more than two years. Perhaps this is because many Californians are acquainted with sprawl, gridlock, and pollution, and moved to Colorado to get away from it. If 21st-century metropolitan Denver comes to look like today's metropolitan Los Angeles, the quality of life for its inhabitants in several important respects—congestion and pollution in particular—will decline.

Nonetheless, Denver has long had an insecurity problem, cringing at the thought that some people call it a "cow town," or that every time the Denver Broncos play on Monday Night Football, there seems to be a blizzard, leading everyone outside Colorado to think that the state has the climate of Siberia. But Denver has a splendid, semi-arid climate, with more than 300 days of sunshine per year, and while it may be transitioning from its cow-town past, it actually has more in common with Atlanta of 20 years ago. Both are white-collar professional cities that serve as regional headquarters for most of the nation's Fortune 500 corporations. While isolated by vast prairies and uninhabited mountains from the rest of the nation, Denver is an oasis of human activity on the edge of buffalo commons.

Like Atlanta, in the long term the new airport will likely stimulate economic growth, immigration of population, international trade, and a more stimulating, cosmopolitan culture. Atlanta began a boom in the mid-1960s that was propelled by its airport reconfiguration and expansion in 1980 and has not slowed down since. It is hardly recognizable from the sleepy southern town it once was. If Denver experiences only half the change that has transformed Atlanta, it will be unrecognizable in the 21st century—culturally, economically, and visually. Denver will be an even more cosmopolitan and prosperous city if it fulfills its dream of becoming an international aviation hub.

Sustainable development

If all the growth that DIA was intended to stimulate actually materializes, the Front Range of Denver could one day look like today's Los Angeles basin, with its enormous pollution and congestion problems. This is but one of the many costs that we have identified in this book.

If we have assessed them correctly, it is highly dubious that DIA fits the model of a transportation project that promotes sustainable development—providing benefits today without jeopardizing the welfare of future generations. But the concept of sustainable development is very broad and includes several different dimensions. On the one hand, development should be environmentally sustainable, which suggests no harm to the natural environment nor the rampant use of resources at rates faster than they can be naturally renewed or economically replaced. On the other hand, development must also be economically sustainable. Projects that provide short-term gain but long-term distress are obviously not desirable. Since so many economic and political decisions today are based on short-term time horizons that have limited long-range utility, this aspect of sustainable development must also be acknowledged.

In light of these concepts, assessment of airport projects like DIA become more problematic. Air transportation is certainly not the most energy-efficient mode of transportation. Plus, aircraft and airports contribute significantly to noise and air pollution. Airports, located on the outskirts of cities, help to create more decentralized spatial patterns of urban development, thus contributing to more ground vehicle miles and even more air pollution emissions. Yet airports and air transportation are important to long-term economic development, especially global trade, which involves long-distance and overseas travel, where other modes are not feasible. Even if large shifts to more energy-efficient modes occur, there will still be a need for air transport and airports.

In the case of DIA, the question becomes whether growth spurred by the airport will turn Denver and the Front Range region into a major megalopolis and, if so, whether the region can accommodate this degree of growth. Though it is not necessarily desirable that Denver become a huge population center, population growth inevitably accompanies economic growth. Only time will tell to what degree DIA will induce growth and to what degree that growth will be permitted to jeopardize sustainability of the Denver region.

What if DIA crashes?

As of this writing, DIA's revenues exceed its costs, even though Continental Airlines has been relieved of $30 million of its $58 million gate obligations on Concourse A (the concourse built for Continental), causing DIA's other airlines to pay between $1 to $1.25 more per departing passenger (Imse 1995a). United's lease contract imposes a ceiling of $20 per passenger in 1992 dollars, beyond which United will not pay. But to date, passenger costs at DIA are several dollars lower than that ceiling. Predictions by airport critic Michael Boyd that the airport would be insolvent within a year of its opening were unfulfilled.

But in the longer term, any doom-and-gloom scenario can materialize, particularly in the airline industry. Prolonged economic recession or depression could radically reduce the number of passengers flying. After a few more years like the disastrous 1989 to 1994 period, the principal tenant, United Airlines (whose balance sheet caused its debt to be downgraded by Wall Street to junk), could find itself in bankruptcy or could pack up and leave Denver, as did Continental Airlines. In 1994, United had net equity of $76 million and debt of $12 million (GAO 1996). Airlines suffer from high fixed costs (new aircraft are ordered in good times and delivered in bad), highly cyclical demand, and chronic excessive capacity (Dempsey 1995). Their fastest growing expense items are taxes and airport rates and charges. Although the latter accounts for about 6 percent of airline costs, airline net profit margins collapsed to subzero levels in the post-deregulation era (Dempsey 1995, p. 22). As noted in Chapter 5, airports have four primary revenue streams: the ticket tax, passenger facility charges, concessions (including parking), and airline rates and charges. It is the airlines that fear that most of the burden must be borne by them.

Other pessimistic scenarios exist. The high-cost, low-efficiency hub-and-spoke distribution system might lose favor as the primary means of distributing passengers. Technological revolutions in teleconferencing or affordable lighter-than-air vehicles small enough to fit in a two-car garage could eventually diminish airport traffic. Technological developments in short-takeoff-and-landing (STOL) aircraft and the noise level of aircraft may make long runways unnecessary. Just as the canals were made obsolete by the railroads, and rail passenger transportation was supplanted by aviation, some new form of transportation or communications may make much essential air transportation unnecessary and the infrastructure built to support it obsolete. Thus, a 21st-century airport may become a 22nd-century museum of ancient transport technology.

Less draconian scenarios of a pessimistic vein might include Denver defaulting on the bonds, or civil liability imposed on the city for fraud in DIA bond financing, or noise violations running riot—any one of which might force the taxpayers of Denver, the state of Colorado, or the United States (in a federal bail-out) to foot the bill. It is unlikely that a bankruptcy judge would shut down an airport and liquidate the property. More likely, the debt would be restructured to something more tolerable while the bondholders would be forced to take only so many cents on the dollar or a repayment schedule spread over 50 years. The electric utility industry remembers the impact of the WPPSS nuclear facility bond default and the ripple effect it had on interest rates for all public utilities. Politically, such a result could be disastrous for other airports across the country. Joseph Schwieterman, head of the infrastructure planning institute at DePaul University, points to the impact the cost overruns at DIA have already had: "One thing the new Denver airport has done is sour the atmosphere for a proposal to build a new airport here in Chicago. Now people say, 'It could be another Denver'" (*High Country News* 1995a).

Such doomsday scenarios are unlikely to materialize. Whatever the inconveniences and costs to the local Denver passenger, DIA is already a national treasure for both the national air transportation system and for the connecting passenger. In the final analysis, we have an airport whose value will be proven in the 21st century. As passenger demand and air traffic grows, DIA will be unique among the nation's airports in its relatively infinite capacity to absorb such growth. As other airports spend billions of dollars for marginal, incremental improvements of capacity, Denver's high costs will become relatively less significant. Inflation alone will, over time, reduce real DIA costs to the airlines and their passengers, so cost will, in the 21st century, be less of a concern.

How did Denver do it?

Many major cities have attempted to build new airports—cities like Chicago, Miami, St. Louis, and San Diego. But typically, a small group of highly organized opponents have scuttled the projects, using the available zoning, environmental, and public relations tools at their disposal either to tie the projects in legal knots or turn the public against the projects. Thus, for two decades, no new airport was built in the United States, the world's largest air transport market. Most have had to settle for modest expansion (new or extended concourses or a new runway here or there), which typically takes a decade or more to build at enormous cost.

With so many other cities stymied by political, legal, and environmental opposition in their attempts to build a new airport, how was Denver able to accomplish what had eluded so many other cities for two decades? Denver airport director Jim DeLong, among the most eloquent of DIA's proponents, offers six reasons:

1. *A visionary mayor.* Denver had a tenacious mayor who was a visionary. "Imagine a Great City," was Federico Peña's mayoral theme. The mayor was a builder of public works projects. Not only was a new airport built, but the city also built a new convention center, library, stadium, and other civic improvements. Peña realized that the window of opportunity (created by the following circumstances) was narrow and put the project on as fast a track as possible.
2. *Strong federal governmental support.* The mood in Washington was one of support for new infrastructure to help relieve a saturated air transport system. With three carriers hubbed at Denver, delays at Stapleton Airport were clogging the national air transportation system. With a metropolitan population of two million people, and situated

at the center of a 1000-mile radius that takes in the Pacific Coast and the Mississippi River, Denver was ideally situated as a major connecting hub airport. On a trip to examine the oil spill of the Exxon *Valdez* in Alaska, DOT Secretary Sam Skinner was flown by helicopter over the new airport site. He couldn't believe that such a vast amount of undeveloped real estate stood so near a major city. He quickly became a proponent. The FAA wanted a demonstration project to stimulate needed airport infrastructure capacity growth and sent George Brewer to Denver to serve as liaison, with orders to "Make it happen!"

3. *Sufficient undeveloped land.* Because the Rocky Mountain Arsenal had geographically blocked development in the northeast quadrant of the metropolitan area, Denver had ample room to build an airport on accessible land that was unencumbered by rivers, mountains, hills, or people, within reasonable driving distance (24 miles) of the city. It was anticipated that takeoffs and landings would impact fewer than 300 people in the *verboten* 65 Ldn range.
4. *An entrepreneurial spirit.* Like Texans (who had built Dallas/Fort Worth Airport two decades earlier), Coloradans are not as risk-averse as are their counterparts elsewhere in the nation. Denver had a tradition of building essential transportation infrastructure to ensure its role as the dominant regional economic capital stretching back to the days of stagecoaches and railroads. Tenacious self-determination and willingness to build it ensured that Denver would be the major city of the Rocky Mountains/Great Plains region rather than Cheyenne, Pueblo, or Salt Lake City. Moreover, because of its isolation in the center of the continent, Denver has always been more cognizant of its reliance on transportation corridors to connect it with the rest of the nation. This dominant mood took a short setback in the 1970s, when antigrowth forces (led by then-Governor Dick Lamm) aborted the scheduled Winter Olympic Games and refused hundreds of millions in federal dollars to build interstate highways. But this temporary setback to the growth philosophy of Colorado was overshadowed by the recession of the 1980s. It is difficult to be antigrowth when there is none.

5. *A poor economy.* Colorado's oil and gas bubble had burst in 1983, causing unemployment to soar and creating a regional economic depression. Denver had the highest office-space vacancy rate of any city in America, save Houston. Only sharp outmigration arrested high unemployment. The people were willing to support and finance public works projects for the projected short-term and long-term employment and economic benefits—to "jump start" the economy. The demand for economic growth trumped green environmental opposition (in fact, environmental opposition was neutralized because closing down Stapleton Airport would offer enormous noise relief to a large constituency of nearby residents). Again, it is difficult to be antigrowth when the people's wallets are empty. Infrastructure investment is a type of public works project most likely to serve as a catalyst for economic growth in the private sector. Early referenda of the electorate in Adams County and Denver diffused the organized opposition.
6. *The airlines couldn't derail it.* United Airlines CEO Stephen Wolf said he never fought anything as vigorously in his life as he fought DIA. Airlines almost always oppose a major expansion in airport capacity because they fear its economic costs, and they fear the new airline competition it might produce. But the airlines' majority-in-interest clauses had expired at Stapleton, and they couldn't use that effective veto to scuttle the project. They attempted to scuttle it by withholding support, in an unsuccessful attempt to dissuade Wall Street from financing the project. But the bonds sold well, even without airline leases, because no one could imagine that local or national political leaders would allow a major city's airport to go "belly up." Unfortunately, this meant the project had to go forward without the airlines' input on the concourses they would occupy. Their belated jump on the departing bandwagon caused major cost escalation at DIA (Brewer 1995; DeLong 1995).[2]

Thus, all the planets were lined up perfectly to allow a new airport to go forward to completion in a relatively compressed time frame and, given what such infrastructure costs elsewhere, at a relatively reasonable price.

2. Many of these points were developed in interviews with Jim DeLong and George Brewer in October 1995.

But not everyone is enamored with DIA. Ed Marston, publisher of the Colorado western-slope biweekly *High Country News*, cynically offered three additional reasons why DIA was built:

7. *An ineffective press.* "DIA was possible, first and foremost, because Colorado lacks an effective press. Colorado's two major newspapers—the *Denver Post* and the *Rocky Mountain News*—and the Front Range television stations acted as cheerleaders for the project. Only the Denver alternative weekly, *Westword*, and columnist Gene Amole of the *Rocky Mountain News* functioned as journalists, asking the questions cheerleaders didn't want to hear. The press didn't fail because it favored the airport. It failed because it refused to give opponents the same coverage and respect as proponents: thereby, it deprived the public of the chance to understand and debate the need for a new airport."
8. *Inadequate federal oversight.* Said Marston, "Although $500 million in federal funds was invested in the airport, none of Colorado's elected national officials made any publicly visible attempts to determine if this would be spent correctly."
9. *An unorganized and unprotected public interest.* "Hand-in-hand with its lack of a bulldog press and more thoughtful elected officials, Colorado also continues to lack an organized collection of individuals and groups that understands and fights for the broad public good. The airport's billions in public pork enabled the development interests to buy off or coerce almost all of those who professionally 'represent' the public" (Marston 1995).

Aviation consultant Michael Boyd, the most articulate of DIA's opponents, added a tenth item to this list:

10. *Greed.* Boyd insisted that DIA was not needed and was principally built to assuage the economic interests of the landowners, the land developers, minority groups, and the corporations that would generate significant revenue streams from the project (*Westword* 1993).

After careful and dispassionate review of all the evidence, the authors have concluded that neither the proponents nor opponents of DIA have captured the full truth about this project by themselves. Many of the arguments on both sides possess validity. We believe the previous ten reasons adequately identify the principal elements leading to the completion of the first major new airport to have been built in the United States in two decades. However, we feel that they can be succinctly summarized by three main factors, as follows:

1. *Growth impetus*. Large projects, such as DIA, get built because large growth coalitions mobilize powerful forces to promote and take advantage of manifest growth opportunities. In the case of Denver, the new airport became such an opportunity, and the monied interests in the city pushed to make it happen. The particular success of the Denver growth coalition in making the airport a reality can be attributed to a keen desire to reinvigorate the local economy during the 1980s, which subsumes the entrepreneurial spirit, poor economy, and greed reasons given above. But growth coalitions exist in every city and country in the world, so other factors were also necessary.
2. *Willing politicians*. Growth coalitions cannot build large public projects on their own. They need to work with elected political leaders who themselves are willing to engage in public-private partnerships for the "good of the community." Otherwise these projects cannot be built. This was most certainly the case for Denver, as politicians like Mayor Federico Peña, Governor Roy Romer, and others were elected on platforms that emphasized economic development, job creation, and vibrant communities. Neither Peña nor Romer set out to build a new airport when elected. They did see the airport, however, as a way to achieve their political goals, while at the same time alleviating the problems of airport capacity and noise at Stapleton. Plus, for Denver, this was a rare opportunity to increase its land area in the wake of annexation restrictions enacted in the 1970s. The personalities and inclinations of individual elected officials in the end do matter, because both Peña and Romer, unlike other politicians elsewhere and in the past, were willing to work with the growth coalition to help make the new airport a reality. Thus, the "visionary mayor" and sufficient undeveloped land reasons can be subsumed into this factor. Yet even the power of the growth coalition and the willing politicians do not entirely explain why DIA got built.

3. *Favorable external circumstances*. For any large project like DIA to succeed, circumstances external to the principal decision-makers and players had to be favorable. In this case, the congested state of the air transportation system in the mid-1980s in the wake of deregulation created a keen desire on the part of the U.S. Department of Transportation and the Federal Aviation Administration to build more airport capacity to alleviate the principal bottlenecks. Stapleton was identified as one of the key bottlenecks, and since Denver had already made substantial progress in planning for a new airport, it was a natural candidate. This strong federal support was crucial in getting the airlines to agree eventually to sign on to the new airport. Even though the major airlines were against the airport, they could not mount an effective opposition, largely because of the industry's volatility after deregulation. Both federal and local politicians had longer periods of tenure than airline CEOs during the time DIA was being planned. Thus, the airlines were not as well organized as they could have been, and the city's lack of coordination with the airlines, though not fatal, created many difficulties. Finally, the media, citizen interest groups, and the public at large were unwilling or unable to stop the project. Thus, the federal government, airlines, media, and public interest reasons can be subsumed into this factor.

We feel that these three factors explain the essence of how and why DIA was built. Whether it was for better or worse is more difficult to say.

The most impressive accomplishments of the Denver International Airport project were that a new airport was built at all, that it reserved sufficient real estate for low-cost future growth and expansion, and that because of its airfield and runway configuration, it is among the most efficient and safest hub airports in the world. These are monumental achievements indeed, despite their compression into a single paragraph. Clearly, DIA is an airport for the 21st century, which will grow and mature over time, and whose value is likely to, like fine wine, improve with age. Nonetheless, the question will remain for years to come whether DIA's monetary and other costs were simply too high a price to pay. Only time will tell whether DIA will promote sustainable development or come to be regarded as a monument to hubris, greed, and reckless ambition.

Lessons learned

In the preceding chapters, we identified numerous specific lessons. Here we attempt to draw broader conclusions about new airport planning and development.

- *We have learned that despite all the obstacles, a major American city with determined political leadership, a strong growth coalition, and a favorable economic environment can build a new airport.*
- *Since most airport megaprojects go over budget, we believe that better planning for predictable and unpredictable contingencies might enhance their success.*
- *We have learned that adopting cutting-edge technologies for essential systems increases the risk of failure.* We believe essential systems should be built with off-the-shelf technology or have backup systems in place.
- *We have learned that there are costs to be incurred in attempting to maximize safety and efficiency in air transportation.* We believe that designing an airport to be an efficient hub may lead to trade-offs that make it less convenient for the local population.
- *We have learned not to believe the overly optimistic projections of demand and cost.* We believe the process would be better served with more honesty and less exaggeration. Means should be identified to shed more light and less heat on the objective facts.
- *We have learned that airline deregulation has created tremendous uncertainty for airport planners.* The enormous public investment in airports is jeopardized by the financial uncertainty in the airline industry. We believe Congress and the DOT should develop policies that will create a more stable environment.
- *We believe that the federal government should take a stronger role in protecting competitors and consumers from the predatory and monopolistic pricing and other practices, respectively, of dominant airlines at concentrated hub airports.*
- *We have learned that politicians are essential for building political consensus, but that they often feel compelled to reward their supporters.* We believe that airports serving large

regions should be built and run by regional airport authorities rather than individual municipalities. We also believe that major airport projects need some measure of federal oversight so that waste and cost overruns are minimized.

- *We have learned that affirmative action can be used to mask political cronyism.* We believe that, used in this way, it undercuts support for and corrupts an otherwise legitimate social goal.
- *We have learned that new approaches to airport design and implementation are required.* Arrangements must be made that permit public examination of a project's assumptions. Flexibility is essential and the organizations that plan and implement the project must be prepared to deal with the unexpected.
- *We have learned that while most foreign airports carefully plan intermodal transportation access, such access is, unfortunately, a very low priority for U.S. airports.* We believe that legislative and policy changes must rectify this oversight.
- *We have learned that peak-period pricing and multiple-use gates, ticket counters, and ground and baggage-handling facilities can reduce capacity saturation at airports and therefore reduce the need for massive infrastructure expansion.*
- *Finally, we believe that more airport capacity will inevitably be needed, and that our governments, federal and local, should help facilitate airport growth and development and the more efficient use of existing airway and airport capacity.*

References

Baca, Stacey. 1995. DIA boon to shuttles. *Denver Post.* June 10.

Berenson, Alex. 1995a. GP Express exiting west slope. *Denver Post.* July 19.

————. 1995b. MarkAir shutdown could spark higher fares. *Denver Post.* August 3.

Brewer, George. 1995. Interview. October 25.

Chandler, David. 1993a. Cash landing! *Westword.* February 17.

————. 1993b. Denver does Dallas? *Westword.* September 22.

DeLong, Jim. 1995. Interview. October 25.

Dempsey, Paul. 1995. Airlines in turbulence: Strategies for survival. *Transportation Law Journal*. 15.

———— and Andrew R. Goetz. 1992. *Airline Deregulation & Laissez-Faire Mythology*. Westport: CT: Quorum Books. 252

Denver International Airport (DIA). 1995a. Cost per passenger.

————. 1995b. How DIA's cost compares to other new airports and expansions.

————. 1996. Press release. June 17.

Denver Post. 1994. Continental plans to cut 1,400 Denver flight-crew jobs. March 4.

————. 1995a. Denver's DIA is a bargain in a global comparison. February 28.

————. 1995b. Comparison. July 16.

Flynn, Kevin and Burt Hubbard. 1995. Webb berates United for $40 fare increase. *Rocky Mountain News*. January 29.

General Accounting Office (GAO). 1996. Denver Airport. February.

Green, Chuck. 1995. DIA debate in the view of history. *Denver Post*. February 26.

Hellman, Wayne. 1994. Fly the Springs. *Gazette Telegraph*. October 9.

High Country News. 1995a. Airport blues: A crash in the making. January 23.

————. 1995b. Airport blues: Denver's bet goes sour. January 23.

Imse, Ann. 1995a. DIA will raise airlines' passenger fees. *Rocky Mountain News*. April 5.

————. 1995b. DIA slipping as Springs airport soars. *Rocky Mountain News*. July 25.

————. 1995c. United disputes air fare reports. *Rocky Mountain News*. August 18.

Knight, Al. 1995. Webb's free-market yearning comes late. *Denver Post*. July 19.

Knox, Don. 1995. Marketplace the loser in MarkAir fiasco. *Rocky Mountain News*. August 3.

Leib, Jeffrey. 1994. DIA market is fare game for Springs. *Denver Post*. July 16.

————. 1995a. DIA projections healthy. *Denver Post*. June 6.

————. 1995b. Denver airfares take 46% climb in a year. *Denver Post*. July 8.

————. 1995c. United VP on air-fare hot seat. *Denver Post*. July 12.

————. 1995d. Travelers bypassing DIA for Springs. *Denver Post*. July 15.

Luzzadder, Dan. 1995. Plane fares out of Denver soar 46%. *Rocky Mountain News* July 9.

Marston, Ed. 1995. An ersatz democracy gets what it deserves. *High Country News*. January 23.

Pavlik, Rick. 1994. New airport to open under budget, on time. *Gazette Telegraph*. October 9.

Peña, Federico. 1991. Memorandum from mayor to community leader. March 4.

Rocky Mountain News. 1995a. DIA fees 2nd highest in U.S. January 29.

————. 1995b. Peña defends his role in new airport, slams media coverage. January 29.

Sahagun, Louis. 1995. Denver airport woes add up to distressing start. *Los Angeles Times*. February 24.

Wall Street Journal. 1995. San Diego considers a floating airport, but will it fly? October 3.

Westword. 1993. High fliers, low behavior. February 17.

Epilogue

Many myths have grown up around DIA. In this book, we have done our best to sort fact from fiction. However, we cannot resist quoting the most peculiar news report concerning DIA.

Among the most bizarre allegations concerning DIA is that the airport is actually a front for a concentration camp for the new world order—a plot by the Trilateral Commission, Bilderberg Conference, and others attempting to create a one-world government. Northeast of the terminal, a complex of five buildings was allegedly constructed 60 to 120 feet below the surface, connected by a web of tunnels and linked to the airport's main terminal. Farther east, two 4½-mile runways allegedly were built and buried in 4 inches of dirt to hide them until the new world order can impose martial law on the United States and establish a one-world government under the United Nations by the year 2000. At that point, the runways would be cleared to deliver political prisoners to the underground complex for processing until they could be flown out again to permanent concentration camps in Alaska and Montana. According to an unidentified man who claimed to have worked on the underground complex, DIA was built simply as a diversion for the underground concentration camp (Flynn 1995).

If there is any remaining confusion as to why DIA was built, now we know. The *Rocky Mountain News* tried to deal with the story with a straight face, writing:

> *A review of aerial videos taken during DIA's many years of construction showed no excavation or building work in the area in question. Bedrock on the east side of DIA ranges from just under the surface to about 100 feet below. Underground construction would have required major blasting and could not have gone unnoticed. And scars on the earth would be visible to anyone flying in or out (Flynn 1995).*

At this writing, no one has been reported missing from the new Denver International Airport.

Finally, during the airport's numerous delays and cost overruns, the acronym DIA gave birth to a new generation of comedians across the country. Here are a selected few of the alternative names for Denver's new airport:

- Denver Imaginary Airport
- Doesn't Include Airplanes
- Delayed Indefinitely Again
- Dumbest Idea Anywhere
- Dinosaur In Action
- Dollar Investment Astounding
- Dopes In Authority
- Damned Inconvenient Airport
- Disaster In Aviation
- Debacle In Action
- Denver's Intractable Airport

Reference

Flynn, Kevin. 1995. "Patriots" call DIA a front. *Rocky Mountain News.* April 30.

Appendix

(Kowalski 1994).

ection study released.

rum produced an eco-
tudy in the mid-1980s that
percent of the state's earned
000 direct jobs and 140,000 in-
, to Stapleton International Air-
stimated that a new, replacement
would generate another 90,000 new
nd would require 10,000 construction
kers to build (Albin 1994). It also pro-
ted the new airport would generate $8.2
llion annually in business revenue by the
year 2010 and $206 million in state and lo-
cal taxes. Planners then spoke of an airport
with 200 gates (Kowalski 1994).

86 Denver Mayor Peña wonders whether the new airport could be opened by 1990 (*Denver Post* 1994b). Peña announces accelerated timetable for 1992 opening of DIA (*Denver Post* 1994a).

September 1984 Denver begins construction of Concourse E at Stapleton (*Denver Post* 1994a).

1984 In skirmishing with Adams County, Denver officials predict a new runway could be completed on the Rocky Mountain Arsenal by 1988 (*Denver Post* 1994b).

January 28, 1985 Denver and Adams County signed a memo[...] randum of agreement to close Stapleton a[...] build a new airport east of the Rocky Mo[...] tain Arsenal. According to an early pla[...] facility would cover 15,000 acres, co[...] with Stapleton's 4600, and have eig[...] ways open by the year 2020, com[...] Stapleton's four (Kowalski 1994)[...]

January 1985 New airport master plan begin[...]

1986 Three-hub price competition [...] Continental, and Frontier Ai[...] International Airport is fie[...] port to experience unpr[...] 1986, the number of p[...] percent, to 34 million[...]

Summer 1986 New airport site se[...]

September 1986 The Colorado Fo[...] nomic impact [...] attributed 10 [...] income, 21,[...] direct jobs [...] port. It e[...] airport [...] jobs a[...] worl[...] jec[...] b[...]

September 9, 19[...]

February 1987	Renovation of Concourse B at Stapleton complete (*Denver Post* 1994a).
June 1987	Concourse E at Stapleton opens (*Denver Post* 1994a).
July 1987	United and Continental Airlines question planning and timing of a new airport and begin withholding surcharge payments. Peña halts Stapleton construction projects and cancels a planned new runway (*Denver Post* 1994a).
August 1987	Airfield technical committee (city, airlines, and FAA) approves airfield design.
January 1988	Denver and Adams County officials conclude a preliminary agreement for a 53-square-mile airport east of the Rocky Mountain Arsenal. The agreement provides a guarantee that residential communities in the flight path will experience no more noise than they did from Stapleton; any violations will cost Denver $500,000. Colorado Governor Roy Romer begins the "oatmeal circuit," promoting the new airport with local Adams County residents in a series of breakfast meetings (Kowalski 1994).
May 17, 1988	Adams County referendum cedes 53 square miles to Denver to build DIA by a vote of 56 to 44 percent (Kowalski 1994).
Fall 1988	Environmental assessment delivered to FAA.
Winter 1988	Conceptual terminal design released.
January 1989	Two thirds of airport land is purchased.
January 1989	New airport master plan completed.
February 1989	FAA completes draft environmental impact statement.
May 16, 1989	Denver referendum to build DIA passes by a vote of 63 to 37 percent.
May 1989	U.S. Federal court rules Denver may use Stapleton concession revenue, but not airline revenue, to pay for new airport (*Denver Post* 1994a).

August 1989	FAA approves environmental impact statement for DIA (*Denver Post* 1994a).
Late 1980s	Property transactions surrounding the site of the new airport involve Silverado Savings & Loan, and a subsidiary of Lincoln Savings & Loan, then owned by Charles Keating (Kowalski 1994). Real estate deals projected to cost taxpayers $24 million (*Denver Post* 1994a). Extensive investigations by federal law enforcement officials result in no significant prosecutions (Kowalski 1994).
Late 1980s	FAA predicts that by the year 2000, Denver would have the second-busiest airport in the nation. A city consultant predicted that DIA would serve as many as 100 million passengers annually (Kowalski 1994).
September 28, 1989	Groundbreaking on DIA.
March 1990	Continental Airlines signs an agreement at Denver to become the first major tenant at DIA, to occupy 30 gates on Concourse A at a cost of $100 million. Continental then operated 160 flights a day out of 24 gates (Kowalski 1994).
May 8, 1990	Denver sells its first bonds on Wall Street, worth $704 million (Kowalski 1994).
August 1990	The U.S. Senate approves $85 million for Denver's new airport (*Rocky Mountain News* 1991).
November 1990	Denver Mayor Federico Peña announces he will not seek re-election.
December 1990	Continental Airlines enters Chapter 11 bankruptcy for the second time in seven years (*Rocky Mountain News* 1991).
January 1991	Peña scales back the new airport by $150 million, from 94 gates to 85, from five jet runways to four, and from three terminal buildings to two (*Rocky Mountain News* 1991).
February 1991	FAA revises estimated Denver passengers for 1995 from 55 to 34 million.

Appendix

DIA chronology

1974	Dallas/Fort Worth International Airport opens.
Mid-1970s	Questions surface whether Stapleton, with its close-in proximity to Denver and its limited room for expansion, would be adequate to accommodate expected traffic growth (Kowalski 1994).
September 1978	Denver Chamber of Commerce convened a special Airport Task Force (Albin 1994).
1979	Denver Regional Council of Governments began a four-year Airport Site Selection Study (Albin 1994). It concludes a phased expansion onto the adjacent 17,000-acre Rocky Mountain Arsenal ("the most polluted piece of ground on Earth") was the best alternative (Albin 1994). One of Stapleton's existing runways already extended a mile onto the Arsenal.
February 1983	Former state legislator Federico Peña, a candidate for mayor of Denver, criticizes the concept of building a new airport: "In terms of access, convenience and land use impacts, development of a new regional airport represents an inferior choice. . . . At present, the commitment and financial resources required to build such a facility do not exist" (Kowalski 1994). He believed expansion of Stapleton onto the Rocky Mountain Arsenal "represents the best long-term option available" (Kowalski 1994).
May 1983	Federico Peña defeats Mayor Bill McNichols, and in June, Peña is inaugurated mayor of Denver (Kowalski 1994).
Summer 1984	Mayor Peña reopens discussions with Adams County regarding a new airport site.

September 1984	Denver begins construction of Concourse E at Stapleton (*Denver Post* 1994a).
1984	In skirmishing with Adams County, Denver officials predict a new runway could be completed on the Rocky Mountain Arsenal by 1988 (*Denver Post* 1994b).
January 28, 1985	Denver and Adams County signed a memorandum of agreement to close Stapleton and build a new airport east of the Rocky Mountain Arsenal. According to an early plan, the facility would cover 15,000 acres, compared with Stapleton's 4600, and have eight runways open by the year 2020, compared to Stapleton's four (Kowalski 1994).
January 1985	New airport master plan begins.
1986	Three-hub price competition among United, Continental, and Frontier Airlines at Stapleton International Airport is fierce, causing the airport to experience unprecedented growth. In 1986, the number of passengers grew by 16 percent, to 34 million (Kowalski 1994).
Summer 1986	New airport site selection study released.
September 1986	The Colorado Forum produced an economic impact study in the mid-1980s that attributed 10 percent of the state's earned income, 21,000 direct jobs and 140,000 indirect jobs, to Stapleton International Airport. It estimated that a new, replacement airport would generate another 90,000 new jobs and would require 10,000 construction workers to build (Albin 1994). It also projected the new airport would generate $8.2 billion annually in business revenue by the year 2010 and $206 million in state and local taxes. Planners then spoke of an airport with 200 gates (Kowalski 1994).
September 9, 1986	Denver Mayor Peña wonders whether the new airport could be opened by 1990 (*Denver Post* 1994b). Peña announces accelerated timetable for 1992 opening of DIA (*Denver Post* 1994a).

February 1987	Renovation of Concourse B at Stapleton complete (*Denver Post* 1994a).
June 1987	Concourse E at Stapleton opens (*Denver Post* 1994a).
July 1987	United and Continental Airlines question planning and timing of a new airport and begin withholding surcharge payments. Peña halts Stapleton construction projects and cancels a planned new runway (*Denver Post* 1994a).
August 1987	Airfield technical committee (city, airlines, and FAA) approves airfield design.
January 1988	Denver and Adams County officials conclude a preliminary agreement for a 53-square-mile airport east of the Rocky Mountain Arsenal. The agreement provides a guarantee that residential communities in the flight path will experience no more noise than they did from Stapleton; any violations will cost Denver $500,000. Colorado Governor Roy Romer begins the "oatmeal circuit," promoting the new airport with local Adams County residents in a series of breakfast meetings (Kowalski 1994).
May 17, 1988	Adams County referendum cedes 53 square miles to Denver to build DIA by a vote of 56 to 44 percent (Kowalski 1994).
Fall 1988	Environmental assessment delivered to FAA.
Winter 1988	Conceptual terminal design released.
January 1989	Two thirds of airport land is purchased.
January 1989	New airport master plan completed.
February 1989	FAA completes draft environmental impact statement.
May 16, 1989	Denver referendum to build DIA passes by a vote of 63 to 37 percent.
May 1989	U.S. Federal court rules Denver may use Stapleton concession revenue, but not airline revenue, to pay for new airport (*Denver Post* 1994a).

August 1989	FAA approves environmental impact statement for DIA (*Denver Post* 1994a).
Late 1980s	Property transactions surrounding the site of the new airport involve Silverado Savings & Loan, and a subsidiary of Lincoln Savings & Loan, then owned by Charles Keating (Kowalski 1994). Real estate deals projected to cost taxpayers $24 million (*Denver Post* 1994a). Extensive investigations by federal law enforcement officials result in no significant prosecutions (Kowalski 1994).
Late 1980s	FAA predicts that by the year 2000, Denver would have the second-busiest airport in the nation. A city consultant predicted that DIA would serve as many as 100 million passengers annually (Kowalski 1994).
September 28, 1989	Groundbreaking on DIA.
March 1990	Continental Airlines signs an agreement at Denver to become the first major tenant at DIA, to occupy 30 gates on Concourse A at a cost of $100 million. Continental then operated 160 flights a day out of 24 gates (Kowalski 1994).
May 8, 1990	Denver sells its first bonds on Wall Street, worth $704 million (Kowalski 1994).
August 1990	The U.S. Senate approves $85 million for Denver's new airport (*Rocky Mountain News* 1991).
November 1990	Denver Mayor Federico Peña announces he will not seek re-election.
December 1990	Continental Airlines enters Chapter 11 bankruptcy for the second time in seven years (*Rocky Mountain News* 1991).
January 1991	Peña scales back the new airport by $150 million, from 94 gates to 85, from five jet runways to four, and from three terminal buildings to two (*Rocky Mountain News* 1991).
February 1991	FAA revises estimated Denver passengers for 1995 from 55 to 34 million.

March 12, 1991	By now $469 million in contracts have been issued and $200 million in work completed; Standard & Poor's drops DIA bond rating to BBB-minus, one grade above junk (Flynn 1991).
March 1991	Mayoral candidate Don Bain urges two-year moratorium on construction at DIA.
May 1991	Wellington Webb elected mayor of Denver; inaugurated in June.
June 25, 1991	Six days before leaving office, Mayor Peña signs a contract with United Airlines guaranteeing that DIA would not open without an automated baggage system (Flynn 1995).
July 1991	United Airlines agrees to lease 45 gates at DIA.
October 1991	Shunning Denver and Colorado's generous offers of financial assistance, United Airlines picks Indianapolis over Denver for the site of its $1 billion maintenance base (*Denver Post* 1994a).
October 1991	Airport resized up to five runways, three concourses, and 94 gates, and now includes an automated baggage system, grading for a sixth runway, expanded parking, commuter buildings, terminal basement expansion, and expanded cargo facilities.
December 1991	United Airlines signs 30-year leases for 42 gates at DIA.
January 1992	Denver decides to extend automated baggage system to entire airport.
February 1992	Airport adds 41 commuter positions.
May 1992	Continental Airlines signs five-year leases for 20 gates.
August 1992	Denver moves venue of cargo terminals from the north to the south side of DIA, closer to interstate highways. Federal Express and Airborne sign long-term leases (*Denver Post* 1994a).
November 1992	DIA roof is completed (*Denver Post* 1994a).
November 1992	Airport engineer Bill Smith dies of brain cancer.

January 1993	Federico Peña appointed U.S. Secretary of Transportation.
February 1993	Philadelphia airport director James DeLong appointed director of DIA.
February 1993	Widespread allegations of cronyism in airport contracting process (*Denver Post* 1994a).
March 1993	Webb pushes back opening date to October 28.
April 1993	United Parcel Service becomes the last cargo carrier to sign a lease at DIA (*Denver Post* 1994a).
April 1993	Continental Airlines emerges from its second Chapter 11 bankruptcy.
September 1993	United Airlines dedicates a $90 million maintenance facility at DIA (*Denver Post* 1994a).
September 1993	Denver agrees to move rental cars out of terminal to separate campus.
October 1993	Webb postpones opening date to December 19.
December 1993	Webb again postpones opening date, this time to March 9, 1994 (*Denver Post* 1994a).
1993	32.6 million passengers fly in and out of Stapleton (Kowalski 1994).
March 1, 1994	Webb again delays opening, this time to May 15 (O'Driscoll 1994).
March 9, 1994	Continental cuts Denver service by 32 daily departures to 107 daily flights, compared with United's 248 (Mahoney 1994).
May 1994	DIA again fails to meet opening date; Mayor Webb sets no new opening date; DIA bonds downgraded to "junk" status (*Denver Post* 1994c).
July 1994	City adopts $50 million backup baggage system (Eddy 1994).
August 1994	DIA bonds downgraded to lowest investment grade rating by Flitch and Moody's (Berenson 1994).
August 23, 1994	Mayor Webb announces DIA will open on February 28, 1995 (Eddy and Leib 1994).
January 1995	Delay and cost overruns have totaled at least $461 million (Flynn 1995).

January 1995	At least a dozen investigations were under way, including the following: • The Denver District Attorney was investigating allegations of shoddy workmanship, minority contracting irregularities, and cronyism in the awarding of contracts. • The SEC was investigating whether the city disclosed sufficient details about DIA's delays and other problems to prospective bond purchasers. It also was investigating the relationship between bond underwriters' campaign contributions and their municipal contracts. • The FAA was investigating $402,000 the city spent defending its affirmative action program for minority contractors. • The DOT was investigating the city's management of the project (*Rocky Mountain News* 1995).
February 1995	Denver International Airport opens with 88 gates at a cost of $5.3 billion.

References

Albin, Robert. 1994. Building Denver International Airport was a 15-year obstacle course. *Denver Post.* February 26.

Berenson, Alex. 1994. Delays hit DIA bonds. *Denver Post.* August 24.

Denver Post. 1994a. History in the making. Special section on DIA. March.

———. 1994b. Opening day pushed back three times. March 1.

———. 1994c. Downgrading of DIA bonds is not an imminent crisis. May 15.

Eddy, Mark. 1994. DIA bag tag: $10 million. *Denver Post.* July 28.

Eddy, Mark and Jeffrey Leib. 1994. City promises DIA will open by February 28. *Denver Post.* August 23.

Flynn, Kevin. 1991. Planners: It'll be cheaper to finish airport. *Rocky Mountain News.* March 17.

———. 1995. Who botched the airport baggage system? *Rocky Mountain News.* January 29.

Kowalski, Robert. 1994. Turbulence marks DIA history. *Denver Post.* March.

Mahoney, Michelle. 1994. Airline changes buffet Denver. *Denver Post.* February 13.

O'Driscoll, Patrick. 1994. DIA is MIA again. *Denver Post.* March 1.

Rocky Mountain News. 1991. Chronology. March 17.

———. 1995. An update on DIA inquiries. January 29.

Index

B

C

E

F

I

J

K

L

M

N

O

Q

R

S

T

U

V

W

Y

About the authors

Paul Stephen Dempsey, Professor of Law and Director, Transportation Law Program at the University of Denver, is Vice Chairman of Frontier Airlines. Andrew R. Goetz, Associate Professor of Geography, is Associate Director, Intermodal Transportation Institute, at the University of Denver. Joseph S. Szyliowicz is Professor of International Studies at the University of Denver's Graduate School and Director of the University's Intermodal Transportation Institute. All three have published extensively on transportation and related topics.